Lecture Notes in Mathematics

1712

Editors:
A. Dold, Heidelberg
F. Takens, Groningen
B. Teissier, Paris

Springer
Berlin
Heidelberg
New York
Barcelona
Hong Kong
London
Milan
Paris
Singapore
Tokyo

Niels Schwartz   James J. Madden

Semi-algebraic Function Rings and Reflectors of Partially Ordered Rings

Springer

Authors

Niels Schwartz
Fakultät für Mathematik und Informatik
Universität Passau
Postfach 25 40
D-94030 Passau, Germany
E-mail: schwartz@fmi.uni-passau.de

James J. Madden
Department of Mathematics
Louisiana State University
Baton Rouge, LA 70803, USA
E-mail: madden@math.lsu.edu

Cataloging-in-Publication Data applied for

Die Deutsche Bibliothek - CIP-Einheitsaufnahme

Schwartz, Niels:
Semi-algebraic function rings and reflectors of partially ordered
rings / Niels Schwartz ; James J. Madden. - Berlin ; Heidelberg ;
New York ; Barcelona ; Hong Kong ; London ; Milan ; Paris ;
Singapore ; Tokyo : Springer, 1999
   (Lecture notes in mathematics ; 1712)
   ISBN 3-540-66460-2

Mathematics Subject Classification (1991): 14P10, 06F25, 13J25, 18A40, 54C30

ISSN 0075-8434
ISBN 3-540-66460-2 Springer-Verlag Berlin Heidelberg New York

The use of general descriptive names, registered names, trademarks, etc. in this
publication does not imply, even in the absence of a specific statement, that such
names are exempt from the relevant protective laws and regulations and therefore
free for general use.

Typesetting: Camera-ready T$_E$X output by the authors
SPIN: 10700220   41/3143-543210 - Printed on acid-free paper

Dedicated to
Sybilla Prieß–Crampe

Preface

Our work lays algebraic foundations for real algebraic geometry via a systematic study of *reduced partially ordered rings* (or, as we say, *reduced porings*). Thus, we address real algebraists and real geometers as well as algebraists who are concerned with partially ordered algebraic structures. Since the class of rings that we study comprehends all rings of continuous functions with values in the real numbers, we hope that our work will also be useful for topologists using and studying such rings.

The real spectrum is a functorial construction that links algebra to geometry by associating with each poring a topological space. This space should perhaps be viewed as a substrate on which geometric structures can be built, since in algebraic geometry many questions are concerned with more than mere topology.

How does one impose a geometric structure? It is a well-established method to think of geometric properties by means of the mappings or transformations that preserve them. For example, one would expect a richer geometric structure to be preserved by fewer maps. In order to impose geometric structures on real spectra, we take a cue from modern algebraic geometry and assign a *poring of functions* to each real spectrum in a canonical way. Then the category that includes exactly those mappings between real spectra that arise from homomorphisms between assigned porings provides a setting for geometrical investigations. In a sense, the geometry resides in and is expressed by the porings.

How is the "structure poring" to be chosen? One obvious choice is the reduced poring that defines the spectrum in the first place. However, *there are other possibilities*, for there exist numerous functorial constructions (forming a subclass of the reflectors referred to in the title of the book) that produce extensions for all reduced porings, but do not change the real spectrum when passing from a ring to its extension. Each such construction yields a category whose objects are all real spectra; the categories differ only as to the morphisms. Every one of these categories contributes to our understanding of the geometry and the topology of real spectra from a particular point of view. Accordingly, one of the very first steps in the development of a real algebra that is specifically adapted to the investigation of real spectra has to be a study of the functorial extension operators of the reduced porings.

The core of our work concerns the construction of extensions that apply uniformly to all reduced porings. These concepts can be expressed most clearly and most naturally in category theoretic terms, using monoreflectors of reduced porings. Monoreflectors are a well-known concept that has been used in different parts of mathematics, e.g., the Stone-Čech compactification of a Tychonoff space or the divisible hull of a torsion-free abelian group. The example most relevant to our work occurs in the Artin-Schreier theory of ordered fields: The construction of the real closure of a totally ordered field is a monoreflector of the category of totally ordered fields. It might be argued, in fact, that this monoreflector is what gives real algebraic geometry its distinctive character. Considering the fundamental role of the real closure operator it is to be expected that its relatives, which we study here, will also be of considerable significance.

There are several mathematical traditions that converge in our work and that have exerted a strong influence on our view of the subject — no one being more important than any of the others. With the following comments we wish to acknowledge our debt to previous work by other authors and to put our own contributions into the proper context.

Algebraic geometry has ancient roots. Rejuvenating forces in this century included the recognition of the potential of abstract algebraic techniques (Zariski being one of the first to exploit these possibilities systematically) and the introduction of schemes (by Grothendieck). Both these developments are closely related. Together they are a model of success to which real algebra and real algebraic geometry may aspire. Our view of real spectra as geometric entities and of a suitably chosen class of porings as a means for studying their geometry is inspired by these methods and their extraordinary power.

The great importance of commutative algebra in general algebraic geometry suggests a similar role in *real* algebraic geometry. However, it is clear from the start that conventional commutative algebra, dealing with rings, modules, *etc.*, is not quite suitable for this task. It is necessary to incorporate the order and betweenness relations that distinguish real phenomena from those observerd in general algebraic geometry. As was clearly articulated by Brumfiel, *cf.* [27], this leads almost automatically to the study of partially ordered algebraic structures — porings in particular.

There is a tradition of studying ordered algebraic structures which has produced a vast amount of literature, much of which was aimed at understanding rings of continuous functions (*cf.* the classical book by Gillman and Jerison, [49], and work of Henriksen, Isbell and others too numerous to mention). This work does not directly address questions arising in real algebraic geometry, but it has influenced our thinking deeply. Isbell's paper [66] on algebras of uniformly continuous functions contains the first explicitly category-theoretic discussion of porings that we know of. This paper also contains the concepts *of n-ary continuous operations* and of the closure of an algebra under certain such operations. These ideas were developed further by Henriksen, Isbell and Johnson ([59]), who described a construction that was subsequently recognized (by Hager) as the strongest essential monoreflector in the category of archimedean f-rings (and thus is an analogue of the real closure operator which is of central importance in our work). The paper [59] set the stage for studies of monoreflectors of archimedean ℓ-groups and f-rings that were undertaken by Hager and his collaborators beginning in the late 1970's. The direction of our work as well as the methods that we use are clearly influenced by what has been done in the context of archimedean f-rings. Several of our results have striking analogies in that area. But as we progress with the development of a real algebra that is adapted to the needs of real algebraic geometry it does not take long to reach a point where the analogies are left behind and we begin to break new ground.

We have tried to keep the exposition as self-contained as possible; the formal prerequisites are modest. What is needed beyond some basic algebra can be found in a number of standard texts. For formally real and real closed fields one may consult [68], Chapter VI, or [97], §§1–5, or [101], Kapitel II, §2. The small amount of model theory that we use can be found in [97] or [98] or [99]. The basic category theory in [69] or on the first few pages of [83] is certainly enough. The real spectrum is discussed in [19], Chapitre 7, and in [73], Kapitel III. Elementary facts about porings are contained in the first chapters of [27]; a good reference for everything lattice-ordered is [17].

Finally, it is our pleasure to acknowledge the help and support that we received from various persons and organizations: Financial support was given generously by the Department of Mathematics of LSU, by the Fakultät für Mathematik und Informatik of the Universität Passau, by the National Science Foundation (grants DMS-9104427 and DMS-9401509), the Deutsche Forschungsgemeinschaft and the European Community through the HCM project Real Algebraic and Analytic Geometry. Various parts of our work were presented at conferences in Baton Rouge (USA), Curacao (Netherlands Antilles), Gainesville (USA),

Luminy (France), Oberwolfach (Germany) and Trento (Italy), and in colloquium and seminar lectures in Gainesville (hosted by J. Martinez), Madrid (hosted by C. Andradas), München (hosted by S. Prieß-Crampe), Paris (hosted by F. Delon, M. Dickmann and D. Gondard), Regensburg (hosted by M. Knebusch), Santander (hosted by T. Recio) and Saskatoon (hosted by S. Kuhlmann and M. Marshall). As the manuscript was approaching its final form we received suggestions and comments, as well as a list of misprints, from M. Knebusch and D. Zhang (Regensburg). M. Wieberneit (Passau) was a great help with the proof-reading.

Contents

Introduction

Our topic is *partially ordered rings (= porings)*. The purpose is to develop algebraic techniques adapted specifically to the needs of real algebraic geometry. The method is to study *monoreflectors* of the category **POR/N** of reduced porings.

The *real spectrum* is a functor that associates a topological space with every (reduced) poring. These spaces are among the most basic geometric objects in real algebraic geometry. As a framework for their investigation one needs a category whose objects are the real spectra. The morphisms between the spectra are specified using maps that arise functorially from homomorphisms between porings.

A monoreflector of the category **POR/N** provides an extension for every reduced poring, which is called a *monoreflection* of the poring. It may be thought of as a *completion* or a *closure* of the poring. The monoreflector determines a subcategory whose objects are the *closed* porings, i.e., the objects that are not changed by the extension. Conversely, this subcategory also determines the monoreflector. We refer to the subcategory as the *monoreflective subcategory* associated with the reflector. The class of monoreflectors carries a lattice order; the lattice is complete with bottom and top elements. Many well-known functorial constructions with porings are monoreflectors. Our systematic study leads to many new constructions and reveals relationships between various different monoreflectors, which we usually phrase in terms of the lattice order.

The real spectrum of a poring and the real spectrum of an extension that is defined by a monoreflector are tied together via the functorial map corresponding to the extension homomorphism. This map is always bijective, frequently it is even a homeomorphism. Thus, each monoreflector yields a category whose objects are all real spetra, but with a different class of morphisms. Each of these categories contributes to the investigation of real spectra from a special point of view. The comparison of different monoreflectors helps to clarify for which purposes a given monoreflector (or the corresponding monoreflective subcategory) is best suited. For example, it may happen that there is a largest monoreflector whose extensions do not change a particular property of reduced porings. Then that is a monoreflector which is particularly well adapted to the study of this property. A case in point is the *real closure reflector* which is the largest monoreflector that preserves the topology of the real spectrum. In this way the analysis of monoreflectors explains the importance of particular constructions in real algebraic geometry.

This introduction is divided into five parts which can be read independently to a certain extent. Parts A and B are motivational: We explain the importance of porings for real algebraic geometry and our reasons for studying monoreflectors of porings as a particularly important class of functorial extensions. In part C we summarize the main results; the contributions of the different sections are described in part D. Finally, part E offers some thoughts on future directions: The framework we set up provides a systematic way of breaking down problems in real geometry into a sequence of smaller and, hopefully, more accessible steps. Thus, over contribution may help to decide where to put the stepping stones if one big leap is too much and, therefore, needs to be replaced by a sequence of smaller steps.

A The most basic geometric entities of general algebraic geometry are *algebraic varieties*, i.e., zero sets of polynomials that are contained in some affine space K^n, where K is any field (cf. [57], Chapter I). Real algebraic geometry was originally concerned with algebraic varieties in $\mathbb{R}^n$. Historically, the origins of real algebraic geometry can be traced back to antiquity. The Greeks studied geometric figures, such as line segments, triangles, conic sections and convex bodies, that were embedded in $\mathbb{R}^2$ or $\mathbb{R}^3$ ([25], Chapter I; [45]). After the introduction of coordinates many of their geometric objects were recognized as *algebraic sets* (i.e., definable by polynomial equalities) or *semi–algebraic* sets (i.e., definable by polynomial inequalities). In the 19th century complex algebraic geometry began to overshadow real algebraic geometry; the real theory was often felt to be deficient, lacking, for example, Hilbert's Nullstellensatz. Today it has been recognized that real algebraic geometry is different from complex algebraic geometry, but not inferior to it. There is even a real Nullstellensatz ([19], Théorème 4.1.4). The main feature that distinguishes real algebraic geometry from complex algebraic geometry is the presence of an *order relation*, or, geometrically speaking, a *betweenness relation*, a topic that is given much thought in [63]. In connection with questions about geometric constructions, eventually these considerations led Hilbert to formulate his 17th Problem ([62]). Artin's solution of this problem ([6]) made decisive use of *formally real* and *real closed fields*, introduced by Artin and Schreier in [7]. One consequence of these new concepts was a much more general view of real algebraic varieties and semi–algebraic sets, namely as subsets of affine spaces R^n over arbitrary real closed fields R.

As for many mathematical structures, a very basic question about algebraic varieties is the *classification problem* (cf. [57], Chapter I, Section 8): How does one decide whether two algebraic varieties are isomorphic or not? The answer to this question clearly depends on the category one is working in, i.e., on the maps that are used to compare the geometric objects. For example, to relate two real algebraic varieties to each other one may use either polynomial maps or analytic maps or continuous maps, just to name a few possibilities. Each choice of maps will lead to a different answer, in general. The class of allowable maps between varieties can be fixed by specifying for each variety the ring of allowable maps into the affine line. Therefore rings play a vital part in algebraic geometry. The recognition of the importance of rings led to an algebraization of algebraic geometry (with Zariski as one chief proponent) and, eventually, to Grothendieck's theory of schemes. In fact, in modern scheme theory rings are the dominant structure. The geometry resides in the *Zariski spectrum* which is derived from the rings. The scheme theoretic philosophy is to view every ring as a ring of functions on its Zariski spectrum. Thus, the structure of a ring must have a profound influence on the geometric object to which it belongs, whether it is a Zariski spectrum or a more classical variety. Exploring these connections is one of the central themes in algebraic geometry, and a difficult one in general. The translation between the algebra (of the ring) and the geometry is not a simple dictionary. As an example we mention that normality of the ring of regular functions of a variety entails that the variety does not have any singularities of codimension 1. The converse of this statement is false ([94], p. 276-277). Also, the full geometric meaning of flatness of an extension of rings is felt to be engimatic on the geometric level ([94], p. 295).

On the level of rings, the counterpart to geometric betweenness relations (distinguishing real algebraic geometry from general algebraic geometry) are order relations. Correspondingly, the rings of functions in real algebraic geometry frequently carry a partial order, or some variation thereof, such as a preorder. Therefore porings are among the basic algebraic tools in real geometry. They are pairs (A, P) consisting of a ring A together with the positive cone P belonging to a partial order. Their importance was recognized early on in the development of the subject, most clearly expressed by Brumfiel in the introduction of his book [27] (see also [73]; [19]). Experience shows that a partial order adds enormously to the structure of a ring, it may even rival the one of multiplication. In fact, in the presence of an order relation multiplication is not even a necessary part of a geometrically meaningful algebraic structure. Abelian lattice–ordered groups or vector–lattices have enough structure to recover the full PL–isomorphism type of an arbitrary PL–space from the vector–lattice of all its PL–functions. Orderings—and lattice-orderings in particular—have rich geometric meaning and are well–behaved algebraic objects at the same time (cf. [16]; [84]). A partial order is a complex structure with subtle relations to the algebra of the underlying ring. To support a partial order a ring has to satisfy strong algebraic conditions, e.g. the reduction modulo the nilradical must be a *real ring* ([73], p. 103). As our understanding of partially ordered rings advances, it grows increasingly different from the theory of (unordered) rings.

With every poring one can associate its *real spectrum* as a geometric object. Originally the real spectrum was introduced in a series of papers by Coste and Coste–Roy as the solution of a topos–theoretic problem ([31]; [35]; [32]; [36]; [34]; [33]; see also [19], Chapitre 7). It is a construction associating a topological space with any ring, without reference to partial orders. However, it is only a minor step to extend the definition to porings. The real spectrum $Sper(A, P)$ of the poring (A, P) is a *spectral space* (in the sense of [64]) which depends functorially on the poring, a situation very similar to the relationship between rings and their Zariski spectra encountered in general algebraic geometry. Thus, in the spirit of scheme theory, one of the principal goals of real algebraic geometry is to *study partially ordered rings and their real spectra*.

Algebraic varieties over arbitrary real closed fields have been studied using *polynomial functions* or *regular functions* ([19]) or *Nash functions* ([19]; [120]) or *continuous semi–algebraic functions* ([39]; [41]; [72]). Each one of these rings of functions can be equipped with a partial order in a natural way. The smallest among them is the ring of polynomial functions. All the other rings can be obtained from the ring of polynomial functions by applying certain *functorial completion operators* which are defined on the category of all porings. Besides those already mentioned, there are many other similar constructions leading to other partially ordered rings of functions. Each one of these constructions contributes in its own special way to the investigation of real spectra. Therefore we propose to study all of them in order to incorporate as many different points of view as possible. In particular, we want to understand the relationship between different constructions so that we are able to better appreciate the overall contribution of each of them.

B   The best understood and most well–behaved completion operators on categories are *reflectors* ([83], Chapter IV, Section 4; [1], Chapter 4; [60], Chapter X). They modify, in most instances that we are interested in: enlarge, a given object to produce a new one which is distinguished by some special features.   We apply these concepts to categories of porings. Formally, reflectors are functors that are left adjoint to inclusion functors of subcategories. A reflector determines the subcategory (which is called a *reflective subcategory*), and vice versa.   The unit of the adjunction is a natural transformation whose components are called *reflection morphisms*; they are morphisms from objects of the ambient category to objects of the reflective subcategory.   Their codomains are called *reflections*.   From this discussion it is clear that a reflector determines the reflection morphisms. But the converse is also true. This observation leads to the description of reflectors wich lends itself most easily to an intuitive understanding: Suppose that $D \subseteq C$ is a full isomorphism-closed subcategory. Assume that for each $C \in ob(\mathbf{C})$ there is a morphism $r_C \colon C \to r(C)$ into a **D**–object such that the following universal mapping property holds:

> If $f \colon C \to D$ is any morphism into a **D**-object then there is a *unique* morphism $\overline{f} \colon r(C) \to D$ with $f = \overline{f} \circ r_C$.

If these conditions are satisfied then one can show (easily) that **D** is a reflective subcategory of **C** and the morphisms r_C are the reflection morphisms.

Universal mapping properties of this type are ubiquitous in many parts of mathematics and are familiar to every mathematican. Category theory and reflectors provide a formal framework for their investigation. Here are a few examples from different areas:

- The category of compact topological spaces is monoreflective in the category of completely regular spaces. The reflection of a space is its Stone–Čech compactification.
- The category of divisible torsion free abelian groups is monoreflective in the category of all torsion free abelian groups. The reflection of a group is its divisible hull.
- The category of Boolean lattices is monoreflective in the category of distributive lattices. The reflection of a distributive lattice is its Booleanization.

- The category of sheaves is reflective in the category of presheaves. The reflection of a presheaf is its associated sheaf.

One classical example of a monoreflective subcategory that is closely connected with our main topic and that we wish to emphasize is the subcategory **RCF** of real closed fields in the category **TOF** of totally ordered fields. By the theoy of Argin and Schreier ([7]; [68], Chapter VI), every totally ordered field (K, T) has a unique totally ordered algebraic extension $\rho(K, T)$ which is real closed. The extension is called the *real closure* of (K, T); it is the reflection of (K, T) in the monoreflective subcategory **RCF**. The universal mapping property is one of the most important results of the Artin-Schreier theory.

The reflection morphisms of many important reflectors are monomorphisms; in this case one speaks of *monoreflectors* and *monoreflective subcategories*. Then the reflection of a given object can be viewed as an extension. One simple, but exceedingly important fact about monoreflectors is that the reflection morphisms are also epimorphisms, i.e., one deals with epimorphic extension.

Monoreflectors can be compared to each other using a partial order. First of all, the inclusion relation defines a partial order on the class of monoreflective subcategories. Via the bijective correspondence between monoreflectors and monoreflective subcategories this partial order can be transferred to the class of monoreflectors. But we shall use the dual of this partial order. Thus, if $\mathbf{D}, \mathbf{E} \subseteq \mathbf{C}$ are monoreflective subcategories with corresponding monoreflectors r and s then we say that s is *stronger* than r, or r is *weaker* than s $(r \leq s)$ if $\mathbf{E} \subseteq \mathbf{D}$. The reason for this partial order is that s is stronger than r if and only if for every $C \in ob(\mathbf{C})$ the extension defined by s is larger than the one defined by r.

We work with these concepts inside the category **PREOR** of *preodered rings*. Most results are about monoreflectors of its subcategory **POR/N** of *reduced porings*. Because **POR/N** itself is a reflective subcategory of **PREOR**, composition of left adjoint functors can be used to extend results about reflectors of **POR/N** to the category **PREOR**. A considerable number of constructions with porings that are reflectors have been used in real algebra for quite a while, e.g. *the ring of regular functions* associated with a ring of polynomial functions on a real algebraic variety ([19], p. 55), the *real closure* of a ring ([108]; [110]; [111]), the *f–ring reflector* and the *piecewise polynomial functions reflector* (both of which are intimately connected with the notorious Pierce–Birkhoff Conjecture, cf. [85]; [115]). We shall add many more reflectors to the repertoire of real algebra.

In classical commutative algebra monoreflectors are conspicuous for their scarcity. It is shown in [86] that the category of rings does not have any nontrivial monoreflectors. On the other hand, the category of *reduced* rings has nontrivial monoreflectors; there is a largest one which is described in [55], (5.6). Otherwise, little is known about monoroflectors of rings. The reason for the lack of interest and results in this direction can be seen by considering epimorphisms between fields. (Recall at this point that the reflection morphisms of every monoreflector are epimorphisms.)

In the category of fields an extension is epimorphic if and only if it is algebraic and purely inseparabel. For example, suppose that $i\colon K \to L$ is a proper Galois extension of fields of characteristic 0. Then every element a in $L \backslash K$ is the root of some irreducible polynomial over K. There is no way to distinguish the different roots in L of this polynomial. Therefore, by Galois theory, there are automorphisms of L over K that permute these roots. As a consequence, every epimorphism between fields of characteristic zero is surjective. The same must be true in any category that contains the category of fields of characteristic 0.

The situation is completely different in the category **TOF** of totally ordered fields. If $i\colon K \to L$ is an algebraic extension of totally ordered fields then there does not exist any nontrivial order preserving automorphism of L over K. For, such an automorphism would have to shuffle around the different roots of the minimal polynomial of every a in L. The total order of L provides a way to distinguish between the roots: one of them is the largest root, another one is the second largest root, and so on. An order preserving automorphism must respect these relations between the different roots. Hence they must stay fixed, the automorphism must be the identity. Thus, the order relation, or any other additional structure on a field that distinguishes the different roots of a polynomial, leads to a great deal of rigidity of algebraic field extensions, i.e., there are no nontrivial automorphisms respecting this additional structure. In category theoretic language: Algebraic extensions are epimorphisms in the category of fields with this additional structure.

This difference between fields and totally ordered fields manifests itself very clearly on the level of reduced rings or reduced porings. The reason is that reduced rings are closely connected with fields, whereas reduced porings are connected with totally ordered fields: Every reduced ring is contained in a direct product of fields; every reduced poring is contained in a product of totally ordered fields. As a consequence, in the category **POR/N** there are many more epimorphisms than in the category of reduced rings, thus creating many more possiblities for the construction and application of monoreflectors.

Given any large class of mathematical objects it is always an interesting and important task to classify its elements. This is one of the main questions we pursue in our investigation of the monoreflectors of the category **POR/N**. A very successful and fruitful project of this type is the work of Hager and his co–authors aimed at monoreflectors of archimedean l–groups ([53]; [52]; [55]; [56]; [54]; [51]; [9]; [10]; [11]; [88]; [12]). In fact, Hager's work was one of the main inspirations for our study of porings. The classification of monoreflectos of **POR/N** is one step in the direction of a classification of porings themselves. In its strongest form this means classification up to isomorphism. Of course, this is an impossible task – isomorphism is far too fine an equivalence relation. As an approximation one may try to use a coarser notion of equivalence, for example the one provided by a reflector: Two objects are considered as equivalent if they have isomorphic reflections. Or, even coarser, one may refer only to certain features of the reflections to distinguish between objects.

In part A we explained how porings are connected with the geometric objects of real algebraic geometry. Therefore, the classification of porings is closely related and contributes to the classification of real algebraic sets, another problem which is hopelessly difficult in its strongest form. Once again, a coarser approach is needed. Corresponding to the different reflectors of porings there are different classes of mappings for comparing real algebraic sets. One such class are the continuous semi–algebraic maps. The classification of real algebraic sets up to continuous semi–algebraic isomorphisms is at least as difficult as the problem of classifying PL–manifolds. By the work of Markov ([90]), this problem is algorithmically unsolvable in dimensions greater than 3. So, even with the weaker notion of equivalence provided by continuous semi–algebraic maps the classification problem is still too hard. However, there is an uncounted number of variations, not all of which are of the same hopeless degreee of difficulty. For example, the algebraic sets under consideration may be from a restricted class and the invariants used for classification may be much coarser than isomorphism classes. We feel that the conceptual framework we offer here will be a convenient and natural setting in which to address such problems.

C   The first step in the systematic investigation of the monoreflectors of the category **POR/N** is to associate with every reduced poring (A, P)

its *complete ring of functions*, $\Pi(A, P)$. This is the direct product of all real closed residue fields at the points of the real spectrum. The construction of complete rings of functions is functorial, but it is not a reflector. However, it provides a framework for the subsequent investigation of reflectors: Given any monoreflector of the category **POR/N**, the reflection of a reduced poring (A, P) is always contained in the ring $\Pi(A, P)$. In fact, many reflectors are constructed by describing the reflection of the poring (A, P) as a partially ordered subring of $\Pi(A, P)$. The procedure is to associate with every reduced poring (A, P) an intermediate ring between (A, P) and $\Pi(A, P)$ and then to show that these extensions are the reflection morphisms belonging to some reflector.

Earlier in this introduction we described partial orders on the class of monoreflective subcategories of **POR/N** and on the class of monoreflectors of **POR/N**. In fact, the monoreflective subcategories, or the monoreflectors, of **POR/N** form a complete lattice. The lattice structure is our main vehicle for relating different reflectors of **POR/N** to each other.

The lattice of monoreflective subcategories of **POR/N** is a *proper class* (Theorem 9D.3). It contains a number of categories of porings that have already been studied before, independently from our work. Among these are the categories **FR/N** of *reduced f–rings*, **RCR** of *real closed rings*, **VNRFR** of *von Neumann regular f–rings*, and **SAFR** of *rings of semi–algebraic functions*. We study the multitude of monoreflectors by pursuing two approaches that are complementary to each other.

- Individual monoreflectors are studied to set up landmarks in the lattice.
- Subclasses of monoreflectors are studied to improve our understanding of the lattice as a whole.

Our main results about individual monoreflective subcategories are concerned with **SAFR** and **RCR** and with their very special positions in the lattice: **SAFR** is the smallest of all monoreflective subcategories of **POR/N** (Theorem 8.10); **RCR** is the smallest monoreflective subcategory of **POR/N** such that all reflection morphisms, which are called *real closures*, preserve the real spectrum as a topological space (Theorem 12.12). The latter result gives an explanation for the great importance that **RCR** is known to have for the investigation of the topology of real spectra; more about this in part E of the introduction. The category **RCR** has other remarkable properties: It is the smallest monoreflective subcategory of **FR/N** such that every reflection morphism is an essential monomorphism (Theorem 15.5). The reduced porings (A, P) for which every intermediate poring between (A, P) and its real closure is an epimorphic extension of (A, P) can be characterized in terms of the real spectrum of (A, P) (Theorem 14.6).

Only very few monoreflectors can be distinguished by such extraordinary extremal properties as **SAFR** and **RCR**. In our systematic approach towards the vast majority of monoreflectors we pursue the question how much information is needed to determine any one of them. A trivial, but rather useless, answer is that every monoreflector is determined by its entire family of reflections. Dealing with a subclass of monoreflectors it is not always necessary to know the reflection of every reduced poring in order to characterize one particular reflector in the class. The most general question in this direction is:

(Q1)   Given a subclass C of the class of all monoreflectors and some class X of reduced porings, when is it true that any two reflectors $r, s \in C$ coincide if $r(A, P) = s(A, P)$ for every object $(A, P) \in X$?

There are many possible answers to this question, depending on the choice of C and X.

One important class of reflectors are the *H–closed* monoreflectors. The fact that the monoreflectos of **POR/N** form a proper class implies that not all of these reflectors can be equally important for studying concrete geometric situations, such as, e.g., varieties or semi–algebraic sets over the field R_0 of real algebraic numbers. Therefore it is frequently appropriate to consider monoreflectors having additional properties. One such property is *H–closedness*: The subcategory $\mathbf{D} \subseteq \mathbf{POR/N}$ is *H*–closed, if, given a poring (A, P) belonging to $\mathbf{D}$ and a *surjective* homomorphism $f : (A, P) \to (B, Q)$ in **POR/N**, the codomain (B, Q) also belongs to $\mathbf{D}$. Theorem 10.6 says that an *H*–closed monoreflector of **POR/N** is completely determined by:

- the reflection of the single polynomial ring $\mathbb{Z}[T_n; n \in \mathbb{N}]$; or
- the reflections of the countably many polynomial rings $\mathbb{Z}[T_1, \ldots, T_n]$, $n \in \mathbb{N}$.

(In this introduction a polynomial ring with coefficients in $\mathbb{Z}$ always carries the sums of squares as its positive cone.) Many, but not all, of the most important monoreflective subcategories of **POR/N** are *H*–closed; examples are **FR/N**, **RCR**, **VNRFR** and **SAFR**. It is a first important consequence of Theorem 10.6 that the *H*–closed monoreflectors of

POR/N are a set of cardinality $2^{\aleph_0}$. Similar to the class of all monoreflectors, they are a complete lattice with the identity functor as the weakest element and the semi–algebraic functions reflector as the strongest element (Theorem 10.7).

Theorem 10.6 may be interpreted as saying that H–closed monoreflectors arise from geometric situations. For, the polynomials in $\mathbb{Z}[T_1, \ldots, T_n]$ can be considered as functions defined on the affine space R_0^n. To determine an H–closed monoreflector it suffices to know, for each $n \in \mathbb{N}$, which ring of semi–algebraic functions on R_0^n serves as the reflection of the polynomial ring $\mathbb{Z}[T_1, \ldots, T_n]$. Therefore H–closed monoreflectors are more easily accessible to our geometric intuition than general monoreflectors. For example, consider the category **SAFR** of rings of semi–algebraic functions: The reflection of the polynomial ring $\mathbb{Z}[T_1, \ldots, T_n]$ is the ring of all semi–algebraic functions on R_0^n. Or, consider the category **RCR** of real closed rings: The reflection of the polynomial ring $\mathbb{Z}[T_1, \ldots, T_n]$ is the ring of continuous semi–algebraic functions on R_0^n. In fact, the real closure reflector is introduced as the H–closed monoreflector having exactly these reflections. In the same way one can define an H–closed reflector of r–times continuously differentiable semi–algebraic functions, for each $r \in \mathbb{N}$, and an H–closed monoreflector of semi–algebraic C^∞–functions (*Nash functions*).

The concept of *implicit operations* leads to another view of H-closed monoreflectors. Suppose that $f \in \mathbb{Z}[T_1, \ldots, T_n]$ is a polynomial function with n variables. Given a poring (A, P), any tuple $(a_1, \ldots, a_n)$ in A can be substituted into f to yield an element $f(a_1, \ldots, a_n) \in A$. Thus, f may be seen as an operation on (A, P), similar to addition and multiplication. If f is a semi–algebraic function with n variables defined over R_0, but not a polynomial function, then one can still consider the substitution, but usually $f(a_1, \ldots, a_n)$ will not be defined as an element of the ring. However, there is always a poring containing (A, P) in which the substitution is defined. Usually this poring is a proper extension of (A, P). The semi–algebraic function defines an operation on this poring. We distinguish between two types of operations on porings: The polynomial functions define *explicit operations* (since they can be evaluated using the basic arithmetic operations), whereas the semi-algebraic functions that are not polynomial give *implicit operations* (their values are given implicitly as roots of polynomials).

The discussion of H-closed monoreflectors shows that the H-closed monoreflective subcategories of **POR/N** can be classified by the implicit operations that are defined everywhere on each of their objects. An example of an implicit operation in one variable is the absolute value of a function. This operation is not everywhere defined, e.g., usually the absolute value of a polynomial $g \in \mathbb{Z}[T]$ is not a polynomial. However, every reduced poring can be enlarged canonically in such a way that each element in the extension has an absolute value. This is the f-ring reflection, which belongs to the monoreflective subcategory $\mathbf{FR/N} \subseteq \mathbf{POR/N}$.

Related to this point of view is Hilbert's 13th Problem ([62]) which can be understood as a question about the expressive power of functional composition. It asks whether certain functions can be obtained as composites of functions belonging to some specified restricted class. For any H–closed monoreflector of $\mathbf{POR/N}$, the functions belonging to the reflections of the polynomial rings $\mathbb{Z}[T_1, \dots, T_n]$, $n \in \mathbb{N}$, are closed under functional composition. In the spirit of Hilbert's 13th Problem (cf. [5]) one may ask the following question:

(Q2) Given some H–closed monoreflector of $\mathbf{POR/N}$, does there exist some natural number $N \in \mathbb{N}$ such that the reflector is already determined by the reflections of the polynomial rings $\mathbb{Z}[T_1, \dots, T_n]$, $n \leq N$, i.e., such that a function belonging to any one of the reflections of the $\mathbb{Z}[T_1, \dots, T_n]$, $n \in \mathbb{N}$, can be obtained as a composite of functions in at most N variables?

Any answer to this question provides a great deal of information about a reflector. Examples show that some important reflectors are determined by the reflection of the single polynomial ring $\mathbb{Z}[T_1]$. We suspect that many reflectors are not that simple. However, at this time we do not have a single example proving our suspicion. So, our results in this regard are only fragmentary. We consider this problem to be one of the most important and most difficult unanswered questions about monoreflectors of porings.

The answer of Theorem 10.6 to question (Q1) is very satisfying. However, one may also go in a different direction by choosing for X some particularly simple class of porings, such as, e.g., the class of totally ordered fields or the class of totally ordered integral domains. We study the question for the case that C is an interval $[\mathbf{E}, \mathbf{D}]_H$ in the lattice of H–closed monoreflective subcategories where $\mathbf{E}$ and $\mathbf{D}$ are monoreflective with $\mathbf{E} \subseteq \mathbf{D}$. The result for the interval $[\mathbf{RCR}, \mathbf{FR/N}]_H$ is typical (Theorem 19.8): Two of its reflectors, r and s, coincide if and only if they agree on the class of all fibre products

$$(A, P_A) \times_{(C, P_C)} (B, P_B)$$

where (A, P_A) and (B, P_B) are totally ordered integral domains and (C, P_C) is a totally ordered ring. (Note that the fibre product is formed in the category of *porings*, but the resulting ring belongs to **FR/N**.) Stronger results are to be expected if **FR/N** is replaced by smaller categories. If the category **CPWPFR** of *rings of continuous piecewise polynomial functions* takes the place of **FR/N** then it suffices to compare the reflectors r and s on totally ordered integral domains (Theorem 20.1). Or, any H–closed monoreflective subcategory between **SAFR** and the category **VNRFR** is completely determined by its intersection with the category of totally ordered fields (Theorem 17.6). The most extreme case is the interval $[\mathbf{SAFR}, \mathbf{RCR}]_H$: It consists just of the two endpoints (Corollary 22.8), hence there is little need for criteria to distinguish between its elements.

Not unexpectedly, these results show that small monoreflective subcategories of **POR/N** are easier to understand than large ones. Although there are plenty of open question about the bottom of the lattice of monoreflective subcategories, the interval $[\mathbf{SAFR}, \mathbf{FR/N}]$ is certainly much better understood than the upper part of the lattice: It is a major challenge for future work to improve our knowledge of the upper part of the lattice. To point out one possible direction, note that the reflections of the polynomial rings $\mathbb{Z}[T_1, \dots, T_n]$ in a reflective subcategory $\mathbf{D} \in [\mathbf{SAFR}, \mathbf{FR/N}]$ always contain the sup–inf–definable functions on R_0^n, hence they contain functions that are not differentiable. One may study monoreflective subcategories whose objects are rings of differentiable functions. Geometrically this means that differential features, such as degrees of tangency, will be taken into account when semi–algebraic sets are compared using such reflectors.

D Altogether we have 23 sections. The first five sections contain introductory material to set up terminology and notation and to remind the reader of a few basic facts. In sections 6 through 15 we are concerned with the categories **SAFR** and **RCR**, with the methods needed for their introduction and with some of the ramifications of these methods. Sections 16 through 22 are devoted to the systematic investigation of some intervals in the lattice of H–closed monoreflectors. Finally, section 23 stands alone: It contains a summary of our knowledge about the lattice of all monoreflectors of **POR/N** and about the lattice of H–closed monoreflectors. Here is a section-by-section review of our results.

In **section 1** we introduce the categories that provide the framework for our investigations, in particular **PREOR** (*preodered rings*), **POR** (*porings*), **POR/N** (*reduced porings*) and **FR/N** (*reduced f–rings*).

In **section 2** we collect the most basic facts about reflective subcategories and reflectors.

Section 3 deals with totally ordered fields and real closed fields.

In **section 4** we modify the original definition of real spectra to obtain a functor on the category **PREOR**.

In **section 5** we use the real spectrum for a characterization of epimorphisms in **POR/N** (Theorem 5.2).

The systematic investigation of functorial extensions of porings starts in **section 6** with the introduction of the *complete ring of functions*, $\Pi(A, P)$, associated with any preordered ring (A, P).

In **section 7** the *rings of semi–algebraic functions reflector* is introduced as a subfunctor of the complete rings of functions functor (section 6). The corresponding category is **SAFR**, the category of *rings of semi–algebraic functions*.

The comparison of reflective subcategories of **POR/N** by inclusion or of the corresponding monoreflectors by the weaker–stronger relation is discussed in **section 8**. The main result is that the monoreflective subcategories of **POR/N** are a complete lattice with largest element **POR/N** and smallest element **SAFR** (Theorem 8.10).

To fill our picture of the class of monoreflective subcategories of **POR/N** with life we need ways to produce examples of reflectors. There are two Characterization Theorems of reflective subcategories (Theorem 8.3, Theorem 8.7) which we lift from the literature. They provide very general methods for finding reflectors. However, the results of these constructions are far from being explicit, i.e., frequently we know that some reflector exists without knowing anything about the reflections of the individual objects.

In **section 9** we discuss four methods that use given reflectors to construct new ones. If the input reflectors are given explicitly, frequently the same is true for the output reflectors. The constructions of section 9 produce so many reflectors that the class of all monoreflectors of **POR/N** is recognized as a *proper class* (Theorem 9D.3).

In **section 10** we study the class of *H–closed* monoreflectors of **POR/N**. The main result says that every such reflector is completely determined by the reflection of the polynomial ring $\mathbb{Z}[T_n; n \in \mathbb{N}]$ (Theorem 10.6). The *H*–closed monoreflective subcategories are a complete lattice of cardinality $2^{\aleph_0}$ with **SAFR** as its smallest element (Theorem 10.7).

After dealing with factor porings in section 10, we turn to *porings of*

quotients in **section 11**. Reflective subcategories, and the corresponding reflectors, are called *quotient–closed* if they contain every poring of
quotients of each of its members. Examples show that many important
monoreflectors are quotient–closed, or even both quotient–closed and H–
closed.

Using the results of the preceding sections about H–closed and
quotient–closed monoreflectors of **POR/N**, the *real closure reflector* is
introduced in **section 12**. The corresponding category is **RCR**, the
category of *real closed rings*. The reflection of the polynomial ring
$\mathbb{Z}[T_1, \dots, T_n]$ is the ring of all *continuous* semi–algebraic functions on
R_0^n. This definition of the real closure reflector is new. According to
Theorem 12.10 it yields the same rings as the construction used in the
literature ([108]; [110]; [111]). The main result is that the real closure is
the strongest reflector which does not change the real spectrum (Theorem
12.12).

Section 13 is motivated by Theorem 10.6 and deals with the question
how many of the reflections of the polynomial rings $\mathbb{Z}[T_1, \dots, T_n]$, $n \in \mathbb{N}$,
are actually needed to determine a given H–closed monoreflector. Exam-
ples show that some reflectors are already determined by the reflection of
the single polynomial ring $\mathbb{Z}[T_1]$.

In **section 14** we return to the investigation of epimorphisms in the
category **POR/N**. The main result is Theorem 14.6: Every intermediate
poring of (A, P) and its real closure is an epimorphic extension of (A, P)
if and only if the real spectrum of the reduced poring (A, P) contains no
nontrivial fans.

Essential monoreflectors are studied in **section 15**. They are defined
by the property that every reflection morphism is an essential monomorphism. The main result (Theorem 15.5) shows another extremal property
of the real closure: It is the strongest essential monoreflector of **FR/N**.

Section 16 serves as preparation for the next section. It contains an
investigation of the monoreflective subcategories of the category **TOF** of
totally ordered fields.

Section 17 is devoted to the H–closed monoreflective subcategories
D $\subseteq$ **VNRFR**. It is shown that every such subcategory is determined
by its intersection with **TOF** (Theorem 17.6) and that the intersection is
an elementary class of totally ordered fields in the model theoretic sense
(Theorem 17.7).

Monoreflectors of the category **TOD** of *totally ordered integral domains* are the topic of **section 18**. The results are applied in the next
two sections.

Section 19 deals with the interval $[\mathbf{RCR}, \mathbf{FR/N}]_H$ in the lattice of H–closed monoreflective subcategories. Each one of the corresponding reflectors is determined by its action on certain fibre products of totally ordered domains (Theorem 19.8).

In **section 20** it is shown that the categories in the interval $[\mathbf{RCR}, \mathbf{CPWPFR}]_H$ are determined by their intersections with **TOD** (Theorem 20.1).

Section 21 contains an investigation of the H–closed and quotient–closed monoreflective subcategories between **RCR** and **CPWRFR**, the category of *rings of continuous piecewise rational functions*. Using the lattice operations they can be related to the interval $[\mathbf{SAFR}, \mathbf{VNRFR}]_H$. Any two of these reflectors can be told apart by looking at their effect on totally ordered fields (Corollary 21.5).

In the lattice of H–closed monoreflective subcategories the intervals $[\mathbf{SAFR}, \mathbf{VNRFR}]_H$ and $[\mathbf{RCR}, \mathbf{CPWRFR}]_H$ are *parallel* to each other in the sense that the lattice operations can be used to move information back and forth between them. Given an H–closed and quotient–closed monoreflector $\mathbf{D} \in [\mathbf{RCR}, \mathbf{CPWRFR}]_H$, the interval $[\mathbf{D} \cap \mathbf{VNRFR}, \mathbf{D}]_H$ lies *transversely* to $[\mathbf{SAFR}, \mathbf{VNRFR}]_H$ and $[\mathbf{RCR}, \mathbf{CPWFRF}]_H$, i.e., if $\mathbf{E}, \mathbf{F} \in [\mathbf{D} \cap \mathbf{VNRFR}, \mathbf{D}]_H$ then $\mathbf{E} \cap \mathbf{VNRFR} = \mathbf{F} \cap \mathbf{VNRFR}$ and $\mathbf{E} \vee \mathbf{RCR} = \mathbf{F} \vee \mathbf{RCR}$. The investigation of a category $\mathbf{E} \in [\mathbf{D} \cap \mathbf{VNRFR}, \mathbf{D}]_H$ involves discontinuous semi–algebraic functions. In particular, discontinuous functions may be contained in the reflection of $\mathbb{Z}[T_1]$. In **section 22** we study the question how the discontinuities of such functions can be distributed on the line R_0. Frequently this distribution can be used to distinguish between different reflectors. The two extreme cases are that $\mathbf{D} = \mathbf{CPWRFR}$ and that $\mathbf{D} = \mathbf{RCR}$. If $\mathbf{D} = \mathbf{CPWRFR}$ then $\mathbf{D} \cap \mathbf{VNRFR} = \mathbf{VNRFR}$, and there are $2^{\aleph_0}$ H–closed monoreflective subcategories in the interval $[\mathbf{VNRFR}, \mathbf{CPWRFR}]_H$ (Theorem 22.14). If $\mathbf{D} = \mathbf{RCR}$ then $\mathbf{D} \cap \mathbf{VNRFR} = \mathbf{SAFR}$, and there is no H–closed monoreflective subcategory properly in between the two, i.e., $\mathbf{SAFR}$ and $\mathbf{RCR}$ are consecutive elements of the lattice (Corollary 22.8).

In **section 23** we step back to take a panoramic view of the lattice of all monoreflectors of $\mathbf{POR/N}$ and of the lattice of H–closed monoreflectors. The results pertaining to the structure of the lattices are summarized, providing another place where the necessarily detailed investigations of the different sections can be understood from a larger perspective. We invite every reader to jump ahead at any time to see, for example, where a particular category is located in the lattices or how the details of a particular section fit into the larger context.

E   To close the introduction we return to our original geometric motivation which can be explained in greater detail now after the disscussion
of our methods and results. The starting point of our investigations was
the observation that real algebraic varieties can be studied using various
different classes of functions. The same is true for real spectra. To explain
precisely what we mean by this we fix a monoreflector $r\colon \mathbf{POR/N} \to \mathbf{D}$.
For any reduced poring (A, P) the reflection morphism $r_{(A,P)}\colon (A, P) \to$
$r(A, P)$ is an epimorphism.   Therefore (Theorem 5.2) the functorial
map between the real spectra is bijective, hence even a homeomorphism
with respect to the constructible (or *patch*) topology. A homomorphism
$f\colon (A, P) \to (B, Q)$ between reduced porings induces the functorial map
$Sper(f)\colon Sper(B, Q) \to Sper(A, P)$ between the real spectra. The allowable maps between the real spectra are exactly those that arise in this way.
The reflector r provides a unique extension $r(f)\colon r(A, P) \to r(B, Q)$ of f,
hence also an allowable map $Sper(r(f))\colon Sper(r(B, Q)) \to Sper(r(A, P))$
between the corresponding real spectra. In general, there exist homomorphisms $g\colon r(A, P) \to r(B, Q)$ that are not of the form $r(f)$. In this case
the map $Sper(g)$ is allowable only as a map between $Sper(r(A, P))$ and
$Sper(r(B, Q))$, but not as a map between $Sper(A, P)$ and $Sper(B, Q)$.
Thus, the class of allowable maps is enlarged by passage to the larger
rings of functions provided by the reflector. Suppose that we want to
study some feature of the real spectrum that remains unchanged when
passing from (A, P) to $r(A, P)$. Then the rings (A, P) and $r(A, P)$ are
equally well suited for the investigation of this feature. For example, the
real spectrum with its constructible topology will never change, whereas
the real spectrum as a spectral space (i.e., with the spectral, or Harrison,
topology) is unchanged if and only if r is weaker than the real closure
reflector (Theorem 12.10). Therefore we may use any monoreflector to
study the constructible topology. On the other hand, the investigation
of the spectral topology requires that we work with reflectors that are
weaker than the real closure.

As an example of how the choice of a reflector changes the relationship
between geometric objects, consider the following semi–algebraic subsets
of $\mathbb{R}^2$:

$$
\begin{aligned}
S_1 &= \{x;\ x_2 = 0\};\\
S_2 &= \{x;\ x_2 = |x_1|\};\\
S_3 &= \{x;\ x_2^3 = x_1^2\}.
\end{aligned}
$$

For $i = 1, 2, 3$, let (A_i, P_i) be the ring obtained by restricting polynomial functions on the plane to functions on the semi–algebraic set S_i, partially ordered by the positive semi–definite functions. Using only polynomial functions to compare the semi–algebraic sets, no two of them are isomorphic. If φ is the reduced f-ring reflector then $\varphi(A_1, P_1)$ is isomorphic to $\varphi(A_2, P_2)$, but not to $\varphi(A_3, P_3)$. Correspondingly, S_1 and S_2 are isomorphic to each other when compared using sup–inf–definable functions, but they are not isomorphic to S_3. Finally, the real closures $\rho(A_i, P_i)$ are all isomorphic to each other, and so are S_1, S_2 and S_3 via continuous semi–algebraic functions.

Intuitively, the example shows exactly what one should expect: The more allowable maps are available, the fewer isomorphism classes there are. Fewer isomorphism types mean that objects are easier to understand because subtle differences tend to disappear. This phenomenon can also be expressed algebraically, using the monoreflective subcategories of **POR/N**. Large classes of allowable maps correspond to large reflections, i.e., to strong reflectors, and to small reflective subcategories. Small subcategories are easier to study than large ones. So, corresponding to the hierarchy of monoreflective subcategories we have a mathematically precise way to measure degrees of difficulty: Let Q be a question about reduced porings which can also be posed for every monoreflective subcategory. If **D** and **E** are two monoreflective subcategories of **POR/N**, then Q is easier for **E** than for **D** if **E** is contained in **D**.

These observations suggest the following strategy for dealing with problems concerning **POR/N**: First study a small monoreflective subcategory of **POR/N** to gain some small piece of easily accessible information, then use this information to extend the result to some larger monoreflective subcategory. Continuing in this way, eventually one may be able to go all the way up to **POR/N**. For example, to study real spectra with their spectral topology, the first and easiest step is to use the monoreflective subcategory **RCR** (Theorem 12.10).

One may interpret Theorem 12.10 as saying that the very essence of the real closure reflector is the preservation of the topology of the real spectrum. Thus, an investigation of real spectra via real closures amounts to an analysis of their topology. In the geometric situation this is a method which has been used for a long time. Mathematicians have always studied algebraic varieties looking at their topological properties, such as connectedness or compactness. For the concrete setting of semi–algebraic sets, Delfs and Knebusch have created a mature theory of algebraic topology ([41]; [72]). Major parts of algebraic topology can even be developed on the very general and abstract level of real spectra ([28];

[29]; [119]). To pursue topological questions about algebraic varieties or real spectra does not only mean that one deals with some of the most obvious properties of the geometric objects, but also that one works on the least difficult level in the precise sense explicated above.

The investigation of real spectra via their topology, i.e., using real closed rings, is itself a major enterprise that will never be finished. But it is also a logical first step towards understanding real spectra with respect to smaller classes of functions. Our theory of monoreflective subcategories of **POR/N** is meant as a first step towards future systematic investigations in this direction.

1 Preordered and partially ordered rings

This work is devoted to the investigation of categories of partially ordered rings. As a first step, the categorial framework needs to be set up by introducing some of the categories that will be used througout. The largest category that will ever occur is the category of preordered rings. The categories of partially ordered rings and reduced partially ordered rings are subcategories thereof. Some of the most basic properties of these categories will be discussed.

Preordered rings are rings with some additional structure. Therefore it is necessary to start with a brief discussion of the underlying rings. All rings are commutative and have a unit element. Most of the rings that we consider are reduced, i.e., without nilpotent elements. The rings together with their homomorphisms form a category denoted by $\mathbf{R}$. The zero ring is included in $\mathbf{R}$. The category $\mathbf{R}$ is a *concrete category* or *construct* ([83], p. 26; [1], Definition 5.1; [60], Definition 12.7) via the underlying sets functor $U = U_{\mathbf{SETS,R}} : \mathbf{R} \to \mathbf{SETS}$ to the category of sets. A functor $F : \mathbf{D} \to \mathbf{C}$ that is left adjoint to a forgetful functor is called a *free functor* ([1], Example 19.4(2)); with every $X \in ob(\mathbf{D})$ it associates the free object over X. The forgetful functor $U_{\mathbf{SETS,R}}$ has a left adjoint functor $F = F_{\mathbf{R,SETS}}$; if $X \in ob(\mathbf{SETS})$ then $F(X)$ is the polynomial ring $\mathbb{Z}[T_x; x \in X]$. The category $\mathbf{R}$ is wellpowered ([83], p. 126; [1], Definition 7.82; [60], Definition 6.27) and co–wellpowered ([83], loc. cit; [1], Definition 7.87, Example 7.90(2); [60], loc. cit.), it is strongly complete ([1], Definition 12.2, Proposition 12.5) and strongly cocomplete (as can be checked by using the dual of [60], Theorem 23.8).

Deviating slightly from the usage in [73], p. 140, Definition 1, and [77], p. 776, we define a *preordering* of the ring A to be a subset P, called the *positive cone*, having the following properties:

(1.1)

    (a) $P + P \subseteq P$;

    (b) $P \cdot P \subseteq P$;

    (c) $\forall x \in A : x^2 \in P$;

    (d) $supp(P) = P \cap -P \subseteq A$ is an ideal.

The ideal $supp(P)$ is called the *support* of P. A ring with a preodering is a *preordered ring* and is denoted by (A, P) or (A, A^+) some similar symbol. Occasionally we also omit the positive cone from the notation if it is clear from the context that we are discussing a preordered ring. The preordered rings are the objects of a category which we denote by **PREOR**. The morphisms from (A, P) to (B, Q) in this category are the ring homomorphisms $f \colon A \to B$ with $f(P) \subseteq Q$. There is a forgetful functor $U_{\mathbf{R},\mathbf{PREOR}} \colon \mathbf{PREOR} \to \mathbf{R}$. Every ring is the underlying ring of some preordered ring since, given a ring A, the entire ring is always a preordering of A.

A preordered ring (A, P) is said to be *partially ordered* if $supp(P) = \{0\}$ (cf. [27], p. 32). Such a ring will be called a *poring*. The positive cone P determines a partial order of A through: $a \leq b$ if and only if $b - a \in P$. This partial order returns P as its set of positive elements. The partial order is *total* if $P \cup -P = A$. The full subcategories of **PREOR** whose objects are the porings or the reduced porings are denoted by **POR** or **POR/N**, resp.

Via the underlying ring functors $U_{\mathbf{R},\mathbf{POR}}$ and $U_{\mathbf{R},\mathbf{POR/N}}$, both **POR** and **POR/N** are concrete categories over **R**. Through composition of forgetful functors they are also concrete over **SETS**. Not every ring is the underlying ring of a poring or a reduced poring. To determine exactly which rings are, recall that there are various notions of reality for rings ([73], p. 103, Definition 1; [77], Definition 2.1). The ring A is *semireal* if $1 + a_1^2 + \ldots + a_n^2 \neq 0$ for all $n \in \mathbb{N}$ and all $a_1, \ldots, a_n \in A$; it is *real* if $a_1^2 + \ldots + a_n^2 = 0$ always implies $a_1 = \ldots = a_n = 0$. We add another item to this list of notions: The ring A is *weakly real* if $a_1^2 + \ldots + a_n^2 = 0$ implies that $a_1^2 = \ldots = a_n^2 = 0$. It follows from the definitions and [73], p. 104, Satz 1, or [77], Lemma 2.9, that A is real if and only if it is weakly real and reduced. The ring $\mathbb{R}[T]/[T^2]$ is weakly real and not reduced, hence it is not real. The zero ring is both real and weakly real, but it is not semireal. If the weakly real ring A is not the zero ring then it is semireal. On the other hand, $\mathbb{R}[T_1, T_2]/(T_1^2 + T_2^2)$ is a semireal ring that is not weakly real. The full subcategories of **R** whose objects are the semireal rings or the weakly real rings or the real rings are denoted by **SRR**, **WRR**, and **RR**, resp. The different notions or reality are extended to ideals by saying that an ideal $I \subseteq A$ is *semireal*, *weakly real*, or *real*, if the factor ring A/I has the respective property.

Suppose that A is a weakly real ring. Then the set

$$P_w(A) = \{a_1^2 + \ldots + a_n^2; \, n \in \mathbb{N}, a_i \in A\}$$

is a positive cone of A; it will referred to as the *weak order*. In particular, A is the underlying ring of a poring. Conversely, let (A, P) be a poring and assume that $a_1^2 + \ldots + a_n^2 = 0$. Then

$$a_i^2 = -\sum_{j \neq i} a_j^2 \in P \cap -P = \{0\}$$

for every $i = 1, \ldots, n$. Thus, we have proved

Proposition 1.2 (cf. [27], Proposition 1.2.1) The ring A underlies some poring if and only if it is weakly real. It is the underlying ring of a reduced poring if and only if it is real. $\qquad\qquad\square$

The following diagram exhibits the inclusion functors and forgetful functors introduced so far:

(1.3)

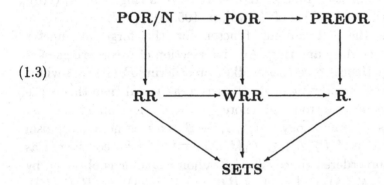

Our next goal is to show that each functor or composition of functors in the diagram has a left adjoint functor. Recall that a composition of left adjoint functors is left adjoint, hence we can proceed in a stepwise fashion.

First of all, the functor $F_{\mathbf{R},\mathbf{SETS}} : \mathbf{SETS} \to \mathbf{R}$ associating with a set X the polynomial ring $\mathbb{Z}[T_x; x \in X]$ and with a set map $f : X \to Y$ the homomorphism $\mathbb{Z}[T_x; x \in X] \to \mathbb{Z}[T_y; y \in Y]$ determined by $T_x \to T_{f(x)}$ is well known to be left adjoint to the forgetful functor $U_{\mathbf{SETS},\mathbf{R}}$.

To find a left adjoint for the inclusion functor $\mathbf{WRR} \to \mathbf{R}$, note that subrings of weakly real rings are weakly real and that the intersection of any set of weakly real ideals is weakly real. Thus, in a ring A there is a smallest weakly real ideal $\sqrt[wr]{(0)}$ which we call the *weakly real radical*. If $f : A \to B$ is any ring homomorphism then there is a unique homomorphism $\overline{f} : A/\sqrt[wr]{(0)} \to B/\sqrt[wr]{(0)}$ such that the following diagram commutes:

$$
\begin{array}{ccc}
A & \xrightarrow{\ f\ } & B \\
\downarrow & & \downarrow \\
A/\sqrt[wr]{(0)} & \xrightarrow{\ \overline{f}\ } & B/\sqrt[wr]{(0)}
\end{array}
$$

The left adjoint functor of the inclusion **WRR** $\to$ **R** associates $A/\sqrt[wr]{(0)}$ with the ring A and $\overline{f}$ with the ring homomorphism f.

It was observed above that a weakly real ring is real if and only if it is reduced. Moreover, one checks easily that $A/Nil(A)$ is weakly real if A is. Therefore the functor **WRR** $\to$ **RR** mapping a weakly real ring A to its reduction $A/Nil(A)$ and a homomorphism $f\colon A \to B$ to the induced homomorphism $\overline{f}\colon A/Nil(A) \to B/Nil(B)$ is left adjoint to the inclusion functor **RR** $\to$ **WRR**. By composition, the left adjoint functor **R** $\to$ **RR** of the inclusion functor **RR** $\to$ **R** maps a ring A to $A/\sqrt[r]{(0)}$, where $\sqrt[r]{(0)}$ is the *real radical* of A ([73], p. 105; [19], p. 77).

To describe the left adjoint functor for the forgetful functor **PREOR** $\to$ **R**, let A be any ring. Any intersection of preorderings of A is a preordering. Hence there is a smallest preordering $P_w(A) \subseteq A$ which is called the *weak preodering* of A. (If A is weakly real then this is the same as the weak order introduced before. So there is no conflict between these two pieces of terminology.) If $f\colon A \to B$ is a ring homomorphism then it is clear that $f(P_w(A)) \subseteq P_w(B)$, i.e., f may be considered as a morphism of preordered rings. The left adjoint functor is obtained by mapping A to $(A, P_w(A))$ and $f\colon A \to B$ to $f\colon (A, P_w(A)) \to (B, P_w(B))$.

Next let $U\colon \mathbf{C} \to \mathbf{D}$ be one of the forgetful functors **POR** $\to$ **WRR**, and **POR/N** $\to$ **RR**. A functor $F\colon \mathbf{D} \to \mathbf{C}$ is defined by sending $A \in ob(\mathbf{D})$ to $(A, P_w(A))$ where $P_w(A)$ is the weak order, i.e, the set of sums of squares in A. If $f\colon A \to B$ is a **D**–morphism then $f(P_w(A)) \subseteq P_w(B)$, i.e., f can also be considered as a **C**–morphism $F(f)\colon (A, P_w(A)) \to (B, P_w(B))$. It is easy to check that this defines a functor which is left adjoint to U.

By composition of left adjoint functors, each of the categories **PREOR**, **POR** and **POR/N** contains a free object over any set S. This free object will be denoted by $F(S)$. Its underlying ring is the polynomial ring $\mathbb{Z}[T_s; s \in S]$; as its preorder or partial order we use the weak one. In fact, since $\mathbb{Z}[T_s; s \in S]$ is a real ring, the weak preorder is a partial order, hence the free objects over S in the three categories coincide. If $S = \{1, \dots, n\}, n \in \mathbb{N}$, then we also use the notation $F(n) = F(S)$.

It remains to define left adjoint functors for the inclusion functors **POR/N** → **POR** and **POR** → **PREOR**. The description is similar to the above definitions of the left adjoint functors of the inclusions **RR** → **WRR** and **WRR** → **R**. First we need the appropriate notion of ideals. This leads us to consider *convex ideals*. Generalizing the definitions of [27], p. 45, and [73], p. 130, Definition 1, we call an ideal I of a preordered ring (A, P) *convex* if $a, b \in P$, $a + b \in I$ always implies that $a, b \in I$. If (A, P) is a poring then this is equivalent to the classical notion of convexity that $a \leq b \leq c$ with $a, c \in I$, $b \in A$ implies $b \in I$ (cf. [17], p. 2; [47], p. 14). It is clear that arbitrary intersections of convex ideals are convex. Some of the most basic properties of convex ideals are collected in

Proposition 1.4 Let (A, P) be a preordered ring.

(a) If $I \subseteq (A, P)$ is a convex ideal then there is a unique partial order $\overline{P}$ on the factor ring $\overline{A} = A/I$ such that the canonical homomorphism $\pi \colon A \to A/I$ is a **PREOR**–morphism and every **PREOR**–morphism $f \colon (A, P) \to (B, Q)$ into a poring with $I \subseteq ker(f)$ factors uniquely as $(A, P) \to (\overline{A}, \overline{P}) \xrightarrow{\overline{f}} (B, Q)$. We call $(\overline{A}, \overline{P})$ the *factor poring modulo I*.

(b) The convex ideals of (A, P) are the kernels of **PREOR**–morphisms from (A, P) into porings.

(c) If $I \subseteq (A, P)$ is convex and $\pi \colon (A, P) \to (\overline{A}, \overline{P})$ is the canonical homomorphism then the map $q \to \pi^{-1}(q)$ is a bijection between the convex ideals of $(\overline{A}, \overline{P})$ and the convex ideals of (A, P) containing I.

Proof (a) We claim that $\overline{P} = \pi(P) \subseteq \overline{A}$ is a partial order. The only property that needs to be checked is that $supp(\overline{P}) = \{0\}$. So, pick $b \in supp(\overline{P})$ and write $b = \pi(a)$, $-b = \pi(a')$ with $a, a' \in P$. Then $a + a' \in ker(\pi) = I$, and convexity implies $a, a' \in I$. We see that $b = 0$. The remaining assertions of (a) are obvious now.   –   **(b)** If $I \subseteq (A, P)$ is convex then (a) shows that I is a kernel as claimed. Conversely, let $f \colon (A, P) \to (B, Q)$ be a **PREOR**–morphism into a poring. Suppose that $a, a' \in P$ with $a + a' \in ker(f)$. Then $f(a) + f(a') = 0$ in B and $f(a), f(a') \in Q$. Thus, $f(a) = f(a') = 0$, i.e., $a, a' \in ker(f)$. This finishes the proof of the convexity of $ker(f)$.   –   **(c)** is trivial.   □

Proposition 1.5 If (A, P) is a preordered ring then $supp(P)$ is the smallest convex ideal of (A, P).

Proof To show that $supp(P)$ is convex, pick $a, b \in P$ with $a + b \in supp(P)$. Then $-a = b + (-a - b) \in P$ and $-b = a + (-a - b) \in P$, hence $a, b \in supp(P)$. Next let $I \subseteq (A, P)$ be any convex ideal and suppose that $a \in supp(P)$. Then $a + (-a) = 0 \in I$ and $a, -a \in P$. By convexity this implies $a \in I$, and we conclude that $supp(P) \subseteq I$. $\square$

Now we are in a position to describe the left adjoint functor of the inclusion **POR** $\to$ **PREOR**. With any preordered ring (A, P) we associate the poring $(A/supp(P), P/supp(P))$ (Proposition 1.4, Proposition 1.5). The canonical ring homomorphism $A \to A/supp(P)$ is also a **PREOR**-morphism. For any **PREOR**-morphism $f : (A, P) \to (B, Q)$ there is a unique **POR**-morphism $\overline{f} : (A/supp(P), P/supp(P)) \to (B/supp(Q), Q/supp(Q))$ such that the diagram

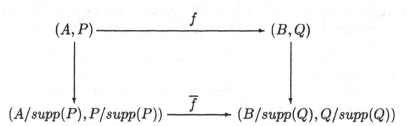

commutes. Associating $\overline{f}$ with f we have obviously defined a functor **PREOR** $\to$ **POR**. One checks easily that it is left adjoint to the inclusion functor.

Finally we describe the left adjoint functor for the inclusion **POR/N** $\to$ **POR**. We consider the nilradical $Nil(A)$ in a poring (A, P). In any ring A the nilradical consists of the nilpotent elements and can be described as the intersection of all prime ideals. In a poring (A, P) the nilradical is the intersection of all *convex* prime ideals ([27], Proposition 2.3.6). In particular, $Nil(A)$ is convex and one can form the factor poring $(A/Nil(A), P/Nil(A))$. This is a reduced poring. Every **R**-morphism maps nilpotent elements to nilpotent elements. Therefore this construction is a functor **POR** $\to$ **POR/N**. It is clearly left adjoint to the inclusion functor **POR/N** $\to$ **POR**.

We saw above that any ring A has a weakest preordering, which we denoted by $P_w(A)$. In general it can happen that $P_w(A) = A$. An example for this phenomenon is the field $\mathbb{C}$ of complex numbers or any finite field ([74], p. 44, Proposition 3.4). But if $P_w(A) \neq A$ then there are always at least two preorderings on A. If A is weakly real then $P_w(A)$ is the weak order. There are rings which carry only one partial order. The field $\mathbb{R}$, or any other real closed field, is a case in point. But frequently there are many different partial orders on a weakly real ring. The smallest one is the weak partial order $P_w(A)$. Zorn's Lemma implies that there are also maximal partial orders. For example, a total order is always a maximal partial order. We shall see below that not every maximal partial order has to be total.

Having taken a first look at the relations between the categories **PREOR**, **POR** and **POR/N**, we now start to study some basic properties of each of them.

Elementary model–theoretic concepts play an important role in our investigations. General references for model theory are [97]; [30]; [98]; [106]. The appropriate language for the first order model theory of porings and preordered rings is the *language of porings*, denoted by $\mathcal{L}_{po}$, which consists of the constants $0, 1$, the function symbols $+, -, \cdot$ and the relation symbol $\leq$. The first order *theory of porings* is given by the axioms of commutative ring theory, the axioms of partial orders, and the axioms of ring–order compatibility. The *theory of reduced porings* is the theory of porings supplemented by the axiom

$$\forall x \, (\, x^2 = 0 \to x = 0).$$

Finally, it is clear that there is also a *theory of preordered rings* using the same language. All these theories are *universal*, i.e., they have sets of axioms consisting only of universally quantified sentences (cf. [106], p. 44).

First we note that monomorphisms are easy to recognize in the three categories **PREOR**, **POR** and **POR/N**:

Proposition 1.6 Let **C** be any one of the categories **POR/N**, **POR**, or **PREOR**. A **C**–morphism $f \colon (A, P) \to (B, Q)$ is a monomorphism if and only if f is injective.

Proof Clearly, injective morphisms are monomorphisms. Now suppose that $f \colon (A, P) \to (B, Q)$ is not injective, say $0 \neq a \in ker(f)$. The homomorphisms $g, h \colon F(1) \to (A, P)$ defined by $g(T) = a$, $h(T) = 0$ (where T is the generator of $F(1)$) are distinct, but $fg = fh$. Thus, f is not a monomorphism.   □

In many categories it is much harder to give a characterization of epimorphisms than of monomorphisms (cf. the collection [107] for the case of
the category **R**). Let $f\colon (A, P) \to (B, Q)$ be a morphism in any one of the
categories **POR/N**, **POR** or **PREOR**. If the underlying ring homomorphism is an epimorphism in **R** then f is clearly epimorphic. It is easy to
see that this sufficient condition is also necessary in the case of **PREOR**.
But this is not true for the other two categories. For example, the in-
clusion $i\colon (\mathbb{Q}, \mathbb{Q}^{\geq}) \to (R_0, R_0^2)$ (with R_0 the field of real algebraic numbers) is epimorphic both in **POR** and in **POR/N**, but the underlying
R–morphism is not an epimorphism. A characterization of **POR/N**–
epimorphisms will be given in section 5, but this requires some preparation. We do not know any characterization of **POR**–epimorphisms
(other than the definition). For **PREOR**, the characterizations of **R**–
epimorphisms in [107] answer the problem.

Following general category theoretic usage, monomorphisms will also
be called *subobjects* or *extensions*. A monomorphism $f\colon (A, P) \to (B, Q)$
of porings or preordered rings is called an *embedding* if $P = f^{-1}(Q)$ and
(A, P) is also called a *sub–preordered ring* or a *subporing* of (B, Q). It
is essential always to bear in mind that subobjects need not be sub-
preordered rings or subporings. (This is similar to the situation one
encounters in topology. If $(X, \mathfrak{T}_X)$ is a topological space and $(Y, \mathfrak{T}_Y)$
is another topological space whose underlying set is a subset of X then
we speak of $(Y, \mathfrak{T}_Y)$ as a subspace of $(X, \mathfrak{T}_X)$ only if $\mathfrak{T}_Y = 2^Y \cap \mathfrak{T}_X$.) A
famous example in **POR** that is concerned with the distinction between
subobjects and subporings is Hilbert's 17th problem (cf. [95]).

Theorem 1.7 The categories **PREOR**, **POR** and **POR/N** are all
well–powered, co–wellpowered, complete and cocomplete. The forgetful
functors (denoted by U) to the category **SETS** preserve limits as well as
direct limits over directed sets.

Proof Using [8], Corollary 1, co–wellpoweredness follows immediately
from the universal axiomatization of the object classes of the three categories.

Wellpoweredness is a consequence of Proposition 1.6: If (A, P) is an
object of any one of the categories and $f\colon (B, Q) \to (A, P)$ is a subobject

then (B, Q) is isomorphic to $(f(B), f(Q))$. The objects (A', P') with $A' \subseteq A$ a subring and $P' \subseteq P$ obviously form a set.

Next we prove that the categories are complete: First of all, it is clear that arbitrary products exist in each of them, and if $\prod_{i \in I}(A_i, P_i)$ is a product then $U(\prod_{i \in I}(A_i, P_i)) = \prod_{i \in I} U(A_i, P_i)$. If $f, g \colon (A, P) \to (B, Q)$ is a parallel pair of morphisms then the equalizer of f and g exists: The underlying set is $E = \{a \in A; f(a) = g(a)\}$. This set is a ring, and the preordered ring or poring $(E, E \cap P)$ is the equalizer. Applying [1], Theorem 12.3 and Proposition 13.14, we conclude that completeness holds and limits are preserved by the forgetful functors U.

For cocompleteness, let $F \colon \mathbf{I} \to \mathbf{PREOR}$ be a diagram in $\mathbf{PREOR}$ (cf. [1], Definition 11.1) and write $F(i) = (A_i, P_i)$, for every $i \in ob(\mathbf{I})$. Composing F with the forgetful functor $U \colon \mathbf{PREOR} \to \mathbf{R}$ we have a diagram in $\mathbf{R}$ with $(UF)(i) = A_i$. Since $\mathbf{R}$ is cocomplete this diagram has a colimit A with canonical maps $f_i \colon A_i \to A$. Suppose that $P \subseteq A$ is the preordering generated by $\bigcup_i f_i(P_i)$. Then every f_i can be considered as a $\mathbf{PREOR}$–morphism $(A_i, P_i) \to (A, P)$. The object (A, P) together with the family of morphisms $(f_i \colon (A_i, P_i) \to (A, P))_i$ is the colimit of F in $\mathbf{PREOR}$, and we are done with this category. Next suppose that $F \colon \mathbf{I} \to \mathbf{POR}$ is a diagram in $\mathbf{POR}$. If (A, P) with morphisms $f_i \colon (A_i, P_i) \to (A, P)$ is the $\mathbf{PREOR}$–colimit of this diagram then $(A/supp(P), P/supp(P))$ with the morphisms

$$(A_i, P_i) \xrightarrow{f_i} (A, P) \to (A/supp(P), P/supp(P))$$

is the $\mathbf{POR}$–colimit. Finally, we consider a diagram $F \colon \mathbf{I} \to \mathbf{POR/N}$. Composing with the inclusion functor $\mathbf{POR/N} \to \mathbf{POR}$ it can also be viewed as a diagram in $\mathbf{POR}$. As such it has a colimit, say (A, P), with morphisms $f_i \colon (A_i, P_i) \to (A, P)$. Evidently, $(A/Nil(A), P/Nil(A))$ together with the morphisms $(A_i, P_i) \xrightarrow{f_i} (A, P) \to (A/Nil(A), P/Nil(A))$ is the colimit in $\mathbf{POR/N}$.

Finally, to check that the forgetful functors preserve direct limits over directed sets, let $\mathbf{C}$ be either $\mathbf{PREOR}$ or $\mathbf{POR}$ or $\mathbf{POR/N}$ and suppose that $G \colon I \to \mathbf{C}$ is a diagram with I a directed set, say $G(i) = (A_i, P_i)$. In the category $\mathbf{R}$ the direct limit $A = \varinjlim A_i$ exists and its underlying set is the direct limit of the *sets* A_i. If $f_i \colon A_i \to A$ are the canonical ring homomorphisms then one checks easily that $(A, \bigcup_{i \in I} f_i(P_i))$ belongs to $\mathbf{C}$ and is the direct limit of G in $\mathbf{C}$. $\square$

Among the most important ring theoretic constructions are the formation
of factor rings and the formation of rings of quotients. We shall briefly
discuss these constructions in the context of the categories **PREOR**,
POR and **POR/N**.

Factor rings modulo convex ideals were already discussed in Proposition 1.4. Now let (A, P) be a preordered ring, let $I \subseteq A$ be any ideal and
$\pi = \pi_I : A \to \overline{A} = A/I$ the canonical homomorphism. One might hope
that $\overline{P} = \pi(P) \subseteq \overline{A}$ is a preordering and π is a **PREOR**–morphism.
However the following example shows that this is not true in general:

Example 1.8 *A partially ordered ring (A, P) with an ideal I such that*
       $P + I/I$ is not a preordering of A/I

Let (A, P) be the polynomial ring $\mathbb{Z}[T_1, T_2]$ with the weak partial order.
We consider the principal ideal $I = (T_1^2 + T_2^2) \subseteq A$ and set $\overline{A} = A/I$, $\overline{P} = P + I/I$. It is obvious that $\overline{P}$ is closed under addition and multiplication
and contains every square of $\overline{A}$. However, $\overline{P} \cap -\overline{P}$ is not an ideal: Denoting
the residue class of $a \in A$ by $\overline{a}$, we have $\overline{T_1^2} = -\overline{T_2^2} \in \overline{P} \cap -\overline{P}$. If
$\overline{P} \cap -\overline{P}$ is an ideal then $\overline{T_1^3} = -\overline{T_1} \, \overline{T_2^2}$ must belong to $\overline{P} \cap -\overline{P}$, i.e.,
there must exist polynomials $F, G \in A$ such that $T_1^3 + F(T_1^2 + T_2^2)$ and
$T_1 T_2^2 + G(T_1^2 + T_2^2)$ are both sums of squares in A. One checks easily that
in a sum of squares the coefficient of any monomial $T_1^i T_2^j$ with i odd and
j even has to be even. Assuming that $T_1^3 + F(T_1^2 + T_2^2)$ is a sum of squares
we conclude that the coefficient of T_1 in F is odd. But this coefficient
coincides with the coefficient of $T_1 T_2^2$ in $T_1^3 + F(T_1^2 + T_2^2)$, and we have
reached a contradiction.   □

Despite the example, it is not unusual at all that $(\overline{A}, \overline{P})$ is a preordered
ring:

Proposition 1.9 Suppose that (A, P) is a preordered ring in which 2 is
invertible. If $I \subseteq A$ is an ideal then $(\overline{A}, \overline{P}) = (A/I, P + I/I)$ is preordered.

Proof It is trivial to check that $\overline{P}$ is closed under addition and multiplication, that $\overline{P}$ contains all squares and that $\overline{P} \cap -\overline{P}$ is additively closed.
It remains to show that $\overline{xy} \in \overline{P} \cap -\overline{P}$ if $\overline{x} \in \overline{A}$ and $\overline{y} \in \overline{P} \cap -\overline{P}$. Suppose
that $\overline{y} = \overline{a} = -\overline{b}$ with $a, b \in P$. Then $a + b \in I$, and the identity

$$ xa + (x^2 + \frac{1}{4})(a + b) = (x + \frac{1}{2})^2 a + (x^2 + \frac{1}{4})b \in P $$

shows that $\overline{xy} = \overline{xa} \in \overline{P}$. (Note that $\frac{1}{4} = (\frac{1}{2})^2 \in A^2 \subseteq P$.) Similarly,

$$xb + (x^2 + \frac{1}{4})(a + b) = (x^2 + \frac{1}{4})a + (x + \frac{1}{2})^2 b \in P$$

implies that $\overline{xy} = -\overline{xb} \in -\overline{P}$. This proves the claim. $\square$

The situation that is most important for our purposes has already been handled: From Proposition 1.4 we know that $(\overline{A}, \overline{P})$ is partially ordered if and only if I is convex. In particular, $(\overline{A}, \overline{P})$ is a reduced poring if and only if I is convex and $I = \sqrt{I}$.

Turning to rings of quotients, we start with a preordered ring (A, P) and a multiplicative subset $S \subseteq A$. As in [27], Proposition 3.1.1, the set

$$P_S = \{\frac{a}{s} \in A_S \, ; \, \exists \, t \in S \colon ast^2 \in P\}$$

is a preordering of A_S. Its support is $supp(P)A_S$. Thus, if (A, P) is a poring or a reduced poring then the same is true for (A_S, P_S); we call (A_S, P_S) a *quotient poring* of (A, P). The canonical ring homomorphism $i_S \colon A \to A_S$ is a **PREOR**–morphism and it has the usual universal mapping property: If $f \colon (A, P) \to (B, Q)$ is any **PREOR**–morphism with $f(S) \subseteq B^\times$ (the group of units of B) then there is a unique **PREOR**–morphism $f_S \colon (A_S, P_S) \to (B, Q)$ with $f = f_S i_S$.

We conclude the section by introducing two more categories of partially ordered rings, the category **FR** of *f-rings* and the category **FR/N** of *reduced f-rings*. By definition, an *f–ring* is a lattice-ordered ring that is isomorphic to a lattice–ordered subring of a direct product of totally ordered rings with componentwise lattice operations ([17], Definition 9.1.1). As general references for *f*–rings we cite: [17], Chapitre 9; [44]; and the survey [58]. Forgetting the lattice–operations, but keeping the partial order we obtain the underlying **POR**–object of an *f*–ring. By **FR** we denote the full subcategory of **POR** whose objects are these porings. We also consider the category **LFR** of *f*–rings with homomorphisms preserving the lattice operations. Such homomorphisms are called *l–homomorphisms*. They are clearly order preserving, hence they are **FR**–morphisms. The following example shows that the converse is not true.

Example 1.10 FR-*morphisms need not be* **LFR**-*morphisms*

Let $A = \mathbb{R}^3$ and $B = \mathbb{R}^2$ with componentwise addition. We define multiplications in both additive groups by

$$A \times A \longrightarrow A : ((a,b,c),(x,y,z)) \to (ax, ay + bx, az + cx),$$
$$B \times B \longrightarrow B : ((a,b),(x,y)) \to (ax, ay + bx).$$

One can be check that both A and B are commutative rings with 1. Lattice orders are defined on both rings by setting

$$(a,b,c) > 0 \quad \text{if and only if} \quad a > 0 \quad \text{or}$$
$$a = 0, \quad b,c \geq 0, \ b + c > 0;$$
$$(a,b) > 0 \quad \text{if and only if} \quad a > 0 \quad \text{or}$$
$$a = 0, \quad b > 0.$$

Once again, it is easy to check that A and B are f-rings with these partial orders; in fact, B is even totally ordered. The homomorphism $f \colon A \to B \colon (a,b,c) \to (a, b + c)$ is clearly order preserving, i.e., f is an **FR**-morphism. But f does not preserve the lattice operations. For, in A we have $(0,1,-1) \vee (0,0,0) = (0,1,0)$ whereas

$$f((0,1,0)) = (0,1),$$
$$f((0,1,-1)) \vee f((0,0,0)) = (0,0) \vee (0,0) = (0,0).$$

$\square$

The full subcategories of **LFR** and **FR** whose objects are the reduced f-rings are denoted by **LFR/N** and **FR/N**. A priori, the same difference that we noted between the morphisms in **FR** and **LFR** must also be expected for **FR/N** and **LFR/N**. However, according to [44], Lemma 2.2, every **FR/N**-morphism preserves the lattice operations, hence is an **LFR/N**-morphism. This means that the inclusion functor **LFR/N** $\to$ **FR/N** is an isomorphism of categories. Therefore these categories are identified, and henceforth we shall use only the notation **FR/N**.

Let $f \colon (A,P) \to (B,Q)$ be an **FR**-morphism. We know from Proposition 1.4 that $ker(f)$ is convex. By [17], Proposition 8.3.4, f is an **LFR**-morphism if and only if $ker(f)$ is an l-ideal, i.e., a convex ideal that is a sublattice of A ([17], Chapitre 8, §2, §3, Chapitre 9, §1). We summarize some of the most frequently used properties of l-ideals in f-rings: The nilradical of an f-ring is an l-ideal ([17], Théorème 9.2.6). So, the reduced poring of an f-ring (A,P) is a reduced f-ring and the canonical homomorphism $\pi_{Nil(A)}$ preserves the lattice operations. An l-ideal

$I \subseteq (A, P)$ is *irreducible* if and only if $(A/I, P/I)$ is totally ordered ([17], Théorème 9.1.5 and §8.4). Any l–ideal containing an irreducible one is irreducible as well. In a reduced f–ring the minimal prime ideals of the ring are exactly the minimal irreducible l–ideals ([17], Théorème 9.3.2). Thus, in every f–ring the minimal prime ideals are l–ideals; if a prime ideal is an l–ideal then the factor poring is totally ordered. Therefore, for a prime ideal convexity is equivalent to being an l–ideal.

Earlier we discussed the set of partial orders on a weakly real ring. Now we show that the partial order of a reduced f–ring is always maximal. A more general discussion concerning the strengthening of partial orders may be found in [117], §2.

Proposition 1.11 Suppose that (A, P) is a reduced f–ring and that $Q \subseteq A$ is a positive cone with $P \subseteq Q$. Then $P = Q$.

Proof Assume that there is some $a \in Q \backslash P$, and write $a = a^+ - a^-$ (where $a^+ = a \vee 0$, $a^- = -a \vee 0$). Since $a^- \in P \subseteq Q$ it follows from

$$-(a^-)^2 = aa^- \in Q$$

(note that $a^+ a^- = 0$) that $(a^-)^2 \in Q \cap -Q = \{0\}$. But then $a = a^+ \in P$, a contradiction. $\qquad\qquad\square$

2 Reflective subcategories

A large number of constructions involving porings can be viewed as, informally speaking, *closure operations*. They have functorial properties and can be studied in terms of *reflectors* and *reflective subcategories* of the categories **PREOR**, **POR** and **POR/N**. In this section we shall discuss the general category theory that we need for these investigations. References for reflective subcategories are [83]; [1]; [60]. Let **C** be a category and let **D** ⊆ **C** be a subcategory. We call **D** a *reflective subcategory* if the inclusion functor has a left adjoint $r: \mathbf{C} \to \mathbf{D}$, which is then called a *reflector*. For each $A \in ob(\mathbf{C})$ the unit $r_A: A \to r(A)$ of the adjunction is called the *reflection morphism* of A. It has the following universal property:

(2.1)   If $f: A \to B$ is any **C**–morphism into a **D**–object then there is a unique **D**–morphism $\bar{f}: r(A) \to B$ such that $f = \bar{f}r_A$.

In practice, the universal property is frequently instrumental in identifying a subcategory as reflective. For, if **D** ⊆ **C** is a subcategory, and if for every $A \in ob(\mathbf{C})$ there are an object $r(A)$ in **D** and a morphism $r_A: A \to r(A)$ which has the universal property (2.1) then **D** is a reflective subcategory of **C** ([83], Chapter IV, §3; [1], Definition 4.16).

Throughout we make the general assumption that **all subcategories are full and isomorphism–closed**. By [83], p. 89, or [1], Proposition 4.20, fullnes of a reflective subcategory is equivalent to idempotency of the corresponding reflector, i.e., if $r: \mathbf{C} \to \mathbf{D}$ is a reflector and if $B \in ob(\mathbf{D})$ then the reflection $r_B: B \to r(B)$ is an isomorphism. In the context of our investigations, a reflection should always be something like a closure operation. Objects are considered as closed when they are not changed by the closure operation. Idempotency of a closure operation means that any object obtained by one applicaton of the closure operation is a closed object. For our purposes the general assumption of fullness is justified since this is the kind of behavior we expect from our constructions. Of course, for different purposes one may also be interested in reflective subcategories which are not full. We give one example showing that such reflective subcategories exist in **POR**. (Of course, at this point – the only one in the entire book – we are suspending the general assumption above.)

Example 2.2 *A reflective subcategory of* **POR** *that is not full and whose reflector is not idempotent*

It was shown in Example 1.10 that the subcategory **LFR** $\subseteq$ **POR** is not full. We claim that this subcategory is reflective: Given any poring (A, P) we form the set of all pairs $\alpha = (I, \overline{P})$ where I is a convex ideal and $\overline{P}$ is a total order of the residue ring such that the canonical homomorphism $\pi_I : A \to A/I$ is a **POR**–morphism. The π_I combine to define a homomorphism $r'_{(A,P)} : (A, P) \to \prod_\alpha (A/I, \overline{P})$. Using the componentwise partial order, the codomain is an f–ring. Let $r(A, P) \subseteq \prod_\alpha (A/I, \overline{P})$ be the smallest subring which is also a sublattice and contains $r'_{(A,P)}(A)$. Then $r(A, P)$ is an f–ring. By restriction of the codomain, $r'_{(A,P)}$ yields the morphism $r_{(A,P)} : (A, P) \to r(A, P)$. In order to prove that **LFR** is reflective it suffices to check the universal property (2.1) for the morphisms $r_{(A,P)}$. So, let $f : (A, P) \to (B, Q)$ be any **POR**–morphism into an f–ring. By definition, there is a representation $h : (B, Q) \to \prod_{i \in I} (B_i, Q_i)$ as a subring and sublattice of a direct product of totally ordered rings. Composing with the projections $p_j : \prod_{i \in I} (B_i, Q_i) \to (B_j, Q_j)$ one obtains homomorphisms $p_j h f : (A, P) \to (B_j, Q_j)$ into totally ordered rings. There is a unique $\alpha(j)$ such that $p_j h f$ factors through $(A/I_{\alpha(j)}, \overline{P}_{\alpha(j)})$. Altogether this leads to a commutative diagram

$$
\begin{array}{ccc}
(A, P) & \xrightarrow{\ f\ } & (B, Q) \\[4pt]
{\scriptstyle r'_{(A,P)}} \downarrow & & \downarrow {\scriptstyle h} \\[6pt]
\prod_\alpha (A/I, \overline{P}) & \xrightarrow{\ g'\ } & \prod_{i \in I} (B_i, Q_i)
\end{array}
$$

where g' is an **LFR**–morphism. The subring $g'^{-1}(h(B)) \subseteq \prod_\alpha (A/I, \overline{P})$ is an f–ring containing $r'_{(A,P)}(A)$. But then it also contains $r(A, P)$, hence g' can be restricted to $g : r(A, P) \to (B, Q)$, which is also an **LFR**–morphism. By construction, $f = g r_{(A,P)}$. It remains to show that g is unique. So, assume that $g_1 : r(A, P) \to (B, Q)$ is another **LFR**–morphism with $f = g_1 r_{(A,P)}$. One checks that

$$
C = \{x \in r(A, P);\ g(x) = g_1(x)\}
$$

is a subring and a sublattice of $r(A, P)$ containing $im(r_{(A,P)})$. From the definition of $r(A,P)$ we conclude that $C = r(A,P)$. But this means that $g = g_1$, and **LFR** is indeed reflective in **POR**.

It was shown in Example 1.10 that **LFR** is not full in **POR**. Thus, it follows from [1], Proposition 4.20, that the reflector $r : \textbf{POR} \to \textbf{LFR}$ is not idempotent. We exhibit an explicit example illustrating this fact: Let (A,P) and (B,Q) be the rings of Example 1.10. Assume that the reflection morphism $r_{(A,P)}\colon (A,P) \to r(A,P)$ is an isomorphism. By the universal property (2.1) the homomorphism $f\colon (A,P) \to (B,Q)$ of Example 1.10 factors as

$$(A,P) \xrightarrow{\ r_{(A,P)}\ } r(A,P) \xrightarrow{\ \overline{f}\ } (B,Q)$$

with $\overline{f}$ an **LFR**–morphism. The isomorphism $r_{(A,P)}$ is clearly an **LFR**–morphism, hence $f = \overline{f}r_{(A,P)}$ is an **LFR**–morphism as well. But it was shown in Example 1.10 that this is not the case. $\qquad\square$

Let X be a class of **C**–morphism. A reflective subcategory $\textbf{D} \subseteq \textbf{C}$ with reflector r is called X–*reflective* if every reflection $r_A\colon A \to r(A)$ belongs to X. If X is the class of epimorphisms or monomorphisms we speak of *epireflectors, monoreflectors, epireflections, monoreflections* and *epireflective* or *monoreflective* subcategories.

Following [55], we shall occasionally use the following description of monorefletors. In a category **C** a *functorial extension operator* is a functor $F\colon \textbf{C} \to \textbf{C}$ together with a natural transformation $t\colon \text{Id}_{\textbf{C}} \to F$ such that each component $t_A\colon A \to F(A)$ of t is a monomorphism and, given a morphism $f\colon A \to B$, then $F(f)$ is the unique morphism $g\colon F(A) \to F(B)$ making the diagram

$$
\begin{array}{ccc}
A & \xrightarrow{\ f\ } & B \\
{\scriptstyle t_A}\downarrow & & \downarrow{\scriptstyle t_B} \\
F(A) & \xrightarrow{\ g\ } & F(B)
\end{array}
$$

commutative. The functorial extension operator is said to be *idempotent* if $t_{F(A)}\colon F(A) \to F(F(A))$ is an isomorphism for all $A \in ob(\textbf{C})$. Examples of idempotent functorial extension operators are obtained from monoreflectors: Let $r\colon \textbf{C} \to \textbf{D}$ be a monoreflector, let F be the composition of r with the inclusion functor $\textbf{D} \to \textbf{C}$. Then F together with the unit of the adjunction is an idempotent functorial extension operator. Conversely, if (F,t) is an idempotent functorial extension operator then let $\mathbf{D} \subseteq \mathbf{C}$ be the full subcategory of all objects A with $t_A : A \to t(A)$ an isomorphism. The objects $F(A)$ belong to this subcategory. Let r be F viewed as a functor into $\mathbf{D}$. Then $\mathbf{D}$ is monoreflective in $\mathbf{C}$ with monoreflector r.

Practically all the reflectors that we shall encounter are epireflectors or monoreflectors. We note the following relationship between epireflectors and monoreflectors:

Proposition 2.3 Every monoreflector is an epireflector.

Proof [83], p. 90, Exercise 3, or [60], Proposition 36.3.   □

We continue with a discussion of a few general properties of reflectors. After that the section concludes with examples. The proofs of most of the following results are simple and are therefore omitted.

Proposition 2.4 Let $r : \mathbf{C} \to \mathbf{D}$ be an epireflector and $f : C \to C'$ a $\mathbf{C}$-morphism. If f is epimorphic in $\mathbf{C}$ then $r(f)$ is epimorphic in $\mathbf{D}$. Suppose that r is a monoreflector. Then f is a $\mathbf{C}$–epimorphism if and only if $r(f)$ is a $\mathbf{D}$–epimorphism, if and only if $r(f)$ is a $\mathbf{C}$–epimorphism.   □

Proposition 2.5 Suppose that $\mathbf{D}$ is reflective in $\mathbf{C}$. A $\mathbf{D}$-morphism $f : C \to D$ is a monomorphism in $\mathbf{D}$ if and only if it is a monomorphism in $\mathbf{C}$.   □

Proposition 2.6 Assume that $\mathbf{D} \subseteq \mathbf{C}$ is epireflective. Then $\mathbf{D}$ is closed under formation of limits existing in $\mathbf{C}$. More precisely, let $F : \mathbf{J} \to \mathbf{D}$ be a diagram and suppose that there is a limit cone $(p_j : C \to F(j))_j$ in $\mathbf{C}$, then C is an object of $\mathbf{D}$ and the cone is a limit cone in $\mathbf{D}$.

Proof [60], Theorem 36.1 (iii).   □

Proposition 2.7 Let $r : \mathbf{C} \to \mathbf{D}$ be a reflector.
(a) If $\mathbf{C}$ is wellpowered then so is $\mathbf{D}$.

(b) If **C** is co–wellpowered and r is a monoreflector then **D** is co–wellpowered.

(c) If **C** is complete and r is an epireflector then **D** is complete.

(d) If **C** is cocomplete then so is **D**.

(e) If **C** has free objects over **SETS** then so does **D**.

Proof (a), (b) and (c) are immediate consequences of Proposition 2.5, Proposition 2.4 and Proposition 2.6, resp. For the proof of (d), first note that reflectors preserve colimits, as do all left adjoint functors ([83], Chapter V, §5; [60] Theorem 27.7). Thus, if $r\colon \mathbf{C} \to \mathbf{D}$ is a reflector and if $F\colon \mathbf{J} \to \mathbf{C}$ is a diagram with colimit cone $(F(j) \to C)_j$ then $(rF(j) \to r(C))_j$ is a colimit cone for the diagram $rF\colon \mathbf{J} \to \mathbf{D}$. Finally, (e) is a consequence of the fact that a composition of left adjoint functors is a left adjoint functor.   □

Let **C** be a complete and wellpowered concrete category over **SETS** with forgetful functor U. Assume that U preserves limits (hence also monomorphisms, cf. [1], Definition 13.1, Proposition 13.5). Then any family of subobjects of any $C \in ob(\mathbf{C})$ has an *intersection* (cf. [1], Proposition 12.5). In fact, if $(m_i\colon C_i \to C)_{i \in I}$ is a family of subobjects with intersection $m\colon D \to C$ then

$$U(m)(U(D)) = \bigcap_{i \in I} U(m_i)(U(C_i)) \subseteq U(C).$$

Now pick some subset $X \subseteq U(C)$ and let $S(X)$ be the class of all subobjects $m\colon C' \to C$ with $X \subseteq U(m)(U(C'))$. The class $S(X)$ contains a smallest object $\langle X \rangle$ (the intersection of all objects belonging to $S(X)$). We call $\langle X \rangle$ the *subobject generated by X*.

In section 1 a number of subcategories of **R** and of **PREOR** were recognized as being reflective (without using the word "reflective"). For future reference we collect the information we have about these subcategories in the following examples.

Example 2.8 *Reflective subcategories of rings*

The subcategories **RR** ⊆ **WRR** and **WRR** ⊆ **R** are epireflective, but not monoreflective. All of these categories have free objects over **SETS**, are wellpowered, complete and cocomplete. The same holds for all their epireflective subcategories. For each of these categories the forgetful functor to **SETS** preserves limits, hence subobjects generated by subsets exist.   □

Example 2.9 *Reflective subcategories of preordered rings*
The subcategories $\mathbf{POR/N} \subseteq \mathbf{POR}$ and $\mathbf{POR} \subseteq \mathbf{PREOR}$ are epireflective, but not monoreflective. They have free objects over $\mathbf{SETS}$; they are wellpowered, co–wellpowered, complete and cocomplete (Theorem 1.7). Thus, if $\mathbf{C}$ is a monoreflective subcategory of any one of these categories then $\mathbf{C}$ has the same properties. For each of these categories the forgetful functor to $\mathbf{SETS}$ preserves limits (Theorem 1.7), hence subobjects generated by subsets exist.   □

We mention at this point that the subcategory $\mathbf{FR/N}$ of reduced f–rings is monoreflective in $\mathbf{POR/N}$. We could prove this by an *ad hoc* method now. But when our techniques have been developed to a certain extent the proof will become very natural and easy. Therefore we defer a detailed treatment to Proposition 6.5.

Suppose that $\mathbf{C} \subseteq \mathbf{PREOR}$ is a subcategory and $r: \mathbf{C} \to \mathbf{D}$ is a reflector. The reflection $r(A, P)$ of a preordered ring (A, P) is also preordered. Its positive cone is uniquely determined by r. Therefore we usually omit it from the notation. Whenever we need to refer to the positive cone then we shall denote it by $r^+(A, P)$.

Example 2.10 *The weak bounded inversion property*
We say that a ring A has the *weak bounded inversion property* if

$$1 + \sum A^2 = \{1 + a_1^2 + \ldots + a_n^2; \ n \in \mathbb{N}, \ a_i \in A\} \subseteq A^\times,$$

where $A^\times$ denotes the group of units of A. Inside $\mathbf{PREOR}$ the full subcategory whose objects have the weak bounded inversion property are denoted by $\mathbf{WBIPREOR}$. For subcategories of $\mathbf{PREOR}$ we use the following generic notation: If $\mathbf{C} \subseteq \mathbf{PREOR}$ then $\mathbf{WBIC} = \mathbf{C} \cap \mathbf{WBIPREOR}$. We shall show that $\mathbf{WBIPREOR} \subseteq \mathbf{PREOR}$, $\mathbf{WBIPOR} \subseteq \mathbf{POR}$ and $\mathbf{WBIPOR/N} \subseteq \mathbf{POR/N}$ are reflective subcategories. If (A, P) is any preordered ring then we form the ring of quotients $(A_{1+\sum A^2}, P_{1+\sum A^2})$. It is claimed that this ring has the weak bounded inversion property: Pick $\frac{a_1}{s_1}, \ldots, \frac{a_n}{s_n} \in A_{1+\sum A^2}$ and form the sum $1 + (\frac{a_1}{s_1})^2 + \ldots + (\frac{a_n}{s_n})^2$. In doing so we may assume that $s_1 = \ldots = s_n = s \in 1 + \sum A^2$, and therefore

$s^2 \in 1 + \sum A^2$. Then

$$1 + (\frac{a_1}{s})^2 + \ldots + (\frac{a_n}{s})^2 = \frac{s^2 + a_1^2 + \ldots + a_n^2}{s^2}$$

Both the numerator and the denominator are elements of $1 + \sum A^2$, hence this is an invertible element of $A_{1+\sum A^2}$. We conclude that $(A_{1+\sum A^2}, P_{1+\sum A^2}) \in ob(\textbf{WBIPREOR})$. The canonical homomorphism $i_A \colon A \to A_{1+\sum A^2}$ is a \textbf{PREOR}–morphism. If $f \colon (A, P) \to (B, Q)$ is any \textbf{PREOR}–morphism into a \textbf{WBIPREOR}–object then the universal property of the formation of rings of quotients in \textbf{PREOR} (cf. section 1) shows that there is a unique morphism $\overline{f} \colon (A_{1+\sum A^2}, P_{1+\sum A^2}) \to (B, Q)$ with $f = \overline{f} i_A$. Using the universal property (2.1) we conclude that the construction is a reflector \textbf{PREOR} $\to$ \textbf{WBIPREOR}. Since the underlying ring homomorphism $i_A \colon A \to A_{1+\sum A^2}$ of the reflection is always epimorphic in \textbf{R}, it follows that i_A is also epimorphic in \textbf{PREOR}, i.e., we have constructed an epireflector.

If (A, P) is a poring or a reduced poring then the same is true for $(A_{1+\sum A^2}, P_{1+\sum A^2})$. Thus, the reflector restricts to reflectors \textbf{POR} $\to$ \textbf{WBIPOR} and \textbf{POR/N} $\to$ \textbf{WBIPOR/N}. Evidently both of them are epireflectors; the second one is even a monoreflector (this is clear since in a real ring no element in $1 + \sum A^2$ can be a zero divisor) whereas the first one is not a monoreflector. This requires an example of a weakly real ring A such that $1 + \sum A^2$ contains a zero divisor. A direct computation proves that $A = \mathbb{Z}[T]/(T^2, 2T)$ is weakly real. The element $1 + 1 \in 1 + \sum A^2$ is a zero divisor. $\square$

Example 2.11 *The bounded inversion property*
A preordered ring (A, P) is said to have the *bounded inversion property* if $1 + P = \{1 + a; a \in P\} \subseteq A^\times$. The category of preordered rings with bounded inversion is denoted by \textbf{BIPREOR}. As in the previous example, for any subcatory $\textbf{C} \subseteq \textbf{PREOR}$ we define $\textbf{BIC} = \textbf{C} \cap \textbf{BIPREOR}$. For any preordered ring (A, P) the set $1 + P \subseteq A$ is multiplicative, hence we can form the preordered ring of quotients (A_{1+P}, P_{1+P}). The canonical ring homomorphism $i_A \colon A \to A_{1+P}$ is a morphism of preordered rings. It is claimed that (A_{1+P}, P_{1+P}) has the bounded inversion property: Suppose that $\frac{a}{s} \in P_{1+P}$ with $s = 1 + p \in 1 + P$ and $ast^2 \in P$ for a suitable $t = 1 + q \in 1 + P$. It is claimed that $1 + \frac{a}{s} \in A_{1+P}^\times$: The numerator of

$$1 + \frac{a}{s} = \frac{s^2 t^2 + ast^2}{s^2 t^2} = \frac{1 + r + ast^2}{s^2 t^2}$$

(with $r = s^2 t^2 - 1 \in P$) belongs to $1 + P$, hence the quotient $\frac{s^2 t^2}{1 + r + ast^2}$ can be formed in A_{1+P}. This proves the claim.

Similar to the previous example it is obvious that the construction is an epireflector **PREOR** $\rightarrow$ **BIPREOR**. It restricts to an epireflector **POR** $\rightarrow$ **BIPOR** and a monoreflector **POR/N** $\rightarrow$ **BIPOR/N**. $\square$

Note that the bounded inversion property implies the weak bounded inversion porperty (trivially), but not vice versa. For example, let (A, P) be the ring $\mathbb{R}[T]$ with the unique total order for which T is positive and infinitely large. Then $(A_{1+\sum A^2}, P_{1+\sum A^2})$ has the weak bounded inversion property, but bounded inversion fails since $1 < T$ and $T \notin A^\times_{1+\sum A^2}$.

3 Totally ordered and real closed fields

The classes of totally ordered fields and of real closed fields are of great importance for each of the categories **PREOR**, **POR** and **POR/N** The corresponding full subcategories of **PREOR** are denoted by **TOF** and **RCF**. The algebraic theory of these fields is treated exhaustively in a large number of references (cf. [19], Chapitre 1; [20], Chapitre 6, §2; [68], Chapter VI; [73], Kapitel I; [97]; [101], Kapitel II, §2); model theoretic aspects are discussed in [30]; [97]; [98]. We have nothing new to add to the existing theory. The purpose of our discussion is to fix some notation and to remind the reader of a few facts.

A field is *formally real* if it can be totally ordered. If it is formally real but has no proper algebraic extension that is also formally real then the field is *real closed*. Real closed fields have a unique total order, the positive cone is the set of squares. A relatively algebraically closed subfield of a real closed field is also real closed. Every totally ordered field (K, T) has a **TOF**–embedding in a real closed field R that is algebraic over K. This embedding is unique up to a unique isomorphism; it is called the *real closure* of (K, T). Using the language of section 2, the content of these remarks can be rephrased by saying that **RCF** is a monoreflective subcategory of **TOF**. Every real closed field contains an isomorphic copy of the field R_0 of real algebraic numbers, i.e., R_0 is the prime field in the class of real closed fields. The class of real closed fields has the *strong amalgamation property*: If $\varphi_1 \colon R \to R_1$ and $\varphi_2 \colon R \to R_2$ are embeddings of a real closed field R into two real closed fields R_1 and R_2 then there exists a real closed field $\overline{R}$ together with embeddings $\psi_1 \colon R_1 \to \overline{R}$, $\psi_2 \colon R_2 \to \overline{R}$ such that $\psi_1\varphi_1 = \psi_2\varphi_2$ and $\psi_1(R_1) \cap \psi_2(R_2) = \psi_1\varphi_1(R)$.

Incidentally, the real closure shows that there is a huge difference between epimorphisms in the categories **R** and **POR**: The inclusion $i \colon (K, T) \to (R, R^2)$ of a totally ordered field in its real closure is always epimorphic in **POR**. In fact, it will be shown in Proposition 16.1 that a homomorphism $f \colon (K, T) \to (K', T')$ of totally ordered fields is an epimorphism in **POR** if and only if it is so in **TOF**. From Artin–Schreier theory it is clear that it is an epimorphism in **TOF** if and only if K' is algebraic over $f(K)$. But the underlying **R**–morphism is an epimorphism if and only if it is an isomorphism.

For the discussion of the model theory of totally ordered and real closed fields we use the language of porings (cf. section 1). In the theory of porings every atomic formula is equivalent to one of the form $f(X_1,\ldots,X_n) = 0$ or $f(X_1,\ldots,X_n) \geq 0$, where $f(X_1,\ldots,X_n)$ is an integer polynomial with variables $X_1,\ldots,X_n$. As a defined relation we also use the symbol $<$, the definition of $f < g$ being $(f \neq g)\&(f \leq g)$. The *theory of totally ordered fields* and the *theory of real closed fields* are obtained from the theory of porings by adjoining the pertinent axioms. The class of real closed fields can also be axiomatized using the language of ring theory since the order relation can be defined by algebraic means. However, without the richer language of porings we do not have elimination of quantifiers (cf. [97], p. 72/73).

Elimination of quantifiers says that for any first order formula $\Theta(X_1,\ldots,X_n)$ (with free variables $X_1,\ldots,X_n$) in the language of porings there is another formula $\Psi(X_1,\ldots,X_n)$ with the same free variables and without quantifiers such that the following formula is a theorem for real closed fields:

$$\forall X_1,\ldots,X_n \ (\Theta(X_1,\ldots,X_n) \leftrightarrow \Psi(X_1,\ldots,X_n)).$$

Since $\mathbb{Z}$ with its natural total order is a prime substructure of the class of real closed fields it follows from quantifier elimination that the theory of real closed fields is complete and that any two real closed fields are elementarily equivalent.

Let $\Theta = \Theta(X_1,\ldots,X_m, Y_1,\ldots,Y_n)$ be a formula in the language of porings with free variables $X_1,\ldots,X_m, Y_1,\ldots,Y_n$. Let R be a real closed field and pick $a_1,\ldots,a_n \in R$. A set of the form

$$S = \{(x_1,\ldots,x_m) \in R^m; \ R \models \Theta(x_1,\ldots,x_m,a_1,\ldots,a_n)\}$$

is said to be *semi–algebraic over R*. By elimination of quantifiers, choose a quantifier–free formula $\Psi = \Psi(X_1,\ldots,X_m, Y_1,\ldots,Y_n)$ that is equivalent to Θ in the theory of real closed fields. The formula Ψ can be written as a disjunction of conjunctions of atomic formulas or negations thereof, i.e., of formulas $f = 0$, $f \geq 0$, $f > 0$ where $f \in \mathbb{Z}[X_1,\ldots,X_m,Y_1,\ldots,Y_n]$. Therefore the semi–algebraic set

$$S = \{(x_1,\ldots,x_m) \in R^m; \ R \models \Psi(x_1,\ldots,x_m,a_1,\ldots,a_n)\}$$

is a finite union of finite intersections of sets of the form

$$\{(x_1,\ldots,x_m) \in R^m; \ f(x_1,\ldots,x_m)\varepsilon 0\}$$

where ε is one of the symbols $=, \geq, >$ and $f \in R[X_1,\ldots,X_n]$.

4 Real spectra of preordered rings

The real spectrum is one of our most important tools. Generel references for the real spectrum of a ring are [19], Chapitre 7, and [73], Kapitel III. We shall present a slight modification of the real spectrum which takes into account partial orders and preorderings. Also, we need to describe the notation we shall use when dealing with the real spectrum.

Let (A, P) be a preordered ring. A *prime cone* of (A, P) is a preordering α having the following properties:

(4.1)

$$
\begin{aligned}
&(a) \quad \alpha \cup -\alpha = A; \\
&(b) \quad -1 \notin \alpha; \\
&(c) \quad supp(\alpha) \text{ is a prime ideal}; \\
&(d) \quad P \subseteq \alpha.
\end{aligned}
$$

The set of prime cones of (A, P) is denoted by $Sper(A, P)$ and is called the *real spectrum* of (A, P). The *real spectrum* of the ring A, denoted by $Sper(A)$, is a spectral space in the sense of [64]. Its elements are the preorderings of A having the properties (a)–(c) of (4.1). Being a spectral space, it comes equipped with two topologies. The *spectral topology* has the sets

$$pos(a) = \{\alpha \in Sper(A); \; -a \notin \alpha\}, \; a \in A,$$

as a subbasis of open sets. It is important to note that this subbasis is not a basis, in general. However, see the discussion of f–rings at the end of this section. A subset of $Sper(A)$ is *constructible* if it belongs to the Boolean algebra of subsets generated by $\{pos(a); a \in A\}$. The constructible sets are the basis of another topology which is called the *constructible topology*. When we talk about *the* topology of a spectral space we always mean the spectral topology. A subset of a spectral space is *proconstructible* if it is closed with respect to the constructible topology. If the topologies of a spectral space are restricted to a proconstructible subset then the subset is a spectral space, too. Since the definition shows immediately that $Sper(A, P) \subseteq Sper(A)$ is a closed subset it follows that $Sper(A, P)$ is proconstructible, hence it is a spectral space (with the induced topologies). If $f\colon (A, P) \to (B, Q)$ is a **PREOR**–morphism then $f^{-1}(\beta)$ is a prime cone of (A, P) for each $\beta \in Sper(B, Q)$. The map $Sper(f)\colon Sper(B, Q) \to Sper(A, P)$ defined in this manner is a morphism of spectral spaces (i.e., it is continuous in both topologies). Altogether, the real spectrum is a functor from **PREOR** to the category of spectral spaces.

The following notation will be used troughout. If $\alpha \in Sper(A, P)$ then

- $supp_{(A,P)}(\alpha) = supp(\alpha)$ is the support of α;
- $A/\alpha = (A/supp(\alpha), \alpha/supp(\alpha))$ is the totally ordered residue domain;
- $\pi_\alpha \colon (A, P) \to A/\alpha$ is the canonical homomorphism;
- $\kappa(\alpha) = \kappa_{(A,P)}(\alpha)$ is the totally ordered quotient field of A/α (also called the *residue field* of A at α);
- $\kappa_\alpha \colon (A, P) \to \kappa(\alpha)$ is the canonical homomorphism;
- $\rho(\alpha) = \rho_{(A,P)}(\alpha)$ is the real closure of $\kappa(\alpha)$, called the *real closed residue field* of A at α;
- $\rho_\alpha \colon (A, P) \to \rho(\alpha)$ is the canonical homomorphism;
- $a(\alpha)$ is the image of $a \in A$ in any one of A/α, $\kappa(\alpha)$ and $\rho(\alpha)$;
- $V(\alpha)$ is the convex hull of A/α in $\kappa(\alpha)$.
- $\overline{V}(\alpha)$ is the convex hull of A/α in $\rho(\alpha)$.

Note that $supp_{(A,P)}$ is not only a map from the real spectrum of (A, P) into the prime spectrum of A, it is even a morphism of spectral spaces.

If $\alpha \in Sper(A, P)$ then the homomorphism

$$\rho_\alpha : (A, P) \xrightarrow{id} (A, \alpha) \to A/\alpha \to \kappa(\alpha) \to \rho(\alpha)$$

is a **PREOR**-morphism into a real closed field. The real spectrum of $\rho(\alpha)$ consists of a single point which is mapped to α by $Sper(\rho_\alpha)$. Let $f \colon (A, P) \to R$ be any **PREOR**-morphism into a real closed field. Then $f^{-1}(R^2) \subseteq A$ is always a prime cone of (A, P). If $f \colon (A, P) \to (B, Q)$ is a **PREOR**-morphism and if $\beta \in Sper(B, Q)$ then f induces an embedding $f/\beta \colon A/f^{-1}(\beta) \to B/\beta$ of totally ordered domains. This extends to a homomorphism $\kappa_{f/\beta} \colon \kappa_{(A,P)}(f^{-1}(\beta)) \to \kappa_{(B,Q)}(\beta)$ of the totally ordered quotient fields and further to a homomorphism $\rho_{f/\beta} \colon \rho_{(A,P)}(f^{-1}(\beta)) \to \rho_{(B,Q)}(\beta)$ of their real closures. For $a \in A$ we note that

$$\rho_{f/\beta}(a(f^{-1}(\beta))) = f(a)(\beta).$$

Let (A, P) be a preordered ring and let $S \subseteq A$ be a multiplicative subset. Since $i_S \colon (A, P) \to (A_S, P_S)$ is a **PREOR**-morphism we have the functorial map $Sper(i_S)$ of the real spectra. It is a routine matter to show that this is a homeomorphism onto the proconstructible subset

$$\{\alpha \in Sper(A, P); \ S \cap supp(\alpha) = \emptyset\}$$

of $Sper(A, P)$. Similarly, if $I \subseteq A$ is a convex ideal then the canonical surjection $\pi_I : (A, P) \rightarrow (A/I, P + I/I)$ yields a homeomorphism onto the closed subset

$$\{\alpha \in Sper(A, P);\ I \subseteq supp(\alpha)\}.$$

The support of every prime cone is convex in (A, P). Therefore $Sper(\pi_{supp(P)})$ is a homeomorphism, and the real spectra of the pre-ordered ring (A, P) and of the poring $(A/supp(P), P/supp(P))$ can be identified canonically.

The constructible subsets of $Sper(A, P)$ are the sets that are closed and open with respect to the constructible topology, or, alternatively, the restrictions of constructible subsets of $Sper(A)$. They can also be described in terms of formulas using the language of porings.

Proposition 4.2 (cf. [33], Definition and Proposition 2.2) Let $\Theta(X_1, \dots, X_n)$ be a formula in the language of porings, with free variables $X_1, \dots, X_n$, and let $a_1, \dots, a_n \in A$. Then the set

$$C = \{\alpha \in Sper(A, P);\ \rho(\alpha) \models \Theta(a_1(\alpha), \dots, a_n(\alpha))\}$$

is constructible. Every constructible subset of $Sper(A, P)$ is of this form.

Proof If C is constructible then $C = C_1 \cup \dots \cup C_r$ where

$$C_i = \{\alpha \in Sper(A, P);\ \forall\, j = 1, \dots, s_i : a_{ij}(\alpha)\varepsilon_{ij}0\}$$

(with $\varepsilon_{ij} \in \{=, \geq, >\}$ and $a_{ij} \in A$). The set can be defined by substituting a_{ij} for X_{ij} in the formula

$$\bigvee_{i=1}^{r} \bigwedge_{j=1}^{s_i} (X_{ij}\varepsilon_{ij}0).$$

Conversely, suppose that C is defined by a formula. By elimination of quantifiers, in the theory of real closed fields the formula $\Theta(X_1, \dots, X_n)$ is equivalent to a quantifier–free formula $\Psi(X_1, \dots, X_n)$. This formula can be rewritten as

$$\bigvee_{i=1}^{r} \bigwedge_{j=1}^{s_i} (f_{ij}(X_1, \dots, X_n)\varepsilon_{ij}0)$$

where $f_{ij} \in \mathbb{Z}[X_1, \dots, X_n]$. Substituting a_k for X_k and defining $a_{ij} = f_{ij}(a_1, \dots, a_n) \in A$ one sees that

$$C = \bigcup_{i=1}^{r} \bigcap_{j=1}^{s_i} \{\alpha \in Sper(A, P); \; a_{ij}(\alpha)\varepsilon_{ij}0\}$$

is constructible. □

There are preordered rings whose real spectrum is empty. This is the case whenever the preordering is the entire ring. In fact, in all other cases the real spectrum is nonempty. For, if $P \subset A$ then $-1 \notin P$. Using Zorn's Lemma we find a preordering that contains P and is maximal with the property of not containing -1. As in [73], proof of Satz 1, it is easy to show that this maximal preordering is a prime cone. In fact, one can prove a bit more than this:

Proposition 4.3 If (A, P) is a preordered ring such that $-1 \notin P$ then there exist convex prime ideals in (A, P), and every one of them is the support of some prime cone.

Proof If $P \subset A$ then the poring $(\overline{A}, \overline{P}) = (A/P \cap -P, P/P \cap -P)$ is not the zero ring. Because of Proposition 1.4 (c) it suffices to deal with $(\overline{A}, \overline{P})$. Since $Nil(\overline{A}) \subset \overline{A}$ is a proper convex ideal there exist convex prime ideals in $\overline{A}$ ([27], Proposition 2.3.6). If $p \subseteq \overline{A}$ is one of these then $(\overline{A}/p, \overline{P} + p/p)$ is a *podomain* (= *partially ordered domain*). According to [27], Proposition 1.6.1 (c), every maximal positive cone of $\overline{A}/p$ that contains $\overline{P} + p/p$ is a total order, hence defines a prime cone $\beta \subseteq \overline{A}$ with $supp(\beta) = p$. □

It was noted in section 1 that the nilradical of a poring is the intersection of all convex prime ideals. Together with the proposition, this implies that every reduced poring can be mapped monomorphically into a direct product of totally ordered fields, say $f: (A, P) \to \prod(K_i, T_i)$. Since f is injective every minimal prime ideal of A is the restriction of some minimal prime ideal of $\prod(K_i, T_i)$ ([21], Chap. II, §2 No.6, Proposition 16). But in the reduced f–ring $\prod(K_i, T_i)$ minimal prime ideals are convex ([17], Théoréme 9.3.2). Then the restriction to A is convex as well. This remark proves the following supplement to Proposition 4.3:

Proposition 4.4 The minimal prime ideals of every poring are convex.$\square$

We conclude the section by pointing out a few special properties of f-rings with regard to the real spectrum. So, suppose that (A, P) is an f-ring. First of all, we note that the subbasis

$$pos(a) = \{\alpha \in Sper(A, P); \; a(\alpha) > 0\}, \; a \in A,$$

of the spectral topology of $Sper(A, P)$ is actually a basis. For, one checks easily that

$$pos(a_1) \cap \ldots \cap pos(a_n) = pos(a_1 \wedge \ldots \wedge a_n).$$

By the remarks at the end of section 1, the support map $supp: Sper(A, P) \to Spec(A)$ is injective and its image is

$$\{p \in Spec(A); \; p \subseteq (A, P) \text{ convex }\},$$

a set which is sometimes called the *Brumfiel spectrum* (cf. [70], p. 208) and is denoted by $Spec_B(A, P)$. The Brumfiel spectrum is a proconstructible subset of the Zariski spectrum, hence it is a spectral subspace of $Spec(A)$. The support is actually a homeomorphism onto the Brumfiel spectrum since $supp(pos(b)) = D(b \vee 0) = \{p \in Spec_B(A, P); b \vee 0 \notin p\}$ (see also [44], Proposition 3.2).

5 Epimorphisms of reduced porings

We pointed out before that recognizing epimorphisms is a problem in many categories. In this section, we shall present a concrete characterization of epimorphisms in **POR/N**. This result is fundamental for everything that follows. In spirit, it is similar to the characterization of **R**–epimorphisms in [78], Proposition 1.5. Our result is not much more than a restatement of [111], Proposition II 2.13, or [112], Theorem 1; only the context is more general.

Suppose that $f_1, f_2 : (A, P_A) \to (B, P_B)$ are two parallel **POR/N**–morphisms. If $f_1 \neq f_2$ then it follows from Proposition 4.4 that there is some **POR/N**–morphism $g : (B, P_B) \to R$ into a real closed field such that $gf_1 \neq gf_2$. Thus, a **POR/N**–morphism $f : (A, P_A) \to (B, P_B)$ is epimorphic if and only if $g_1 f = g_2 f$ implies $g_1 = g_2$ for all **POR/N**–morphisms $g_1, g_2 : (B, P_B) \to R$ into real closed fields. Or, in other workds, the real closed fields form a *cogenerating class* in **POR/N** (cf. [83], p. 123). Before proving the main result it is useful to settle the following special case:

Lemma 5.1 If $\alpha \in Sper(A, P)$ then $\rho_\alpha : (A, P) \to \rho(\alpha)$ is epimorphic in **POR/N**.

Proof Recall that ρ_α is the composition

$$(A, P) \to A/\alpha \to \kappa(\alpha) \to \rho(\alpha)$$

of canonical homomorphisms. Since $(A, P) \to A/\alpha$ is surjective it is epimorphic. The inclusion of A/α in its quotient field is epimorphic even in **R**, hence it is epimorphic in **POR/N**. Thus ρ_α is epimorphic if and only if so is the inclusion $\kappa(\alpha) \to \rho(\alpha)$. But this follows from the uniqueness of real closures *up to unique isomorphism* together with the fact that the real closed fields are a cogenerating class in **POR/N**. $\square$

Theorem 5.2 The **POR/N**–morphism $f : (A, P) \to (B, Q)$ is an epimorphism if and only if the following two conditions hold:

(i) the functorial map $Sper(f)$ is injective;

(ii) for   each β $\in$ $Sper(B,Q)$ the   homomorphism
$\rho_{f/\beta} \colon \rho_{(A,P)}(f^{-1}(\beta)) \to \rho_{(B,Q)}(\beta)$ is an isomorphism.

Proof   First assume that f is epimorphic. If (i) fails then there are
$\beta, \gamma \in Sper(B,Q)$ with $\beta \neq \gamma$ and $f^{-1}(\beta) = \alpha = f^{-1}(\gamma)$. By the strong
amalgamation property there is a commutative diagram

$$
\begin{array}{ccccc}
 & \xrightarrow{\ \rho_{f/\beta}\ } & \rho_{(B,Q)}(\beta) & \xrightarrow{\ \psi_\beta\ } & \\
\rho_{(A,P)}(\alpha) & & & & R \\
 & \xrightarrow{\ \rho_{f/\gamma}\ } & \rho_{(B,Q)}(\gamma) & \xrightarrow{\ \psi_\gamma\ } &
\end{array}
$$

of real closed fields. The homomorphisms

$$
A \xrightarrow{\ f\ } B \xrightarrow{\ \rho_\beta\ } \rho_{(B,Q)}(\beta) \xrightarrow{\ \psi_\beta\ } R,
$$
$$
A \xrightarrow{\ f\ } B \xrightarrow{\ \rho_\gamma\ } \rho_{(B,Q)}(\gamma) \xrightarrow{\ \psi_\gamma\ } R
$$

coincide. Since f is epimorphic it follows that the homomorphisms

$$
B \xrightarrow{\ \rho_\beta\ } \rho_{(B,Q)}(\beta) \xrightarrow{\ \psi_\beta\ } R,
$$
$$
B \xrightarrow{\ \rho_\gamma\ } \rho_{(B,Q)}(\gamma) \xrightarrow{\ \psi_\gamma\ } R
$$

also agree. But this is impossible since $\beta \neq \gamma$; thus we have reached a contradiction, proving (i). For the proof of (ii) we also use the strong amalgamation property. Supposing that (ii) fails we have some $\beta \in Sper(B,Q)$
such that $\rho_{f/\beta} \colon \rho_{(A,P)}(f^{-1}(\beta)) \to \rho_{(B,Q)}(\beta)$ is a transcendental extension.
There exists a commutative diagram

$$
\begin{array}{ccccc}
 & \xrightarrow{\ \rho_{f/\beta}\ } & \rho_{(B,Q)}(\beta) & \xrightarrow{\ \psi_1\ } & \\
\rho_{(A,P)}(f^{-1}(\beta)) & & & & R \\
 & \xrightarrow{\ \rho_{f/\beta}\ } & \rho_{(B,Q)}(\beta) & \xrightarrow{\ \psi_2\ } &
\end{array}
$$

of real closed fields such that

$$
\psi_1(\rho_{(B,Q)}(\beta)) \cap \psi_2(\rho_{(B,Q)}(\beta)) = \psi_1 \rho_{f/\beta}(\rho_{(A,P)}(f^{-1}(\beta))).
$$

In particular, this means that $\psi_1 \neq \psi_2$. On the other hand,

$$
\psi_1 \rho_\beta f = \psi_1 \rho_{f/\beta} \rho_{f^{-1}(\beta)} = \psi_2 \rho_{f/\beta} \rho_{f^{-1}(\beta)} = \psi_2 \rho_\beta f
$$

implies that $\psi_1 = \psi_2$ because $\rho_\beta f$ is an epimorphism (Lemma 5.1). Thus, we have a contradiction, and (ii) has been proved.

Now suppose that both (i) and (ii) are satisfied. Pick two homomorphisms $g_1, g_2 \colon (B, Q) \to R$ into a real closed field such that $g_1 f = g_2 f$. If $Sper(R) = \{\gamma\}$ then $Sper(g_1)(\gamma) = Sper(g_2)(\gamma)$ because $Sper(f)$ is injective and $Sper(g_1 f)(\gamma) = Sper(g_2 f)(\gamma)$. Set $\beta = Sper(g_1)(\gamma)$. Now g_1 and g_2 factor as $g_1 = h_1 \rho_\beta$, $g_2 = h_2 \rho_\beta$ with $h_1, h_2 \colon \rho(\beta) \to R$. Since $\rho_{f/\beta} \rho_{f^{-1}(\beta)}$ is an epimorphism (Lemma 5.1 and hypothesis (ii)) we learn from

$$h_1 \rho_{f/\beta} \rho_{f^{-1}(\beta)} = h_1 \rho_\beta f = h_2 \rho_\beta f = h_2 \rho_{f/\beta} \rho_{f^{-1}(\beta)}$$

that $h_1 = h_2$. Because ρ_β is an epimorphism (Lemma 5.1) this implies that $g_1 = g_2$, and the proof is finished. $\qquad\square$

Remark 5.3 Suppose that $\mathbf{C}$ is a category and $\mathbf{D} \subseteq \mathbf{C}$ is a subcategory. A $\mathbf{D}$–morphism which is epimorphic in $\mathbf{C}$ is always epimorphic in $\mathbf{D}$. But the converse is false, in general. For example, in any subcategory of $\mathbf{POR/N}$ the conditions (i) and (ii) of the theorem are sufficient for a morphism to be epimorphic. But they are not always necessary, e.g., in the category of archimedean porings there are epimorphisms which do not satisfy (i) and (ii). On the other hand, in every *full* subcategory of $\mathbf{POR/N}$ that *contains all real closed fields*, the conditions characterize epimorphisms. For instance, $\mathbf{FR/N}$, $\mathbf{BIPOR/N}$ and $\mathbf{WBIPOR/N}$ are such subcategories. The same is true for almost all other subcategories that we are going to encounter. $\qquad\square$

In this work we are mostly interested in functorial constructions involving epimorphisms of porings. An investigation of epimorphisms emphasizing other aspects may be found in [117].

6 Functions and representable porings

With any preordered ring (A, P) we associate its *complete ring of functions* on the real spectrum, defined as

$$\Pi(A, P) = \prod_{\alpha \in Sper(A,P)} \rho(\alpha).$$

We shall see that the construction of the complete ring of functions is functorial, but it is not a reflector. These rings are important for our purposes since practically every reflection of (A, P) that we encounter in this work will be contained in $\Pi(A, P)$. When we start to compare various reflectors the task will be greatly facilitated by the fact that all reflections of (A, P) are contained in one ring. As in [3], p. 30, the elements of $\Pi(A, P)$ will be called *functions*. If f is a function then $f(\alpha)$ is the α–th component of f. The ring $\Pi(A, P)$ is von Neumann regular. The set

$$\Pi^+(A, P) = \{f \in \Pi(A, P); \ \forall \, \alpha \colon f(\alpha) \geq 0\}$$

is a positive cone. The poring $(\Pi(A, P), \Pi^+(A, P))$ is an f–ring, even a reduced f–ring since it is von Neumann regular. The positive cone consists exactly the squares of $\Pi(A, P)$. In fact, this is the only positive cone of $\Pi(A, P)$. For, if Q is any positive cone then Q contains all squares, i.e., $\Pi^+(A, P) \subseteq Q$. Since $\Pi^+(A, P)$ is the positive cone of a reduced f–ring it does not have any proper refinement (Proposition 1.11), hence $\Pi^+(A, P) = Q$. Notationally we do not distinguish between $\Pi(A, P)$ as a ring and as a poring. Note that this does not create any problems when dealing with the real spectrum. For, any prime cone α of the *ring* $\Pi(A, P)$ contains the squares, hence is a prime cone of the *poring* $\Pi(A, P)$. So, the real spectra of the ring $\Pi(A, P)$ and of the poring $\Pi(A, P)$ agree.

For every $\alpha \in Sper(A, P)$ the homomorphism $\rho_\alpha \colon (A, P) \to \rho(\alpha)$ is a **PREOR**–morphism. Altogether these morphisms define a **PREOR**–morphism $\Delta = \Delta_{(A,P)} \colon (A, P) \to \Pi(A, P)$. If $f \colon (A, P) \to (B, Q)$ is a **PREOR**–morphism then $\Pi(f) \colon \Pi(A, P) \to \Pi(B, Q)$ defined by

$$(\Pi(f)(a))(\beta) = \rho_{f/\beta}(a(f^{-1}(\beta)))$$

is a **PREOR**–morphism and $\Pi(f)\Delta_{(A,P)} = \Delta_{(B,Q)}f$. Thus, the construction of complete rings of functions is a functor $\Pi \colon \mathbf{PREOR} \to \mathbf{POR/N} \subseteq \mathbf{PREOR}$ and $\Delta \colon Id \to \Pi$ is a natural transformation. The homomorphism $\Delta_{(A,P)} \colon (A, P) \to \Pi(A, P)$ is injective if and only if (A, P) is a reduced poring (Proposition 4.4).

None of the properties of the functor Π and the natural transformation Δ exhibited so far give any indication that we are not dealing with a reflector. But we shall show this momentarily. First note that the construction of Π is not idempotent. For, suppose that (A, P) is a preordered ring with countable real spectrum. Such a ring is, for example, $(\mathbb{R}[T], P)$ with $P = \{a \in \mathbb{R}[T]; \; \forall z \in \mathbb{Z} \colon a(z) \geq 0\}$ (its real spectrum is the 2-point compactification of the discrete space $\mathbb{Z}$). Then $\Pi(A, P)$ is a countable direct product of real closed fields. The maximal ideals of $\Pi(A, P)$ correspond to the ultrafilters on the set $Sper(A, P)$. Each residue field modulo a maximal ideal is real closed. Hence the real spectrum of $\Pi(A, P)$ may be identified with $Spec(\Pi(A, P))$. Because this is an uncountable set, $\Pi(\Pi(A, P))$ is a direct product of uncountably many real closed fields. But then $\Pi(A, P)$ and $\Pi(\Pi(A, P))$ cannot be isomorphic. The same ring can be used to show that Π is not a reflector: Assume for the moment that Π is a reflector with a natural transformation τ whose components are the reflection morphisms. Every real closed field belongs to the reflective subcategory corresponding to Π (since Π does not change real closed fields). Thus, for each $\alpha \in Sper(A, P)$ the canonical homomorphism $\rho_\alpha \colon (A, P) \to \rho(\alpha)$ factors uniquely as shown by the diagram:

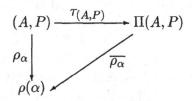

Let $\varphi(\alpha)$ denote the point of $Sper(\Pi(A, P))$ that is defined by $\overline{\rho_\alpha}$. Thus, we have a map $\varphi \colon Sper(A, P) \to Sper(\Pi(A, P))$. For cardinality reasons, φ cannot be surjective. So, pick $\gamma \in Sper(\Pi(A, P)) \backslash im(\varphi)$ and let $\beta \in Sper(A, P)$ be the prime cone determined by the homomorphism

$$(A, P) \xrightarrow{\ \tau_{(A,P)}\ } \Pi(A, P) \xrightarrow{\ \rho_\gamma\ } \rho(\gamma)$$

Then there is a unique homomorphism $i \colon \rho(\beta) \to \rho(\gamma)$ such that the outer square and the upper triangle in the diagram

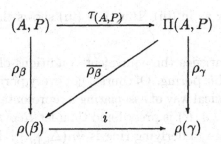

are commutative. The lower triangle is not commutative since $\gamma \notin im(\varphi)$. Now $i\rho_\beta = \rho_\gamma \tau_{(A,P)}$ is a homomorphism from (A,P) into a real closed field. By the universal property of the reflection $\tau_{(A,P)}$ it factors *uniquely* through $\Pi(A,P)$. However, we already know that

$$i\overline{\rho_\beta}\tau_{(A,P)} = i\rho_\beta = \rho_\gamma \tau_{(A,P)}$$

with $i\overline{\rho_\beta} \neq \rho_\gamma$. This contradiction shows that Π is not a reflector.

We noted earlier that $\Delta_{(A,P)}$ is injective if and only if (A,P) is a reduced poring. Even if $\Delta_{(\Delta,P)}$ is injective, usually it will not be an embedding. This will be illustrated by Examples 6.3 and 6.4 below. Here are some useful criteria to decide whether it is an embedding:

Proposition 6.1 For the poring (A,P) the following conditions are equivalent:

(i) $\Delta_{(A,P)}$ is an embedding.
(ii) (A,P) is a subporing of a product of totally ordered fields.
(iii) An element $a \in A$ belongs to P if it satisfies an identity $a^{2e} - ba + c = 0$, with $e \in \mathbb{N}$ and $b,c \in P$.

Proof The implication **(i)** $\Rightarrow$ **(ii)** is trivial.   –   **(ii)** $\Rightarrow$ **(iii)**   By hypothesis there is a family $(K_i, T_i)_{i \in I}$ of totally ordered fields such that $(A,P) \subseteq \prod(K_i, T_i)$ is a subporing. Pick $a = (a_i)_i \in A$ and suppose that $a^{2e} - ba + c = 0$ with $b,c \in P$. For every $i \in I$ we have $b_i \geq 0$ and $c_i \geq 0$. If $a_i < 0$ then $a_i^{2e} - b_i a_i + c_i > 0$ in K_i, which is impossible because of the identity. We conclude that $a_i \geq 0$ for every $i \in I$, hence $a \in P$.   –   **(iii)** $\Rightarrow$ **(i)**   If $a \in A$ is nilpotent then there is some $1 \leq e \in \mathbb{N}$ such that $a^{2e} = (-a)^{2e} = 0$. This identity implies that $a, -a \in P$, hence $a = 0$. Thus the ring A is reduced, and Proposition 4.4 shows that $\Delta_{(A,P)}$ is injective. The Nichtnegativstellensatz ([73], p. 143, Theorem 5) applies to the present situation: If $a(\alpha) \geq 0$ for every $\alpha \in Sper(A,P)$ then a

satisfies an identity as in (iii). But then (iii) shows that $a \in P$.   □

Any poring that satisfies the equivalent conditions of Proposition 6.1 is called a *representable poring*. Of course, not every poring is representable. But there is a canonical way of associating a representable poring with any preordered ring. If (A, P) is preordered then let $Rep(A, P) \subseteq \Pi(A, P)$ be the *subporing* whose underlying ring is $im(\Delta_{(A,P)})$. For any **PREOR**–morphism $f \colon (A, P) \to (B, Q)$ the homomorphism $\Pi(f) \colon \Pi(A, P) \to \Pi(B, Q)$ restricts to a homomorphism $Rep(f) \colon Rep(A, P) \to Rep(B, Q)$. The homomorphisms obtained from $\Delta_{(A,P)}$ and $\Delta_{(B,Q)}$ by restricting the codomains to $Rep(A, P)$ and $Rep(B, Q)$ are denoted by $Rep_{(A,P)}$ and $Rep_{(B,Q)}$. Then $Rep(f)Rep_{(A,P)} = Rep_{(B,Q)}f$. Since $Rep_{(A,P)}$ is surjective, hence epimorphic, the homomorphism $Rep(f)$ is uniquely determined by this property. The full subcategory of **PREOR** whose objects are the representable porings is denoted by **REPPOR**. It is a subcategory of **POR/N**. The construction of the representable poring $Rep(A, P)$ is a functor **PREOR** $\to$ **REPPOR**; Δ is a natural transformation from $Id_{\textbf{PREOR}}$ to Rep followed by the inclusion functor **REPPOR** $\to$ **PREOR**.

Proposition 6.2 The restriction of Rep to a functor **POR/N** $\to$ **REPPOR** is a monoreflector with reflections $Rep_{(A,P)}$.

Proof Most of the work has already been done. We only need to check the universal property (2.1) for the reflection morphisms. So, let $f \colon (A, P) \to (B, Q)$ be a **POR/N**–morphism into a representable poring. We have a commutative diagram

$$
\begin{array}{ccc}
(A, P) & \xrightarrow{\ f\ } & (B, Q) \\
{\scriptstyle Rep_{(A,P)}} \downarrow & & \downarrow {\scriptstyle Rep_{(B,Q)}} \\
Rep(A, P) & \xrightarrow{\ Rep(f)\ } & Rep(B, Q)
\end{array}
$$

in which $Rep_{(B,Q)}$ is an isomorphism, hence $f = Rep_{(B,Q)}^{-1} Rep(f) Rep_{(A,P)}$. Since $Rep_{(A,P)}$ is epimorphic, $Rep_{(B,Q)}^{-1} Rep(f)$ is the only homomorphism $\overline{f} \colon Rep(A, P) \to (B, Q)$ such that $f = \overline{f} Rep_{(A,P)}$. Thus, the universal property (2.1) is satisfied, and **REPPOR** is indeed reflective in **POR/N**. It is monoreflective since the reflection homomorphisms are clearly injective. □

Note that the real rings are exactly the underlying rings of representable porings. In general, a real ring may support many representable porings. This can be seen by looking at fields: If A is a formally real field then any positive cone P of a partial order is an intersection of positive cones of total orders ([97], Corollary 1.6), hence (A, P) is representable. The real spectrum of (A, P) is its space of orderings ([97], §6). For every positive cone P, $Sper(A, P) \subseteq Sper(A)$ is a closed subset. In the extreme case that A is an SAP–field (cf. [97], §9) there is actually a bijective correspondence between positive cones of A and closed subsets of $Sper(A)$.

We finish the general discussion of representable porings by giving two examples of reduced porings that are not representable, together with their representable reflections.

Example 6.3 *Hilbert's 17th problem*
Let $(A, P_w(A))$ be the polynomial ring over $\mathbb{R}$ in the two variables T_1, T_2 with its weak order. Hilbert was the first to observe that $(A, P_w(A))$ is not representable ([61]). The positive cone of $Rep(A, P_w(A))$ consists of the positive semidefinite polynomials. A positive semidefinite polynomial that is not in $P_w(A)$ was explicity exhibited by Motzkin ([93], p. 217), namely

$$1 + T_1^2 T_2^4 + T_1^4 T_2^2 - 3T_1^2 T_2^2.$$

Hilbert's observation led to his 17th problem. Solving the problem, Artin showed that the positive cone of $Rep(A, P_w(A))$ is $A \cap P_w(qf(A))$ ([6]).□

Example 6.4 *A partially ordered integral domain that is not representable*
Let $A = \mathbb{R}[T]$ and define a positive cone P by $\sum_{i=m}^{n} a_i T^i > 0$ (where $a_m \neq 0$) if and only if $a_m > 0$ and $m = 0$ or $m \geq 2$. Since $T \notin P$, but $T^3 \in P$ this cannot be a representable poring. It is easy to check that $Rep(A, P)$ is the ring A together with the total order defined by $\sum_{i=m}^{n} a_i T^i > 0$ (where $a_m \neq 0$) if and only if $a_m > 0$. □

In section 2. we postponed the proof that **FR/N** is a monoreflective subcategory of **POR/N**. Now we shall prove this using the complete rings of functions.

Proposition 6.5 The subcategory **FR/N** is monoreflective in **REP-POR**, hence also in **POR/N**.

Proof First of all, recall that reduced f–rings are representable by their very definition. The method of proof is to describe the morphism that will eventually be the reflection morphism of $(A, P) \in ob(\textbf{REPPOR})$ and then to check the universal property (2.1). Via $\Delta_{(A,P)}$, (A, P) is embedded into $\Pi(A, P)$. The set of sub–f–rings of $\Pi(A, P)$ containing $im(\Delta_{(A,P)})$ is nonempty (since $\Pi(A, P)$ belongs to the set). Let $\varphi(A, P)$ be the intersection of all these sub–f–rings. One checks easily that $\varphi(A, P)$ is a sub–f–ring of $\Pi(A, P)$, hence it is the smallest one containing $im(\Delta_{(A,P)})$. The morphism obtained from $\Delta_{(A,P)}$ by restricting the codomain to $\varphi(A, P)$ is denoted by $\varphi_{(A,P)}$. We claim that $\varphi_{(A,P)}$ has the universal property (2.1). So, let $f: (A, P) \to (B, Q)$ be any **REPPOR**–morphism into an **FR/N**–object. The functor Π and the natural transformation Δ supply us with the following commutative diagram:

$$
\begin{array}{ccc}
(A, P) & \xrightarrow{\;\;f\;\;} & (B, Q) \\
{\scriptstyle \Delta_{(A,P)}} \downarrow & & \downarrow {\scriptstyle \Delta_{(B,Q)}} \\
\Pi(A, P) & \xrightarrow{\;\;\Pi(f)\;\;} & \Pi(B, Q)
\end{array}
$$

Proposition 1.11 implies that $\Delta_{(B,Q)}$ is an isomorphism onto a sub–f–ring. Then $\Pi(f)^{-1}(im(\Delta_{(B,Q)}))$ is a sub–f–ring of $\Pi(A, P)$ containing $im(\Delta_{(A,P)})$, i.e., it contributes to the intersection defining $\varphi(A, P)$. This observation gives us the commutative diagram

$$
\begin{array}{ccc}
(A, P) & \xrightarrow{\;\;f\;\;} & (B, Q) \\
{\scriptstyle \varphi_{(A,P)}} \downarrow & \nearrow {\scriptstyle \overline{f}} & \\
\varphi(A, P) & &
\end{array}
$$

where $\overline{f}(c) = \Delta_{(B,Q)}^{-1}(\Pi(f)(c))$. It remains to check whether $\overline{f}$ is unique. Suppose that $f': \varphi(A, P) \to (B, Q)$ is another homomorphism with $f = f'\varphi_{(A,P)}$. The subset

$$
C = \{c \in \varphi(A, P); \; \overline{f}(c) = f'(c)\} \subseteq \varphi(A, P)
$$

is a subring and a sublattice (by [44], Lemma 2.2), hence it is a sub–f–ring of $\varphi(A, P)$ and also of $\Pi(A, P)$. Since it contains $im(\Delta_{(A,P)})$ it follows that $\varphi(A, P) \subseteq C$. But then $\varphi(A, P) = C$, and this shows that $\overline{f} = f'$. Having finished the proof of the universal property (2.1), we conclude that **FR/N** is monoreflective in **REPPOR** and in **POR/N**.$\square$

The fact that the reflections of all **REPPOR**– or **POR/N**–objects in **FR/N** can be described explicity (as done in the proof) strengthens our result considerably. The monoreflector **POR/N** $\to$ **FR/N** will be denoted by φ throughout.

7 Semi–algebraic functions

With every preordered ring (A, P) we associate a subring $\sigma(A, P)$ of $\Pi(A, P)$. It is called the *ring of semi–algebraic functions* over (A, P), its elements are the *semi–algebraic functions*. This ring has already appeared in the literature, implicitly ([108], §3; [110], Chapter III, §1) and explicitly as the *ring of constructible sections* ([111], Corollary I 2.6; [117], §2) or as the *ring of definable functions* ([3], p. 47). It has been defined in terms of real spectra and also in model theoretic terms. We give a new model theoretic description. The category formed by the rings of semi–algebraic functions is a monoreflective subcategory of **POR/N**. In fact, in the next section we shall show that it is the *smallest* monoreflective subcategory of **POR/N**. Because of this property the category plays a key part and deserves a careful investigation. The present section contains an introduction of semi–algebraic functions via implicit operations and a discussion of the most basic ring theoretic and category theoretic properties of rings of semi–algebraic functions.

In the special case that (A, P) is the weakly ordered ring $\mathbb{Z}[X_1, \ldots, X_n]$ the extension we propose to construct will be the ring of all semi–algebraic functions $R_0^n \to R_0$. Recall that a function $f : R_0^n \to R_0$ is *semi–algebraic* if there is a formula $\Theta_f = \Theta_f(X_1, \ldots, X_n, Y)$ in the language of porings with free variables $X_1, \ldots, X_n, Y$ such that the graph of f is the semi–algebraic set

$$\{(x, y) \in R_0^n \times R_0;\ R_0 \models \Theta_f(x_1, \ldots, x_n, y)\}.$$

Note that the formula Θ_f is by no means unique. If Θ is a formula that is equivalent to Θ_f in the theory of real closed fields then Θ_f and Θ define the same semi–algebraic subset of R_0^{n+1}, hence both define the graph of f. Conversely, if Θ is another formula which also defines the graph of f then Θ and Θ_f are equivalent in the theory of real closed fields. For, to start with, the equivalence holds in the real closed field R_0. By completeness of the theory of real closed fields it is then also true in every real closed field.

A formula $\Theta = \Theta(X_1, \ldots, X_n, Y)$ defines the graph of a semi–algebraic function $R_0^n \to R_0$ if and only if the following sentences belong to the theory of real closed fields

(a) $\forall\, X_1, \ldots, X_n \,\exists\, Y\; \Theta(X_1, \ldots, X_n, Y);$

(7.1) (b) $\forall\, X_1, \ldots, X_n \,\forall\, Y, Y'(\Theta(X_1, \ldots, X_n, Y)\; \&\; \Theta(X_1, \ldots, X_n, Y')$
$\to Y = Y').$

(The second condition was studied in [8], p. 472, under the name *univalence*.) A formula satisfying conditions (7.1) will be called a *functional formula*. From the preceding remarks we know that there is a bijective correspondence between semi–algebraic functions $R_0^n \to R_0$ and equivalence classes of functional formulas $\Theta(X_1, \ldots, X_n, Y)$.

An example of a functional formula is

$$(Y^3 - X_1 Y + X_2 = 0)\; \&\; \forall\, Z(Z^3 - X_1 Z + X_2 = 0 \to Z \leq Y);$$

its meaning is: For any choice of parameters x_1 and x_2 the unique y satisfying the formula is the largest root of the polynomial $Y^3 - x_1 Y + x_2$.

The polynomial ring $\mathbb{Z}[X_1, \ldots, X_n]$ with the weak partial order is denoted by $F(n)$. This is the free object on the set $\{1, \ldots, n\}$ in each of the categories **PREOR**, **POR** and **POR/N** (section 1). The ring of semi–algebraic functions $R_0^n \to R_0$ is denoted by $\sigma(F(n))$, and $\sigma_{F(n)}\colon F(n) \to \sigma(F(n))$ is the canonical homomorphism. If $f \in \sigma(F(n))$ and $g_1, \ldots, g_n \in \sigma(F(k))$ are semi–algebraic functions then the composition

$$f(g_1, \ldots, g_n)\colon R_0^k \xrightarrow{(g_1, \ldots, g_n)} R_0^n \xrightarrow{f} R_0$$

is semi–algebraic as well. Thus, f defines maps

$$\sigma(F(k))^n \longrightarrow \sigma(F(k))\colon (g_1, \ldots, g_n) \to f(g_1, \ldots, g_n)$$

for all $k \in \mathbb{N}$, i.e., f yields an n–ary *operation* on all rings of semi–algebraic functions. This operation is denoted by $\omega = \omega_f$. Examples of n–ary operations are provided by polynomials in $\mathbb{Z}[X_1, \ldots, X_n]$. These are called *explicit operations*, all other operations will be called *implicit operations*. Note that, starting from the polynomial ring $F(n)$, the ring $\sigma(F(n))$ can be constructed via operations. If ω is any k–ary operation and if $P_1, \ldots, P_k \in F(n)$ then $\omega(P_1, \ldots, P_k) \in \sigma(F(n))$. All semi–algebraic functions are obtained in this way. This observation will be used to extend the notion of semi–algebraic functions far beyond the original geometric context. We proceed in several steps.

Suppose that R is some real closed field. Every n–ary operation ω defines a map $\omega_R\colon R^n \to R$. If $f = \omega_{R_0}\colon R_0^n \to R_0$ is the semi–algebraic map belonging to ω then ω_R is the semi–algebraic map obtained from f by extension of the base field (cf. [40], §4; note that R_0 is considered as a subfield of R). Thus, ω defines an n–ary operation on R. The operation can be extended immediately to products of real closed fields: Let $(R_i)_{i \in I}$ be an arbitrary family of real closed fields. For each $i \in I$ let $\omega_i \colon R_i^n \to R_i$ be the n–ary operation on R_i defined by ω. Then we define

$$\omega_{\prod R_i} \colon (\textstyle\prod R_i)^n \to \textstyle\prod R_i \colon ((x_{1i})_i, \ldots, (x_{ni})_i) \to (\omega_i(x_{1i}, \ldots, x_{ni}))_i.$$

In particular, this applies to the complete ring of functions associated with any preordered ring.

If (A, P) is preordered then we denote the map

$$(A, P)^n \xrightarrow{\Delta^n_{(A,P)}} (\Pi(A, P))^n \xrightarrow{\omega_{\Pi(A,P)}} \Pi(A, P)$$

by $\omega_{(A,P)}$ and call it an n–ary operation on (A, P). This concept of an n–ary operation does not provide a map $(A, P)^n \to (A, P)$ as one may expect from an operation. For example, if ω is the unary operation "identity", i.e., the operation belonging to the semi–algebraic map $R_0 \to R_0 \colon x \to x$, then $\omega_{(A,P)}(a) = \Delta_{(A,P)}(a) \in \Pi(A, P)$. Similarly, if ω is the binary operation "+", i.e., the operation belonging to the semi–algebraic map $R_0^2 \to R_0 \colon (x_1, x_2) \to x_1 + x_2$, then $\omega_{(A,P)}(a_1, a_2) = \Delta_{(A,P)}(a_1) + \Delta_{(A,P)}(a_2) \in \Pi(A, P)$. If the operation is *explicit* then $im(\omega_{(A,P)}) \subseteq im(\Delta_{(A,P)})$; if the operation is *implicit* then this is not the case, in general. Thus, implicit operations provide us with new functions. We make the following

Definition 7.2 If (A, P) is a preordered ring then a function $a \in \Pi(A, P)$ is called a *semi–algebraic function* over (A, P) if there are some $n \in \mathbb{N}$, some n–ary operation ω and elements $a_1, \ldots, a_n \in A$ such that $a = \omega_{(A,P)}(a_1, \ldots, a_n)$. $\square$

The set of semi–algebraic functions over (A, P) is denoted by $\sigma(A, P)$. There is a canonical map $\sigma_{(A,P)} \colon A \to \sigma(A, P)$ which is obtained from the identity operation, i.e., $\sigma_{(A,P)}(a) = \Delta_{(A,P)}(a)$. In other words, $\sigma_{(A,P)}$ is $\Delta_{(A,P)}$ with the codomain restricted to $\sigma(A, P)$.

Note that a semi–algebraic function $a \in \sigma(A, P)$ may have many different representations, say

$$\begin{aligned} a &= \omega_{(A,P)}(a_1, \ldots, a_n) \\ &= \omega'_{(A,P)}(a'_1, \ldots, a'_p). \end{aligned}$$

For example, if $(A, P) = \mathbb{Z}$ then $\sqrt{2} \in R_0$ is a semi–algebraic function over (A, P). Let ω be the unary operation belonging to the constant semi–algebraic map $R_0 \to R_0 : x \to \sqrt{2}$; let ω' be the binary operation belonging to the semi–algebraic map

$$R_0^2 \to R_0 : (x_1, x_2) \to \begin{cases} \sqrt{x_1 x_2} & \text{if } x_1 x_2 > 1 \\ 0 & \text{otherwise.} \end{cases}$$

Then $\sqrt{2} = w_{(A,P)}(0) = w'_{(A,P)}(2, 1)$.

Our first result deals with the functorial behavior of operations:

Proposition 7.3 Suppose that ω is an n–ary operation. If $f : (A, P) \to (B, Q)$ is a morphism of preordered rings then the diagram

$$
\begin{array}{ccc}
A^n & \xrightarrow{\;f^n\;} & B^n \\
{\scriptstyle \omega_{(A,P)}}\big\downarrow & & \big\downarrow{\scriptstyle \omega_{(B,Q)}} \\
\Pi(A, P) & \xrightarrow{\;\Pi(f)\;} & \Pi(B, Q)
\end{array}
$$

is commutative in the category **SETS**.

Proof If $\beta \in Sper(B, Q)$ then it suffices to show that

$$p_{f/\beta}(p_{f^{-1}(\beta)}\omega_{(A,P)}(a_1, \ldots, a_n)) = p_\beta \omega_{(B,Q)}(f(a_1), \ldots, f(a_n))$$

for all $(a_1, \ldots, a_n) \in A^n$ (where $p_{f^{-1}(\beta)} : \Pi(A, P) \to \rho(f^{-1}(\beta))$ and $p_\beta : \Pi(B, Q) \to \rho(\beta)$ are the projections). Because of

$$p_{f^{-1}(\beta)}\omega_{(A,P)}(a_1, \ldots, a_n) = \omega_{\rho(f^{-1}(\beta))}(a_1(f^{-1}(\beta)), \ldots, a_n(f^{-1}(\beta))),$$

$$p_\beta \omega_{(B,Q)}(f(a_1), \ldots, f(a_n)) = \omega_{\rho(\beta)}(f(a_1)(\beta), \ldots, f(a_n)(\beta))$$

the claim follows immediately from the univalence of any functional formula defining ω, or from the most elementary properties of semi–algebraic maps under extension of the base field ([40], §4). $\qquad\square$

Because of Proposition 7.3 we can simplify the notation concerning operations. If ω is an n–ary operation then any particular instance $\omega_{(A,P)} : A^n \to \Pi(A, P)$ of the operation is denoted just by ω. Thus, notationally we treat these operations in exactly the same way as addition and multiplication.

Corollary 7.4 If $f : (A,P) \to (B,Q)$ is a **PREOR**–morphism then $\Pi(f)$ restricts to a map $\sigma(f) : \sigma(A,P) \to \sigma(B,Q)$. The diagram

$$
\begin{array}{ccc}
A & \xrightarrow{\ f\ } & B \\[2pt]
\sigma_{(A,P)} \downarrow & & \downarrow \sigma_{(B,Q)} \\[2pt]
\sigma(A,P) & \xrightarrow{\ \sigma(f)\ } & \sigma(B,Q)
\end{array}
$$

is commutative.

Proof If $a = \omega_{(A,P)}(a_1, \dots, a_n) \in \sigma(A,P)$ then

$$\Pi(f)(a) = \omega(f(a_1), \dots, f(a_n))$$

by Proposition 7.3, i.e., $\Pi(f)(a) \in \sigma(B,Q)$. Thus, $\Pi(f)$ can be restricted as claimed. Commutativity of the diagram follows from the fact that $\sigma_{(A,P)}$ and $\sigma_{(B,Q)}$ are both defined by the identity operation. $\qquad\square$

It follows immediately from the functorial properties of Π that $\sigma(gf) = \sigma(g)\sigma(f)$ if $f \colon (A, P_A) \to (B, P_B)$ and $g \colon (B, P_B) \to (C, P_C)$ are **PREOR**–morphisms.

Having defined what individual semi–algebraic functions are, now we turn to the investigation of the sets $\sigma(A, P)$. It will turn out that they are von Neumann regular rings with a large number of special properties. We start with two auxiliary results:

Lemma 7.5 For any preodered ring (A, P) and any n–ary operation ω, $\omega : (\Pi(A,P))^n \to \Pi(A,P)$ maps $\sigma(A,P)^n$ into $\sigma(A,P)$.

Proof Choose elements $a_1, \dots, a_n \in \sigma(A,P)$ and operations ω_i such that $a_i = \omega_i(a_{i1}, \dots, a_{ir_i})$ with $a_{ij} \in A$. Pick functional formulas $\Theta_i = \Theta_i(X_{i1}, \dots, X_{ir_i}, Y_i)$ defining the operations $\omega_1, \dots, \omega_n$; suppose that ω is defined by $\Theta = \Theta(X_1, \dots, X_n, Y)$. Then the formula

$$
\begin{aligned}
H \ &= \ H(X_{11}, \dots, X_{nr_n}, Y) \\
&\equiv \ \exists X_1, \dots, X_n \ \Big(\Theta(X_1, \dots, X_n, Y) \ \& \ \bigwedge_{i=1}^{r} (\Theta_i(X_{i1}, \dots, X_{ir_i}, X_i))\Big)
\end{aligned}
$$

is functional; let ω' be the corresponding operation. Then

$$\omega(a_1, \ldots, a_n) = \omega'(a_{11}, \ldots, a_{nr_n}) \in \sigma(A, P),$$

and the proof is finished.   □

Lemma 7.6

(a) For any pair of semi–algebraic functions $f, g \in \sigma(A, P)$ the set $\{\alpha \in Sper(A, P); f(\alpha) = g(\alpha)\}$ is constructible.
(b) If $Sper(A, P) = C_1 \cup \ldots \cup C_r$ is a partition into constructible subsets and if $f_1, \ldots, f_r \in \sigma(A, P)$ then there is a unique semi–algebraic function $f \in \sigma(A, P)$ such that $f(\alpha) = f_i(\alpha)$ for all $\alpha \in C_i$.

Proof (a) Suppose that $f = \omega(a_1, \ldots, a_m)$ and $g = \omega'(b_1, \ldots, b_n)$ where ω and ω' are given by functional formulas Θ and Θ'. Then

$$\begin{aligned}
&\{\alpha \in Sper(A, P); \ f(\alpha) = g(\alpha)\} \\
&= \{\alpha \in Sper(A, P); \ \rho(\alpha) \models \ \exists Y \ (\Theta(a_1(\alpha), \ldots, a_m(\alpha), Y) \\
&\hspace{4cm} \& \ \Theta'(b_1(\alpha), \ldots, b_n(\alpha), Y))\}.
\end{aligned}$$

This set is constructible by Proposition 4.2.

(b) For each $i = 1, \ldots, r$ there are an n_i–ary operation ω_i and elements $a_{ij}, j = 1, \ldots, n_i$ such that $f_i = \omega_i(a_{i1}, \ldots, a_{in_i})$. Pick functional formulas $\Theta_i(X_{i1}, \ldots, X_{in_i}, Y)$ defining the operations. Moreover, each one of the constructible sets C_i can be described as

$$C_i = \{\alpha \in Sper(A, P); \ \rho(\alpha) \models H_i(b_{i1}(\alpha), \ldots, b_{ik_i}(\alpha))\}$$

with some formula $H_i(Y_{i1}, \ldots, Y_{ik_i})$ and suitable $b_{ij} \in A$ (Proposition 4.2). The formula $\bigvee_{i=1}^{r} (H_i \ \& \ \Theta_i)$ is functional. If the corresponding operation is denoted by ω then

$$f = \omega(a_{ij}, b_{ik}; i = 1, \ldots, r, j = 1, \ldots, n_i, k = 1, \ldots, k_i). \qquad \square$$

Proposition 7.7 $\sigma(A, P)$ is a von Neumann regular sub–f–ring of $\Pi(A, P)$. The induced partial order of $\sigma(A, P)$ is its only partial order; the positive cone consists of the squares. The map $\sigma_{(A,P)}: (A, P) \to \sigma(A, P)$

is a **PREOR**–morphism.

Proof Let α, β, γ and δ be the operations corresponding to the binary semi–algebraic maps "addition", "multiplication", "supremum" and "infimum". Using Lemma 7.5 it follows immediately that $\sigma(A, P) \subseteq \Pi(A, P)$ is a subring and a sublattice, hence $\sigma(A, P)$ is a sub–f–ring. To show that $\sigma_{(A,P)}$ is a **PREOR**–morphism, pick some $a \in P$ and note that $a(\alpha) \geq 0$ for all $\alpha \in Sper(A, P)$. Thus, $\sigma_{(A,P)}(a) = \Delta_{(A,P)}(a) \geq 0$. For von Neumann regularity, suppose that $a = \omega(a_1, \dots, a_n) \in \sigma(A, P)$ and $f \colon R_0^n \to R_0$ is a corresponding semi–algebraic function. The function

$$R \colon R_0^n \to R_0 \colon x \to \begin{cases} 0 & \text{if } f(x) = 0 \\ f(x)^{-1} & \text{if } f(x) \neq 0 \end{cases}$$

is also semi–algebraic, and $f^2 g = f$. Let ω' be the n–ary operation corresponding to g. Then $b = \omega'(a_1, \dots, a_n) \in \sigma(A, P)$ is a semi–algebraic function with $a^2 b = a$, proving von Neumann regularity.

It remains to check that the positive cone of $\sigma(A, P)$ consists of the squares. (Uniqueness of the partial order is an automatic consequence of this property.) So, pick $0 \leq a = \omega(a_1, \dots, a_n) \in \sigma(A, P)$, let $f \colon R_0^n \to R_0$ be the semi–algebraic function corresponding to ω. If ω' is the n–ary operation belonging to $f \vee 0$ then also $a = \omega'(a_1, \dots, a_n)$. Therefore we may assume that $f \geq 0$. In $\sigma(F(n))$ there exists a unique semi–algebraic function $g \geq 0$ such that $g^2 = f$. Let ω' be the corresponding operation and define $b = \omega'(a_1, \dots, a_n) \in \sigma(A, P)$. Then $a = b^2$, and the claim has been proved. $\qquad\square$

From Proposition 7.7 in connection with Corollary 7.4 and the remark following its proof we learn that σ is a functor **PREOR** $\to$ **POR/N** and that the homomorphisms $\sigma_{(A,P)}$ are the components of a natural transformation from the identity functor of **PREOR** to the composition of the functor σ and the inclusion functor **POR/N** $\to$ **PREOR**. The functor σ together with the natural transformation will be studied more closely below. But first we collect more information about the rings $\sigma(A, P)$. Notationally we deal with $\sigma(A, P)$ exactly as with $\Pi(A, P)$, i.e., we do not distinguish between the ring and the poring. The positive cone of $\sigma(A, P)$ is denoted by $\sigma^+(A, P)$.

Corollary 7.8   For the ring $\sigma(A, P)$ the support map $supp\colon$

$Sper(\sigma(A,P)) \to Spec(\sigma(A,P))$ is a homeomorphism.

Proof Since $\sigma(A,P)$ is an f–ring the map is a homeomorphism onto the image. The image contains all minimal prime ideals (Proposition 4.4). Because $\sigma(A,P)$ is von Neumann regular every prime ideal is minimal, i.e., *supp* is also surjective. $\square$

Proposition 7.9 The functorial map $Sper(\sigma_{(A,P)}) \colon Sper(\sigma(A,P)) \to Sper(A,P)$ is bijective. It is a homeomorphism with respect to the constructible topology of $Sper(A,P)$.

Proof First note that the spectral topology of $Sper(\sigma(A,P))$ coincides with the constructible topology since $\sigma(A,P)$ is von Neumann regular. So we consider only constructible topologies in this proof. We define a map $\tau : Sper(A,P) \to Sper(\sigma(A,P))$ by sending $\alpha \in Sper(A,P)$ to the prime cone defined through the homomorphism

$$\sigma(A,P) \overset{i}{\hookrightarrow} \Pi(A,P) \xrightarrow{p_\alpha} \rho(\alpha)$$

(where p_α is the projection onto the α–th component). It is clear that $Sper(\sigma_{(A,P)})\tau = id_{Sper(A,P)}$. Moreover, Lemma 7.6 (a) implies that τ is continuous. Both spaces are compact, hence τ is a homeomorphism onto a closed subspace of $Sper(\sigma(A,P))$. If $im(\tau) \subset Sper(\sigma(A,P))$ then there is some $\beta \in Sper(\sigma(A,P))\backslash im(\tau)$. By the remarks at the end of section 4 we find some $a \in \sigma(A,P)$ such that $a(\beta) > 0$ and $a(\tau(\alpha)) = 0$ for all $\alpha \in Sper(A,P)$. But then $a = 0$ in $\Pi(A,P)$, hence also $a = 0$ in the subring $\sigma(A,P)$. This contradiction finishes the proof. $\square$

From Corollary 7.8 and Proposition 7.9 we have a rather clear picture of the real spectrum and the prime spectrum of $\sigma(A,P)$. To complete this picture we shall now determine the residue fields at the prime (= maximal) ideals of $\sigma(A,P)$.

Proposition 7.10 Let α be a prime cone of (A,P), let $\beta \subseteq \sigma(A,P)$ be the prime cone corresponding to α (Proposition 7.9), and define $q = supp(\beta)$. Then the canonical homomorphism

$$\sigma(A,P)/q \longrightarrow \rho(\alpha)$$

is an isomorphism. In particular, $\rho_{\sigma_{(A,P)}/\beta} : \rho(\alpha) \to \rho(\beta)$ is an isomorphism.

Proof Since $\rho(\alpha)$ is the real closure of $\kappa_{(A,P)}(\alpha)$, for every $a \in \rho(\alpha)$ there is a formula $\Theta = \Theta(X_1, \dots, X_m, X)$ in the language of porings and there are $a_1, \dots, a_m \in A$ such that the following statements are true in $\rho(\alpha)$:

- $\exists X \, [\Theta(a_1(\alpha), \dots, a_m(\alpha), X)$
  $\&\forall X'(\Theta(a_1(\alpha), \dots, a_m(\alpha), X') \to X = X')];$
- $\Theta(a_1(\alpha), \dots, a_m(\alpha), a).$

In general the formula Θ will not be functional. However, it can be modified to obtain a functional formula which still defines a: Let $\Psi = \Psi(X_1, \dots, X_m, Y)$ be the formula

$$\exists X, X' \, [\Theta(X_1, \dots, X_m, X)$$
$$\& \, \Theta(X_1, \dots, X_m, X') \, \& \, X \neq X' \, \& \, Y = 0]$$
$$\lor \quad \forall X \, [\neg\Theta(X_1, \dots, X_m, X) \, \& \, Y = 0]$$
$$\lor \quad \forall X, X' \, [\{ \, \Theta(X_1, \dots, X_m, X) \, \& \, \Theta(X_1, \dots, X_m, X') \to X = X'\}$$
$$\&\Theta(X_1, \dots, X_m, Y)].$$

One checks that Ψ is functional and that

$$\rho(\alpha) \vDash \Psi(a_1(\alpha), \dots, a_m(\alpha), a).$$

Therefore, if ω is the operation defined by Ψ then $\omega(a_1, \dots, a_m) \in \sigma(A, P)$ and $\omega(a_1, \dots, a_m)(\alpha) = a$. Thus, we have shown that the canonical homomorphism

$$\sigma(A, P) \subseteq \prod(A, P) \xrightarrow{p_\alpha} \rho(\alpha)$$

is surjective; this finishes the proof. $\square$

Corollary 7.11 If $f : (A, P) \to (B, Q)$ is an epimorphism in **POR/N** then $\sigma(f)$ is an epimorphism as well.

Proof According to Proposition 7.9 the map $Sper(\sigma(f))$ is injective; by Proposition 7.10 the homomorphisms $\rho_{\sigma(f)/\beta}$, $\beta \in Sper(\sigma(B, Q))$ are isomorphisms. The claim follows from Theorem 5.2. $\square$

Proposition 7.12 If (A, P) is a reduced poring then $\sigma_{(A,P)}$ is an epimorphism in the category **POR/N**. If (A, P) is a preordered ring and

$f, g: \sigma(A, P) \to (B, Q)$ are homomorphisms into a reduced poring such that $f\sigma_{(A,P)} = g\sigma_{(A,P)}$ then $f = g$.

Proof First suppose that $(A, P) \in ob(\mathbf{POR/N})$. Because of the Propositions 7.9 and 7.10 we can apply the criterion of Theorem 5.2 to see that $\sigma_{(A,P)}$ is epimorphic in $\mathbf{POR/N}$. Now let $(A, P) \in ob(\mathbf{PREOR})$, let $\pi: (A, P) \to (\overline{A}, \overline{P})$ be the reflection in $\mathbf{POR/N}$. Note that $\Pi(\pi): \Pi(A, P) \to \Pi(\overline{A}, \overline{P})$ is an isomorphism. Since π is surjective it follows immediately that $\sigma(\pi): \sigma(A, P) \to \sigma(\overline{A}, \overline{P})$ is an isomorphism as well. Because of $f\sigma(\pi)^{-1}\sigma_{(\overline{A},\overline{P})}\pi = g\sigma(\pi)^{-1}\sigma_{(\overline{A},\overline{P})}\pi$ and surjectivity of π we see that $f\sigma(\pi)^{-1}\sigma_{(\overline{A},\overline{P})} = g\sigma(\pi)^{-1}\sigma_{(\overline{A},\overline{P})}$. The homomorphisms in this identity are all $\mathbf{POR/N}$–morphisms. Since $\sigma_{(\overline{A},\overline{P})}$ is epimorphic in $\mathbf{POR/N}$ we conclude that $f\sigma(\pi)^{-1} = g\sigma(\pi)^{-1}$. This implies that $f = g$, which ends the proof. $\square$

The proposition implies that for any $\mathbf{PREOR}$–morphism $f: (A, P) \to (B, Q)$ the homomorphism $\sigma(f)$ is the unique morphism $g: \sigma(A, P) \to \sigma(B, Q)$ with $g\sigma_{(A,P)} = \sigma_{(B,Q)}f$. For, let g be any such homomorphism. Then $\sigma(f)\sigma_{(A,P)} = \sigma_{(B,Q)}f = g\sigma_{(A,P)}$, hence $g = \sigma(f)$.

Now we take a closer look at properties of the functor σ. First we study the question under what conditions the morphism $\sigma(f)$ associated with a $\mathbf{PREOR}$–morphism or a $\mathbf{POR/N}$–morphism $f: (A, P) \to (B, Q)$ is an isomorphism. We recall the following criterion for a homomorphism $h: M \to N$ of von Neumann regular rings to be an isomorphism:

(7.13) h is an isomorphism if and only if $Spec(h) : Spec(N) \to Spec(M)$ is bijective and for all $q \in Spec(N)$ the canonical homomorphism $M/h^{-1}(q) \to N/q$ is an isomorphism.

(If h is an isomorphism then the conditions are trivial. For the converse, note that the conditions say that $Spec(h) : Spec(N) \to Spec(M)$ is an isomorphism of schemes, hence $h : M \to N$, being the homomorphism between the rings of global sections, is an isomorphism.)

Lemma 7.14 Let $f : (A, P) \to (B, Q)$ be a $\mathbf{PREOR}$–morphism such that $Sper(f)$ is bijective and $\rho_{f/\beta} : \rho(f^{-1}(\beta)) \to \rho(\beta)$ is an isomorphism for every $\beta \in Sper(B, Q)$. Then $\sigma(f) : \sigma(A, P) \to \sigma(B, Q)$ is an isomorphism.

Proof In the commutative diagram

$$
\begin{array}{ccc}
(A,P) & \xrightarrow{\ f\ } & (B,Q) \\
\Delta_{(A,P)} \downarrow & & \downarrow \Delta_{(B,Q)} \\
\Pi(A,P) & \xrightarrow{\ \Pi(f)\ } & \Pi(B,Q)
\end{array}
$$

the homomorphism $\Pi(f)$ is an isomorphism. Since $\sigma(f)$ is a restriction of $\Pi(f)$ we know that it is injective. Since $Sper(f)$ is bijective (by hypothesis) it is a homeomorphism with respect to the constructible topology, hence Proposition 7.9 implies that $Sper(\sigma(f))$ is a homeomorphism. The same is true for the functorial map $Spec(\sigma(f))$ of the prime spectra (Corollary 7.8). Finally, the hypothesis about the real closed residue fields together with Proposition 7.10 shows that the homomorphisms $\sigma(A,P)/\sigma(f)^{-1}(q) \to \sigma(B,Q)/q$ induced by $\sigma(f)$ are isomorphisms for all $q \in Spec(B)$. Thus, (7.13) implies that $\sigma(f)$ is an isomorphism. □

Corollary 7.15 Let $f : (A,P) \to (B,Q)$ be a **POR/N**–epimorphism. Then $Sper(f)$ is surjective if and only if $\sigma(f)$ is an isomorphism.

Proof If $\sigma(f)$ is an isomorphism then surjectivity of $Sper(f)$ follows from Proposition 7.9. Conversely, suppose that $Sper(f)$ is surjective. Now Theorem 5.2, Proposition 7.9, Proposition 7.10 and Lemma 7.14 show that $\sigma(f)$ is an isomorphism. □

In an important special case these results make it particularly easy to check whether a **POR/N**–morphism of the form $\sigma(f)$ is an isomorphism:

Proposition 7.16 Suppose that $(A,P) \in ob(\mathbf{POR/N})$ and that $\sigma_{(A,P)}$ factors in **POR/N** as gf with $f: (A,P) \to (B,Q)$ and $g : (B,Q) \to \sigma(A,P)$. Then the following conditions are equivalent:

(a) f is an epimorphism.
(b) $Sper(g)$ is surjective.
(c) $Sper(f)$ is injective.
(d) $\sigma(f)$ is an isomorphism.

Proof Because of $gf = \sigma_{(A,P)}$ and Proposition 7.9 we know that $Sper(f)$

is surjective. Since $Sper(\sigma_{(A,P)})$ is bijective, injectivity of $Sper(f)$ is equivalent to surjectivity of $Sper(g)$, i.e., (b) and (c) are equivalent. – (a) $\Rightarrow$ (c) Theorem 5.2. – (a) $\Rightarrow$ (d) Corollary 7.15. – It remains to show that both (c) and (d) imply (a): If (c) or (d) holds then $Sper(f)$ is bijective. Proposition 7.10 shows that $\rho_{f/\beta}\colon \rho(f^{-1}(\beta)) \to \rho(\beta)$ is an isomorphism for every $\beta \in Sper(B,Q)$. The proof is finished by applying Theorem 5.2.   □

As another consequence of Lemma 7.14 we note that the construction of $\sigma(A,P)$ is idempotent:

Corollary 7.17 For any preordered ring (A,P) the homomorphism $\sigma_{\sigma(A,P)}\colon \sigma(A,P) \to \sigma(\sigma(A,P))$ agrees with $\sigma(\sigma_{(A,P)})$ and is an isomorphism.

Proof It follows from Lemma 7.14 that $\sigma(\sigma_{(A,P)})$ is an isomorphism. Proposition 7.12 together with commutativity of

$$
\begin{array}{ccc}
(A,P) & \xrightarrow{\;\sigma_{(A,P)}\;} & \sigma(A,P) \\[2pt]
{\scriptstyle\sigma_{(A,P)}}\big\downarrow & & \big\downarrow{\scriptstyle\sigma_{\sigma(A,P)}} \\[2pt]
\sigma(A,P) & \xrightarrow{\;\sigma(\sigma_{(A,P)})\;} & \sigma(\sigma(A,P))
\end{array}
$$

shows that $\sigma(\sigma_{(A,P)}) = \sigma_{\sigma(A,P)}$.   □

The preordered rings for which $\sigma_{(A,P)}$ is an isomorphism will be called *rings of semi–algebraic functions*. In **PREOR** the full subcategory whose objects are all rings of semi–algebraic functions is denoted by **SAFR**. Many of the results about the rings of semi–algebraic functions we have obtained so far are summarized in

Theorem 7.18 The category **SAFR** is a reflective subcategory of **PREOR**. The reflector is the functor σ, the reflection of an object (A,P) is the homomorphism $\sigma_{(A,P)}\colon (A,P) \to \sigma(A,P)$. The reflector σ restricts to a monoreflector **POR/N** $\to$ **SAFR**, which will also be denoted by σ.

Proof Corollary 7.17 shows that σ is a functor into the subcategory **SAFR** of **PREOR**. If $f : (A, P) \to (B, Q)$ is a **PREOR**–morphism into a **SAFR**–object then $f = \sigma_{(B,Q)}^{-1}\sigma(f)\sigma_{(A,P)}$. If $g : \sigma(A, P) \to (B, Q)$ is any other homomorphism with $f = g\sigma_{(A,P)}$ then it follows from Proposition 7.12 that $g = \sigma_{(B,Q)}^{-1}\sigma(f)$. Thus, the universal property (2.1) has been checked, and σ is indeed a reflector. Since **SAFR** $\subseteq$ **POR/N** it is clear that σ can be restricted as claimed. Moreover, if (A, P) is a reduced poring then $\sigma_{(A,P)}$ is $\Delta_{(A,P)}$ with restricted codomain. Because $\Delta_{(A,P)}$ is injective the same is true for $\sigma_{(A,P)}$, i.e., σ is a monoreflector on **POR/N**. $\square$

We finish the discussion of the category **SAFR** in this section with a characterization of the objects of **SAFR** among all reduced porings. For this purpose it is useful to have the following particularly simple characterization of epimorphisms in **SAFR**. First note that the category **SAFR** contains all real closed fields. Therefore the criterion of Theorem 5.2 can be used to check whether a **SAFR**–morphism is epimorphic in this category.

Proposition 7.19 In **SAFR**, the epimorphisms are the surjective homomorphisms.

Proof Surjective homomorphisms are clearly epimorphic. Conversely, suppose that $f: (A, P) \to (B, Q)$ is an epimorphism in **SAFR**. Since f is a homomorphism of reduced f–rings it preserves the lattice operations. Therefore $(f(A), f(P)) \subseteq (B, Q)$ is a sub–f–ring; let i be the inclusion homomorphism. It is claimed that these rings coincide, i.e., that i is an isomorphism. By Theorem 5.2, Corollary 7.8 and Proposition 7.9 the prime spectra are homeomorphic via the functorial map. Both rings are von Neumann regular and the residue fields $f(A)/i^{-1}(q)$ and B/q are isomorphic for any prime ideal $q \subseteq B$. Thus, the hypotheses of (7.13) are satisfied, and i is an isomorphism. $\square$

In any category **C**, an object C is called *epicomplete* if every monomorphic and epimorphic morphism $f: C \to D$ is an isomorphism. As a consequence of the proposition we can determine the class of epicomplete objects of **POR/N**:

Corollary 7.20 In **POR/N** an object is epicomplete if and only if it belongs to **SAFR**.

Proof Suppose that $f \colon (A, P) \to (B, Q)$ is an epimorphic extension in **POR/N** and that (A, P) is a ring of semi–algebraic functions. Then

$$(A, P) \xrightarrow{\;\;f\;\;} (B, Q) \xrightarrow{\;\sigma(B,Q)\;} \sigma(B, Q)$$

is an epimorphic extension in **SAFR**, hence it is bijective (Proposition 7.19). This implies that $\sigma_{(B,Q)}f$ is an isomorphism. But then f, being an initial epimorphic factor of an isomorphism, is an isomorphism as well. We conclude that (A, P) is epicomplete. Conversely, if (A, P) is epicomplete then the epimorphic extension $\sigma_{(A,P)} \colon (A, P) \to \sigma(A, P)$ is an isomorphism, hence (A, P) belongs to **SAFR**.   □

Since semi–algebraic functions and rings of semi–algebraic functions have occurred in the literature before we conclude the section by discussing the relationship of the present definition with the definition that exists in the references. The name *constructible sections* (instead of semi–algebraic functions) was used in [111], p. 8, Proposition I 2.5 (see also [108]; [110]). Both definitions are clearly similar to each other. A discussion with A. Prestel helped us to establish the formal connection. If $a = \omega(a_1, \dots, a_n) \in \sigma(A, P)$ and ω is defined by the functional formula $\Theta(X_1, \dots, X_n, X)$ then $\Theta(a_1, \dots, a_n, X)$ is a formula in the language $L_{po}(A)$ of porings with constant symbols added for all elements of A. For every $\alpha \in Sper(A, P)$ the following conditions are satisfied:

$$
\begin{aligned}
&\rho(\alpha) \models \Theta(a_1(\alpha), \dots, a_n(\alpha), a(\alpha)), \\
(7.21) \quad &\rho(\alpha) \models \exists X [\Theta(a_1(\alpha), \dots, a_n(\alpha), X) \;\&\; \\
&\qquad\qquad \forall Y (\Theta(a_1(\alpha), \dots, a_n(\alpha), Y) \to X = Y)].
\end{aligned}
$$

By [111], Proposition I 2.5, a is a constructible section. Conversely, if a is a constructible section then there is a formula $\Theta(a_1, \dots, a_n, X)$ in the language $\mathcal{L}_{po}(A)$ (with X as its only free variable) such that (7.21) holds. We obtain the formula $\Theta(X_1, \dots, X_n, X)$ in the language $\mathcal{L}_{po}$ by replacing every occurence of a constant symbol $a_i \in A$ by a variable X_i. In general this formula is not functional. However, a modification of Θ

helps. A formula $\phi(X_1, \ldots, X_n, X)$ is defined by:

$$[\exists Y \exists Z (Y \neq Z \& \Theta(X_1, \ldots, X_n, Y) \& \Theta(X_1, \ldots, X_n, Z))$$
$$\to X = 0]$$
$$\vee \quad [\forall Y (\neg\Theta(X_1, \ldots, X_n, Y)) \to X = 0]$$
$$\vee \quad [\Theta(X_1, \ldots, X_n, X) \& \forall Y (\Theta(X_1, \ldots, X_n, Y) \to X = Y)].$$

This is a functional formula. If ω is the operation defined by ϕ then one checks easily that $a = \omega(a_1, \ldots, a_n)$.

It is frequently useful to have a description of $\sigma(A, P)$ available which uses only the real spectrum. We restate the definition of [111], p. 8, in the following form: Let $A[T]$ be the polynomial ring in one variable, let $\pi \colon Sper(A[T]) \to Sper(A)$ be the functorial map. We define $Sper((A, P)[T]) = \pi^{-1}(Sper(A, P))$ and restrict π to the surjective map

$$\pi' \colon Sper((A, P)[T]) \to Sper(A, P).$$

If we consider both spaces with their constructible topologies then a semi–algebraic function is a section of π' which is a homeomorphism onto a closed and open subspace of $Sper((A, P)[T])$.

Rings of semi–algebraic functions have been used to study model theoretic properties of reduced f–rings and von Neumann regular f–rings (cf. [79]; [82], §6; [123], §4; [122]), such as existence of model companions etc. The real closed commutative regular f–rings of [79] are our rings of semi–algebraic functions. In [79], Theorem 1, a unique real closure is associated with any von Neumann regular f–ring (A, P). The real closure is the same as our homomorphism $\sigma_{(A,P)} \colon (A, P) \to \sigma(A, P)$. It is important to keep in mind that this notion of a real closure applies only to von Neumann regular f–rings. In section 12 we shall discuss a far more general notion of real closures.

8 Comparing reflectors

The class of all subcategories of a category $\mathbf{C}$ is partially ordered by inclusion. This partial order restricts to a partial order of the classes of reflective or epireflective or monoreflective subcategories. By the unique correspondence between reflective subcategories and reflectors there is also a partial order on the class of reflectors. The investigation of these partially ordered classes is at the core of our work. We shall use the partial order to establish relations between reflective subcategories of **PREOR**, **POR** and **POR/N** and thus bring some degree of order to the multitude of constructions involving preordered or partially ordered rings. In this section we lay the foundations for all applications using the partial order between reflectors in later sections.

First of all we show how the partial order of reflective subcategories of a category $\mathbf{C}$ can be recognized at the level of reflectors.

Proposition 8.1 Let $\mathbf{D}, \mathbf{E} \subseteq \mathbf{C}$ be reflective subcategories with reflectors r and s, resp. Given any functor t from $\mathbf{C}$ into a subcategory, let $\bar{t}$ be the composition with the inclusion functor into $\mathbf{C}$. Consider the following conditions:

(a) $\mathbf{E} \subseteq \mathbf{D}$.
(b) There is a natural equivalence $\bar{r}\bar{s} \to \bar{s}$.
(c) For every $C \in ob(\mathbf{C})$ there is a morphism $\varphi_C : r(C) \to s(C)$ such that $\varphi_C r_C = s_C$.
(d) There is a natural equivalence $\bar{s} \to \bar{s} \circ \bar{r}$ whose component at C is $s(r_C)$.

Between these conditions the following implications hold: (a) $\Leftrightarrow$ (b) $\Rightarrow$ (c) $\Leftarrow$ (d). If r is an epireflector then the conditions are all equivalent.

Proof (a) $\Rightarrow$ (b) If $C \in ob(\mathbf{C})$ then $s(C)$ is in $\mathbf{D}$, hence $id_{s(C)}$ factors as

$$
\begin{array}{ccc}
s(C) & \xrightarrow{\ r_{s(C)}\ } & r(s(C)) \\
 & \searrow{\scriptstyle id_{s(C)}} & \downarrow{\scriptstyle \overline{id_{s(C)}}} \\
 & & s(C)
\end{array}
$$

Since $r_{s(C)}$ is an isomorphism we conclude that $\overline{id_{s(C)}}$ is an isomorphism
as well. It is clear that the family $(\overline{id_{s(C)}})_C$ is a natural transformation.
(b) $\Rightarrow$ **(a)** Let $\tau\colon \overline{rs} \to \overline{s}$ be a natural equivalence. If $C \in ob(\mathbf{E})$ then
$s_C\colon C \to s(C)$ and $\tau_C\colon r(s(C)) \to s(C)$ are isomorphisms. Thus C is iso-
morphic to an object of $\mathbf{D}$. Because $\mathbf{D}$ is isomorphism-closed this implies
$C \in ob(\mathbf{D})$.
(a) $\Rightarrow$ **(c)** Define φ_C to be the unique morphism $r(C) \to s(C)$ such that
$\varphi_C r_C = s_C$.
(d) $\Rightarrow$ **(c)** In the diagram

$$
\begin{array}{ccc}
C & \xrightarrow{r_C} & r(C) \\
{\scriptstyle s_C}\Big\downarrow & & \Big\downarrow{\scriptstyle s_{r(C)}} \\
s(C) & \xrightarrow{s(r_C)} & s(r(C))
\end{array}
$$

the arrow $s(r_C)$ is an isomorphism by assumption. Therefore $\varphi_C =
s(r_C)^{-1}s_{r(C)}$ has the desired property.

Now we assume that r is an epireflector and we propose to show that
(c) $\Rightarrow$ **(a)**. We pick $C \in ob(\mathbf{E})$ and note that $\varphi_C r_C$ is an isomorphism.
Now r_C is an initial epimorphic factor of an isomorphism, hence r_C is an
isomorphism. But then $C \in ob(\mathbf{D})$.

Finally, let r and s both be epireflectors. Then **(a)** – **(c)** $\Rightarrow$ **(d)**: First
of all, the family of morphisms $(s(r_C))_C$ is always a natural transforma-
tion (without any special hypotheses). It is only necessary to show that
the components are isomorphisms. Let φ_C be the morphism of condition
(c). In the diagram

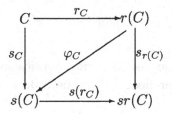

the upper triangle commutes by the choice of φ_C, the lower triangle commutes because r_C is epimorphic. There is a unique morphism
$\overline{\varphi_C}\colon sr(C) \to s(C)$ with $\varphi_C = \overline{\varphi_C}s_{r(C)}$. Then, using the universal property of s_C and $s_{r(C)}$ and

$$
\begin{aligned}
\overline{\varphi_C}s(r_C)s_C &= \overline{\varphi_C}s_{r(C)}r_C = \varphi_C r_C = s_C = id_{s(C)}s_C, \\
s(r_C)\overline{\varphi_C}s_{r(C)} &= s(r_C)\varphi_C = s_{r(C)} = id_{sr(C)}s_{r(C)},
\end{aligned}
$$

we conclude that $\overline{\varphi_C}$ and $s(r_C)$ are mutually inverse isomorphisms. □

Since reflectors and reflective subcategories are in a one–to–one correspondence the partial order "inclusion" can be transferred from the class of reflective subcategories of $\mathbf{C}$ to the class of reflectors. If $r\colon \mathbf{C} \to \mathbf{D}$ and $s\colon \mathbf{C} \to \mathbf{E}$ are reflectors and $\mathbf{E} \subseteq \mathbf{D}$ then we say that s is *stronger* than r, or r is *weaker* than s, and write $r \leq s$. The weaker–stronger relation is a partial order on the class of reflectors of $\mathbf{C}$. It is the opposite partial order of the one transferred from the class of reflective subcategories. If s is stronger than r then $\mathbf{E}$ is a reflective subcategory of $\mathbf{D}$; the reflector is $s|_\mathbf{D}$. If $\mathbf{E}$ is epireflective or monoreflective in $\mathbf{C}$ then the same is true for $\mathbf{E}$ in $\mathbf{D}$.

Corollary 8.2 (cf. [60], Exercise 37 C(a)) As in Proposition 8.1, let $\mathbf{D}$, $\mathbf{E} \subseteq \mathbf{C}$ be reflective subcategories with reflectors r and s. Suppose that the subcategory $\mathbf{E}$ is monoreflective. If $\mathbf{E} \subseteq \mathbf{D}$ then $\mathbf{D}$ is also monoreflective. Moreover, after identifying $s(C) = sr(C)$ via the isomorphism $s(r_C)$, we have $s_C = s_{r(C)}r_C$, hence $s_{r(C)}\colon r(C) \to s(C)$ is a subobject.

Proof By Proposition 8.1 every reflection $s_C\colon C \to s(C)$ factors as $\varphi_C r_C\colon C \to r(C) \to s(C)$. If s_C is monomorphic then so is r_C. The rest is trivial. □

Under suitable hypotheses about the category $\mathbf{C}$ it is possible to characterize epireflective subcategories as those subcategories that are closed under certain constructions. We shall state two such results, Theorem 8.3 and Theorem 8.7. Although proofs may be found in, or extracted from, the literature ([1], Corollary 16.9; [52], Theorem 1.2; [60], Theorem 37.2) we include proofs here to improve the readability of the present work. Comments about the origins of the Characterization Theorems may be found in [52] after the statement of Theorem 1.2.

Theorem 8.3 (First Characterization Theorem) Let $\mathbf{C}$ be a complete category which is both wellpowered and co–wellpowered, let $\mathbf{D} \subseteq \mathbf{C}$ be a full and isomorphism–closed subcategory. Then $\mathbf{D}$ is epireflective in $\mathbf{C}$ if and only if the class of $\mathbf{D}$–objects is closed under the formation of products and extremal subobjects.

Being *closed under the formation of products* means that any product in
C all of whose components are **D**–objects actually belongs to **D**. For the
notion of an extremal subobject, suppose that C is an object of **C**. A
subobject $m \colon C' \to C$ is *extremal* if in any factorization $m = fe$, e is an
isomorphism if it is epimorphic. We say that **D** is *closed under extremal
subobjects* if every extremal subobject of a **D**–object belongs to **D**.

   Note that the equalizer of a pair of parallel morphisms is always extremal ([1], Corollary 7.63). Before proving Theorem 8.3 we illustrate the
notion of extremal subobjects with two examples.

Example 8.4 *Extremal subobjects in the category* **TOF** *of totally
       ordered fields*
In the category **TOF** every injective morphism (i.e., every morphism
whose codomain is not the zero field) is a subobject. A morphism
$f \colon (K, T_K) \to (L, T_L)$ is an epimorphism if and only if L is algebraic
over $f(K)$. Thus, f is an extremal subobject if and only if $f(K)$ is algebraically closed in L. In particular, real closed fields are always extremal
subobjects. $\square$

Example 8.5 *Extremal subobjects in the category of torsion free abelian
       groups*
A homomorphism $f \colon G \to H$ between torsion free abelian groups is a
subobject if and only if it is injective. It is an epimorphism if and only
if the divisible hull of $f(G)$ in H is all of H. This implies that f is an
extremal subobject if and only if $f(G) \subseteq H$ is a pure subgroup ([48], p.
76). In particular, divisible groups are always extremal subobjects. $\square$

Proof of Theorem 8.3 First we suppose that **D** $\subseteq$ **C** is epireflective
with reflector r. To show that **D** is closed under formation of products,
pick any family $(D_i)_{i \in I}$ of **D**-objects. Since **C** is complete the product
$C = \prod_{i \in I} D_i$ exists in **C**. Let $\pi_i \colon C \to D$ be the projections. The universal
property (2.1) applied to the reflection morphism $r_C \colon C \to r(C)$ and
the projection maps yields morphisms $\overline{\pi_i} \colon r(C) \to D_i$ with $\pi_i = \overline{\pi_i} r_C$.
Altogether these morphisms define $\overline{\pi} \colon r(C) \to C = \prod_{i \in I} D_i$ such that $\overline{\pi_i} =$

$\pi_i\bar{\pi}$. From $\pi_i\bar{\pi}r_C = \bar{\pi}_i r_C = \pi_i = \pi_i \circ id_C$ it follows that $\bar{\pi}r_C = id_C$. By hypothesis r_C is an epimrophism. Being an initial epimorphic factor of an isomorphism it is even an isomorphism. Since $\mathbf{D}$ is isomorphism–closed we conclude that $C \in ob(\mathbf{D})$.

Next we prove that $\mathbf{D}$ is closed under extremal subobjects. So, pick any $D \in ob(\mathbf{D})$ and let $m: C \to D$ be an extremal subobject in $\mathbf{C}$. By the universal property (2.1) of reflection morphisms there is unique morphism $\bar{m}: r(C) \to D$ such that $m = \bar{m}r_C$. Because r_C is an epimorphism (since r is an epireflector) and m is extremal it follows that r_C is an isomorphism, hence $C \in ob(\mathbf{D})$, as claimed.

For the converse suppose that $\mathbf{D}$ is closed under the formation of products and extremal subobjects. First we consider two parallel morphisms $a, b: A \to B$ in $\mathbf{D}$. Since $\mathbf{C}$ is complete, the equalizer $c: E(a,b) \to A$ exists in $\mathbf{C}$. Equalizers are extremal subobjects, hence it is an equalizer in $\mathbf{D}$. We see that $\mathbf{D}$ has products and equalizers, hence $\mathbf{D}$ is complete ([83], Chapter V, Section 2, Theorem 1; [60], Theorem 23.8). In particular: If $m_j: A_j \to D, j \in J$, is a family of subobjects in $\mathbf{D}$ then the intersection exists in $\mathbf{D}$.

Let $f: C \to D$ be a morphism from a $\mathbf{C}$-object to a $\mathbf{D}$-object. Since $\mathbf{C}$ is well-powered there is a set $m_j: A_j \to D, j \in J$, of subobjects in $\mathbf{C}$ representing the class of all $\mathbf{C}$-subobjects $m': D' \to D$ with $D' \in ob(\mathbf{D})$ such that f factors through D'. Thus, f can be written as $m_j c_j$ with morphisms $c_j: C \to A_j$ for every $j \in J$. Let $m: A \to D$ be the intersection of the subobjects $m_j: A_j \to D$ and write $m = m_j a_j$, $a_j: A \to A_j$. Then f factors uniquely through m, say $f = mc$, $c: C \to A$.

Claim c is an epimorphism in $\mathbf{C}$.

Proof Suppose that $g, h: A \to B$ are $\mathbf{C}$-moprhisms with $gc = hc$. If $e: E(g,h) \to A$ is the equalizer of g and h then c factors uniquely as $c = ed$, $d: C \to E(g,h)$. Since $A \in ob(\mathbf{D})$ and e is extremal it follows that $E(g,h) \in ob(\mathbf{D})$ as well. Thus, there is some $k \in J$ such that $me: E(g,h) \to A \to D$ is equivalent to $m_k: A_k \to D$, i.e., there is an isomorphism $u: E(g,h) \to A_k$ with $m_k u = me = m_k a_k e$. Since m_k is a monomorphism this implies $u = a_k e$. Thus, a_k is a final monomorphic factor of an isomorphism, and we conclude that a_k is an isomorphism, hence so is e. But if e is an isomorphism then $g = h$, and c is an epimorphism.

Starting with any $C \in ob(\mathbf{C})$ we now construct a morphism $r_C: C \to r(C)$ into a $\mathbf{D}$-object. To start with, let $f_i: C \to D_i$ be a set of representatives for the class of **C**-epimorphisms whose codomain belongs to **D** (observe that **C** is co-wellpowered). Let $f\colon C \to D = \prod_{i\in I} D_i$ be the morphism into the product defined by the f_i's. Then $D \in ob(\mathbf{D})$ and we can apply the previous discussion to this morphism $f\colon C \to D$: Choose a set $m_j\colon A_j \to D$, $j \in J$, of **C**-subobjects with domain in **D** representing the class of all **C**-subobjects $m'\colon D' \to D$, $D' \in ob(\mathbf{D})$, sucht hat f factors through m'. We define $m\colon r(C) \to D$ to be the intersection of the subobjects $m_j\colon A_j \to D$. Let $r_C\colon C \to r(C)$ be the unique morphism with $f = mr_C$. Then $r(C) \in ob(\mathbf{D})$ and r_C is an epimorphism (by the claim).

We finish the proof by showing that r_C has the universal property (2.1) of reflection morphisms: Let $g\colon C \to G$ be any morphism into a **D**-object. As before, there is a set $n_k\colon G_k \to G$, $k \in K$, of **C**-subobjects with $G_k \in ob(\mathbf{D})$ representing the class of **C**-monomorphisms $n'\colon G' \to G$ such that $G' \in ob(\mathbf{D})$ and such that there exists $g'\colon C \to G'$ with $g = n'g'$. Let $n\colon F \to G$ be the intersection of these subobjects. Then $F \in ob(\mathbf{D})$ and $g = nh$, $h\colon C \to F$. The Claim says that h is an epimorphism. Hence there is some $l \in I$ and there is an isomorphism $v\colon D_l \to F$ with $vf_l = h$. If $\pi_l\colon D = \prod_{i\in I} D_i \to D_l$ is the projection then

$$g = nh = nv\pi_l mr_C,$$

i.e., $nv\pi_l m\colon r(C) \to G$ is a morphism providing the desired factorization of g. It is unique because r_C is an epimorphism. Altogether, r_C satisfies the universal property (2.1), and the proof is finished. $\square$

Remark 8.6   In the category **POR/N**, every monomorphism $f\colon (A, P) \to (B, Q)$ can be written as the composition of two extensions $g\colon (A, P) \to (B', Q')$ and $k\colon (B', Q') \to (B, Q)$ with k extremal. To prove this claim we first choose a set $f_i\colon (A, P) \to (A_i, P_i)$, $i \in I$, representing the isomorphism classes of epimorphic extensions of (A, P) (recall that **POR/N** is co–wellpowered). Let $J \subseteq I$ be the subset of those $i \in I$ for which f factors as $k_i f_i$ with $k_i\colon (A_i, P_i) \to (B, Q)$ a monomorphism. For $i, j \in J$ we define $i \leq j$ if there is a (necessarily unique) homomorphism $f_{ji}\colon (A_i, P_i) \to (A_j, P_j)$ such that $f_j = f_{ji}f_i$. This relation is a partial order on J. Since **POR/N** is cocomplete the diagram $D\colon J \to \mathbf{POR/N}$ given by $D(i) = (A_i, P_i)$ and $D(i \leq j) = f_{ji}$ has a colimit which we denote by (B', Q'); let $g_i\colon (A_i, P_i) \to (B', Q')$ be the canonical maps. Since every f_i is an epimorphism it follows from $k_i f_i = f = k_j f_j = k_j f_{ji} f_i$ that

$k_i = k_j f_{ji}$ whenever $i \leq j$. Therefore there exists a unique morphism $k\colon (B', Q') \to (B, Q)$ with $kg_i = k_i$ for all $i \in J$. The universal property of colimits implies that the canonical homomorphism $g\colon (A, P) \to (B', Q')$ is an epimorphism. It is also an extension since $f = kg$. One checks easily that k is injective and that $Q' = k^{-1}(Q)$. To show that k is extremal, let $k = le$ with $e\colon (B', Q') \to (B'', Q'')$ an epimorphism and $l\colon (B'', Q'') \to (B, Q)$ any morphism. We define $\overline{B''} = l(B'') \subseteq B$ and $\overline{Q''} = \overline{B''} \cap Q$. If $\pi\colon (B'', Q'') \to (\overline{B''}, \overline{Q''})$ is the canonical homomorphism then $l = \bar{l}\pi$ with $\bar{l}\colon (\overline{B''}, \overline{Q''}) \to (B, Q)$ injective. Now k also factors as $\bar{l}(\pi e)$ with πe an epimorphic extension and $\bar{l}$ monomorphic. Thus, $\pi eg\colon (A, P) \to (\overline{B''}, \overline{Q''})$ can be identified with one of the $f_i, i \in J$. But then πe must be an isomorphism. Because e is an initial epimorphic factor of an isomorphism it follows that e is also an isomorphism. We conclude that k is an extremal subobject.

The proof of the existence of an extremal subobject $B', Q')$ of (B, Q) containing (A, P) does not give us any information about the elements of B belonging to B'. This remains an important open problem. A related question is to find practical criteria to decide whether a given subobject is extremal.   □

If $\mathbf{C}$ satisfies the hypotheses of Theorem 8.3 and $\mathbf{D}_i$, $i \in I$, is any class of epireflective subcategories then it follows immediately that their intersection is epireflective as well. Thus, the class of all epireflective subcategories is a complete lattice. If X is any class of $\mathbf{C}$-objects then there is a smallest epireflective subcategory containing X. This category can be described more constructively: Let $\mathbf{S_eP}(X)$ denote the full subcategory of $\mathbf{C}$ whose objects are those $\mathbf{C}$-objects that are extremal subobjects of products of members of X. Since we are looking at a category that is complete and wellpowered it follows from [60], Propositions 34.2, 34.4, that $\mathbf{S_eP}(X)$ is an epireflective subcategory of $\mathbf{C}$. Obviously, it is the smallest one containing the class X. We shall call it the *epireflective subcategory generated by* X.

We have just seen that an epireflective subcategory can be specified by a class of objects. Another useful way of creating epireflective subcategories is through a class of epimorphisms. This is the content of the other Characterization Theorem. Before stating it we recall some terminology. Let $\mathcal{E}$ be some class of morphisms in the category $\mathbf{C}$. An object I of $\mathbf{C}$ is *$\mathcal{E}$-injective* ([53], Abstract) or is *$\mathcal{E}$-extendible* ([52], §6) if for any morphism $(e\colon C \to C') \in \mathcal{E}$ and any $\mathbf{C}$-morphism $f\colon C \to I$ there is a morphism $f' \colon C' \to I$ with $f = f'e$. We write $Inj(\mathcal{E})$ for the class of $\mathcal{E}$–injective objects.

Theorem 8.7 (Second Characterization Theorem) Let **C** be a complete, wellpowered and co–wellpowered category, let $\mathbf{D} \subseteq \mathbf{C}$ be a full and isomorphism–closed subcategory. Then **D** is epireflective if and only if there is a class of epimorphisms $\mathcal{E}$ such that $Inj(\mathcal{E})$ is the class of **D**–objects.

Proof First suppose that **D** is epireflective. Then we choose $\mathcal{E}$ to be the class of all reflection morphisms $r_C \colon C \to r(C)$, $C \in ob(\mathbf{C})$. By hypothesis $\mathcal{E}$ consists of epimorphisms. Pick $D \in ob(\mathbf{D})$ and let $f \colon C \to D$ be any **C**-morphism. By the universal property (2.1) there is a morphism $\overline{f} \colon r(C) \to D$ such that $f = \overline{f}r_C$. Thus, $D \in Inj(\mathcal{E})$, and the inclusion $ob(\mathbf{D}) \subseteq Inj(\mathcal{E})$ has been proved. For the reverse inclusion, let $D \in Inj(\mathcal{E})$ and consider the identity morphism id_D. There exists a morphism $f \colon r(D) \to D$ sucht that $id_D = fr_D$. Then r_D is an isomorphism, being an initial epimorphic factor of an isomorphism. Since **D** is isomorphism–closed we conclude that $D \in ob(\mathbf{D})$, proving $Inj(\mathcal{E}) = ob(\mathbf{D})$.

Now suppose that $ob(\mathbf{D}) = Inj(\mathcal{E})$. We check that $Inj(\mathcal{E})$ is closed under products and under extremal subojects. Together with Theorem 8.3 this will prove that **D** is epireflective. Closedness under products is immediate from the notion of $\mathcal{E}$-injectivity and from the universal mapping property of products. So, we only have to deal with extremal subobjects. Suppose that $a \colon B \to A$ is an extremal subobject in **C** with $A \in Inj(\mathcal{E})$. Pick any morphism $e \colon C \to D$ in $\mathcal{E}$ and any morphism $f \colon C \to B$. Then there exists some $g \colon D \to A$ such that $af = ge$. In A let $m_i \colon A_i \to A$, $i \in I$, be a set of subobjects representing the class of all subobjects $m \colon A' \to A$ with the property that there exists some $g' \colon D \to A'$ with $g = mg'$. (Recall that **C** is wellpowered.) Since **C** is complete the intersection $m_0 \colon A_0 \to A$ of all the subobjects $m_i \colon A_i \to A$ exists. Factoring $g = m_0g_0$, $g_0 \colon D \to A_0$, we know from the proof of Theorem 8.3 that g_0 is an epimorphism. Thus, the composition $g_0e \colon C \to A_0$ is an epimorphism as well.

Inside A pick a set $n_j \colon B_j \to A, j \in J$, of subobjects representing the class of all subobjects $n' \colon B' \to A$ such that both m_0 and a factor through n'. As before, let $n_0 \colon B_0 \to A$ be the intersection of these subobjects and consider the commutative diagram:

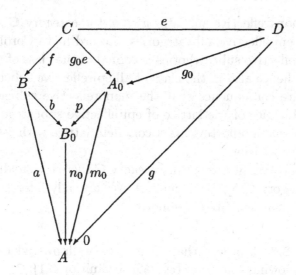

It is claimed that b is an epimorphism. To prove this, let $u, v : B_0 \to U$ be morphisms with $ub = vb$. If $w : E(u,v) \to B_0$ is the equalizer then $b = wb'$, $b' : B \to E(u,v)$. Because of $bf = pg_0e$ we see that $upg_0e = vpg_0e$, even: $up = vp$ (as g_0e is an epimorphism). Thus, p factors as $p = wp'$, $p' : A_0 \to E(u,v)$. Together this shows that the subobject $n_0w : E(u,v) \to A$ is equivalent to one of the subobjects $n_j : B_j \to A$:

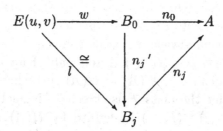

Since $n_j{}'$ is a final monomorphic factor of the isomorphism l it follows that $n_j{}'$ is an isomorphism, hence so is w. But then $u = v$, and b is an epimorphism.

Now $a = n_0 b$ is a factorization of an extremal subobject with an initial epimorphic factor. This factor is an isomorphism by the definition of extremal subobjects. But then

$$(b^{-1}pg_0)e = b^{-1}(bf) = f$$

is the desired factorization of f, i.e., B is $\mathcal{E}$–injective. $\qquad\square$

The class of monoreflective subcategories of the category $\mathbf{C}$ is contained in the class of epireflective subcategories. According to Corollary 8.2 the class of monoreflective subcategories is cofinal in the class of all reflective subcategories, hence also in the class of all epireflective subcategories. It is an immediate consequence that the monoreflective subcategories are a complete sublattice of the lattice of epireflective subcategories. Correspondingly, the monoreflectors are a complete lattice with respect to the weaker–stronger relation.

Under certain conditions on the category $\mathbf{C}$ there is a smallest monoreflective subcategory ([55], Theorem 2.2). We are only interested in a special case which can be stated as follows:

Proposition 8.8 Suppose that $\mathbf{C}$ is a co–wellpowered category and satisfies the following condition (cf. [55], Definition 3.1):

(8.9) For every directed small diagram $F\colon J \to \mathbf{C}$ such that each $F(i \to j)$ is monomorphic, the direct limit $\lim_{\to} F$ exists and the canonical morphisms $F(i) \to \lim_{\to} F$ are monomorphisms.

Then there is a smallest monoreflective subcategory of $\mathbf{C}$.

Proof (cf. [55]) Let $\mathcal{F}$ be the class of all functorial extension operators in $\mathbf{C}$. The elements of $\mathcal{F}$ are denoted by (F, t_F). One notes that $t_{F,A}\colon A \to F(A)$ is always an epimorphism. The class $\mathcal{F}$ is partially ordered by: $(T, t_F) \le (G, t_G)$ if and only if for each $A \in ob(\mathbf{C})$ there is a monomorphism $m_A\colon F(A) \to G(A)$ such that $t_{G,A} = m_A t_{F,A}$. With this partial order the class $\mathcal{F}$ is directed. Namely, if the *composition* $(H, t_H) = (G, t_G) \circ (F, t_F)$ is defined by $H(A) = G(F(A))$ and $t_{H,A} = t_{G,F(A)} \circ t_{F,A}$ then $(F, t_F) \le (H, t_H)$ and $(G, t_G) \le (H, t_H)$. By co–wellpoweredness, each class $(t_{F,A})_{F \in \mathcal{F}}$ of epimorphisms has a *set* of representatives. Let $\mathcal{F}_A \subseteq \mathcal{F}$ be a *set* such that $\{t_{F,A}; F \in \mathcal{F}_A\}$ represents these epimorphisms. Since this is a directed partially ordered set, condition (8.9) shows that the direct limit $L(A) = \lim_{\to} F(A)$ exists; let $l_A\colon A \to L(A)$ be the canonical monomorphism. One checks that $(L, l) \in \mathcal{F}$. From the construction it follows that this is the largest functorial extension operator of $\mathbf{C}$. Hence it is idempotent and belongs to a monoreflector. Clearly, this is the strongest monoreflector of $\mathbf{C}$. $\square$

One can check that the hypotheses of Proposition 8.8 are satisfied by the category **POR/N**, hence there is a smallest monoreflective subcategory in **POR/N**. But the proposition does not tell us what the objects of this subcategory are. However, we know enough about **POR/N** already to clarify this point:

Theorem 8.10 The subcategory **SAFR** is the smallest monoreflective subcategory of **POR/N**.

Proof We know that **SAFR** ⊆ **POR/N** is monoreflective (Theorem 7.18). Every monoreflective subcategory of **POR/N** must contain every epicomplete object ([55], Corollary 5.2(a)). Since **SAFR** consists exactly of the epicomplete objects (Corollary 7.20) the proof is finished.   □

Corollary 8.11 The class of monoreflective subcategories of **POR/N** is a complete sublattice of the lattice of epireflective subcategories.   □

The next result shows an extremal property of **SAFR** as a reflective subcategory of **PREOR**. It is possible to obtain Theorem 8.10 as a corollary of this result. However, the above proof of Theorem 8.10 is so simple that it is impossible to make it shorter.

Theorem 8.12 The category **SAFR** is the smallest reflective subcategory of **PREOR** containing all real closed fields.

Proof We know that **SAFR** is reflective in **PREOR** and that all real closed fields are contained in **SAFR**. Now let $r\colon \textbf{PREOR} \to \textbf{D}$ be a reflector and suppose that **D** contains every real closed field. It is claimed that **SAFR** ⊆ **D**, i.e., that $r_{(A,P)}$ is an isomorphism if $(A, P) \in ob(\textbf{SAFR})$.

The composition

$$\sigma_{r(A,P)}r_{(A,P)}\colon (A,P) \to r(A,P) \to \sigma(r(A,P))$$

is a homomorphism of von Neumann regular f–rings. The main step in the proof is to show that $\sigma_{r(A,P)}r_{(A,P)}$ is an isomorphism. This is done by using (7.13). First recall that the real spectra of $r(A,P)$ and $\sigma(r(A,P))$ are canonically bijective to each other and that the corresponding real closed residue fields are isomorphic (Proposition 7.9, Proposition 7.10). So it suffices to show that $Sper(r_{(A,P)})$ is a bijection and that corresponding real closed residue fields of (A,P) and $r(A,P)$ are isomorphic.

For any $\alpha \in Sper(A,P)$ the universal property of the reflection $r_{(A,P)}$ yields a uniqe homomorphism $\overline{\rho_\alpha} \colon r(A,P) \to \rho(\alpha)$ such that $\rho_\alpha = \overline{\rho_\alpha} r_{(A,P)}$. The prime cone of $r(A,P)$ defined by $\overline{\rho_\alpha}$ is denoted by α'. Thus, $r_{(A,P)}^{-1}(\alpha') = \alpha$, $\rho_{\alpha'} = \overline{\rho_\alpha}$ and $\rho_{r_{(A,P)}/\alpha'} \colon \rho(\alpha) \to \rho(\alpha')$ is an isomorphism. To check the hypotheses of (7.13) it suffices to prove that $Sper(r_{(A,P)})$ is injective. So, pick $\beta, \gamma \in Sper(r(A,P))$ with

$$r_{(A,P)}^{-1}(\beta) = \alpha = r_{(A,P)}^{-1}(\gamma)$$

and suppose that $\beta \not\subseteq \gamma$. There is some $a \in r(A,P)$ with $a(\beta) \le 0$ and $a(\gamma) > 0$. The amalgamation property of real closed fields provides a commutative diagram

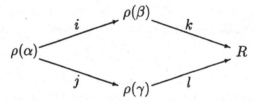

of real closed fields (where i and j are induced by $r_{(A,P)}$). Then

$$k\rho_\beta r_{(A,P)} = ki\rho_\alpha = lj\rho_\alpha = l\rho_\gamma r_{(A,P)}.$$

The universal property of $r_{(A,P)}$ implies that $k\rho_\beta = l\rho_\gamma$. On the other hand, $k\rho_\beta(a) \le 0$ and $l\rho_\gamma(a) > 0$. This contradiction shows that $Sper(r_{(A,P)})$ is injective and, hence, (7.13) proves that $\sigma_{r(A,P)} r(A,P)$ is an isomorphism.

Because of

$$id_{r(A,P)} r(A,P) = r_{(A,P)} = r_{(A,P)}(\sigma_{r(A,P)} r(A,P))^{-1} \sigma_{r(A,P)} r(A,P)$$

the universal property of $r_{(A,P)}$ implies that

$$id_{r(A,P)} = r_{(A,P)}(\sigma_{r(A,P)} r(A,P))^{-1} \sigma_{r(A,P)}.$$

But then $r_{(A,P)}$ is an isomorphism with inverse $(\sigma_{r(A,P)} r(A,P))^{-1} \sigma_{r(A,P)}$. $\square$

Corollary 8.13 A full subcategory $\mathbf{D} \subseteq \mathbf{POR/N}$ is monoreflective if and only if it contains all real closed fields and is closed under the formation of products and extremal subobjects.   □

If (A, P) is any totally ordered intermediate ring of $\mathbb{Z}$ and R_0 then the category of all (A, P)–algebras in **PREOR** is denoted by **(A,P)PREOR**.
The following generic notation will be used: If $\mathbf{C} \subseteq \mathbf{PREOR}$ is a subcategory then $\mathbf{(A,P)C} = \mathbf{C} \cap \mathbf{(A,P)PREOR}$. We shall show that
$\mathbf{(A,P)POR/N}$ is a monoreflective subcategory of **POR/N**. There are
three different tools available for doing this: the universal property (2.1),
Theorem 8.3, and Theorem 8.7. We choose to use the universal property since this includes the determination of the reflections of all reduced
porings. And this is a piece of information we want to have in any event.

First of all, note that the canonical homomorphism $\mathbb{Z} \to (A, P)$ is an
epimorphism in **POR/N** (Theorem 5.2). Every real closed field R contains R_0 as a subfield, hence $R_0 \subseteq \Pi(B, Q)$ for any reduced poring (B, Q).
Thus, $(A, P) \subseteq \Pi(B, Q)$ is a subobject as well. We define $r(B, Q)$ to be
the subring of $\Pi(B, Q)$ generated by A and $im(\Delta_{(B,Q)})$, together with
the partial order generated by P and $\Delta_{(B,Q)}(Q)$. Let $r_{(B,Q)} : (B, Q) \to$
$r(B, Q)$ be the morphism $\Delta_{(B,Q)}$ with restricted codomain. Obviously,
$r_{(B,Q)}$ is injective and $r(B, Q)$ is an (A, P)–algebra. It is claimed that
$r_{(B,Q)}$ satisfies the universal property (2.1) with respect to the subcategory $\mathbf{(A,P)POR/N} \subseteq \mathbf{POR/N}$.

Claim 1   The **POR/N**–morphism $r_{(B,Q)}$ is an epimorphism.
Proof Suppose that $f, g : r(B, Q) \to (B', Q')$ are **POR/N**–morphisms
with $fr_{(B,Q)} = gr_{(B,Q)}$. Inside $r(B, Q)$ the set

$$E = \{a \in r(B, Q); \ f(a) = g(a)\}$$

is a subring containing $im(r_{(B,Q)})$. We equip E with the restriction of
the partial order of $r(B, Q)$ and obtain the reduced poring (E, Q_E). The
homomorphism $\mathbb{Z} \to (B', Q')$ has a unique extension $(A, P) \to (B', Q')$,
hence $f|_A = g|_A$. Therefore E contains A and $im(r_{(B,Q)})$; since $r(B, Q)$
is generated by A and $im(r_{(B,Q)})$ we conclude that $E = r(B, Q)$. This
means that $f = g$.   □

Claim 2 If $f : (B, Q) \to (B', Q')$ is any homomorphism into an object
of **(A,P)POR/N** then there is some $\overline{f} : r(B, Q) \to (B', Q')$ such that
$f = \overline{f}r_{(B,Q)}$.
Proof The homomorphism $\Pi(f) : \Pi(B, Q) \to \Pi(B', Q')$ is a morphism belonging to **(A,P)POR/N**. Note that $\Pi(f)(im(r_{(B,Q)})) =$

$\Delta_{(B',Q')}(im(f))$ and that $A \subseteq \Pi(f)^{-1}(im(\Delta_{(B',Q')}))$ (because (B',Q') is an (A,P)–algebra). Since $r(B,Q)$ is generated by A and $im(r_{(B,Q)})$ it follows that $\Pi(f)$ can be restricted to a ring homomorphism $\overline{f} : r(B,Q) \to B'$ with $f = \overline{f} r_{(B,Q)}$. Because of $f(Q) \subseteq Q'$ and $P \subseteq Q'$ it is clear that $\overline{f}$ is also order preserving, i.e., $\overline{f}$ is the desired map. $\square$

We summarize the content of this discussion in

Proposition 8.14 If (A,P) is a totally ordered intermediate ring of $\mathbb{Z}$ and R_0 then the category **(A,P)POR/N** is a monoreflective subcategory of **POR/N**. The reflection of any reduced poring (B,Q) in **(A,P)POR/N** is the subring of $\Pi(B,Q)$ generated by $\Delta_{(B,Q)}(B)$ and the image of A in $\Pi(B,Q)$ together with the weakest partial order containing $\Delta_{(B,Q)}(Q)$ and the image of P. $\square$

Let $\mathbf{D} \subseteq \mathbf{POR/N}$ be any monoreflective subcategory with reflector $r: \mathbf{POR/N} \to \mathbf{D}$. Let (A,P) be some totally ordered ring between $\mathbb{Z}$ and R_0. Then **(A,P)D** is a monoreflective subcategory of **POR/N**. From Theorem 8.10 and Corollary 8.2 it follows that $r(\mathbb{Z})$ is a totally ordered intermediate ring of $\mathbb{Z}$ and R_0. Suppose that (B,Q) belongs to **D**. By the universal property of reflections the homomorphism $f: \mathbb{Z} \to (B,Q)$ has a unique extension $\overline{f} : r(\mathbb{Z}) \to (B,Q)$. Thus, (B,Q) is an $r(\mathbb{Z})$–algebra, and we see that $\mathbf{D} = \mathbf{r(\mathbb{Z})D}$. Evidently, $r(\mathbb{Z})$ is the largest totally ordered intermediate ring (A,P) of $\mathbb{Z}$ and R_0 such that $\mathbf{D} = \mathbf{(A,P)D}$.

9 Constructing reflectors

In the previous section it was shown that the class of all monoreflectors of **POR/N** is a complete lattice with respect to the stronger–weaker relation. In this section we study four methods to construct new reflectors from one or two given reflectors. According to the four constructions, the section is subdivided into four parts. The constructions allow us to move around in the lattice, to relate different parts of the lattice to each other and to explore small parts of it, thus greatly enhancing our overall understanding of the lattice and of individual reflectors.

The first two constructions are derived directly from the lattice structure. It is their purpose to determine, as explicitly as possible, the reflection of any $(A, P) \in ob(\mathbf{POR/N})$ under the supremum and the infimum of two reflectors $r \colon \mathbf{POR/N} \to \mathbf{D}$ and $s \colon \mathbf{POR/N} \to \mathbf{E}$. In the worst case, both $r \vee s(A, P)$ and $r \wedge s(A, P)$ cannot really be determined explicitly. On the other hand, the nicest conceivable situation is that $r \vee s(A, P) = r(s(A, P))$ or $= s(r(A, P))$ and that $r \wedge s(A, P) = r(A, P) \cap s(A, P)$ (the intersection is formed inside $\sigma(A, P)$, cf. Theorem 8.10, Corollary 8.2). Fortunately, it is not at all unusual for two reflectors to be that nice. The third construction is a variation of the description of $r \wedge s(A, P)$ as $r(A, P) \cap s(A, P)$. If the reflector s is replaced by a subfunctor F ([50], 0.1.7.1) of the complete ring of functions functor Π (section 6) then one can form the intersection $r(A, P) \cap F(A, P)$ in $\Pi(A, P)$. For example, $F(A, P) = \prod_{\alpha \in Sper(A,P)} \kappa_{(A,P)}(\alpha)$ is such a subfunctor. Sufficient conditions for the intersection to be a reflector are given in Proposition 9C.1. For the fourth construction we assume that the reflector s is properly stronger than r. Using auxiliary subcategories of **POR/N** it is frequently possible to find many intermediate reflectors between r and s. Using this method it will be shown that the class of monoreflectors of **POR/N** is a proper class (Theorem 9D.3).

A The supremum of two reflectors

Let $\mathbf{C}$ be a wellpowered, co–wellpowered, complete and cocomplete category satisfying condition (8.9), let $r \colon \mathbf{C} \to \mathbf{D}$ and $s \colon \mathbf{C} \to \mathbf{E}$ be two epireflectors. Then $\mathbf{D} \cap \mathbf{E} \subseteq \mathbf{C}$ is epireflective (by Theorem 8.3) with reflector $r \vee s \colon \mathbf{C} \to \mathbf{D} \cap \mathbf{E}$. Writing down the reflective subcategory belonging to $r \vee s$ is no problem. However, in applications one frequently works with the reflections of the members of $\mathbf{C}$. Then it is necessary to know them as explicitly as possible. Therefore we attempt to describe $r \vee s(C)$ for any $C \in ob(\mathbf{C})$. The most favorable situation is dealt with in the next proposition. The easy proof is left to the reader.

Proposition 9A.1 Suppose that $r \colon \mathbf{C} \to \mathbf{D}$ and $s \colon \mathbf{C} \to \mathbf{E}$ are epireflectors. If $s(D) \in ob(\mathbf{D})$ for each $D \in ob(\mathbf{D})$ then the $r \vee s$–reflection of $C \in ob(\mathbf{C})$ is $C \xrightarrow{r_C} r(C) \xrightarrow{s_{r(C)}} s(r(C))$. $\qquad\qquad\square$

If the proposition is not applicable then it is natural to consider the sequence

$$(9A.2) \quad C \xrightarrow{r_C} r(C) \xrightarrow{s_{r(C)}} sr(C) \xrightarrow{r_{sr(C)}} rsr(C) \longrightarrow \cdots .$$

The direct limit exists since $\mathbf{C}$ is cocomplete; denote it by $t(C)$ and let $t_C : C \to t(C)$ be the canonical homomorphism. For any $\mathbf{C}$–morphism $f : C \to C'$ the universal property of direct limits provides a morphism $t(f) : t(C) \to t(C')$ such that t becomes a functor $\mathbf{C} \to \mathbf{C}$. The canonical homomorphisms t_C are the components of a natural transformation $Id \to t$. We assume now that both r and s are monoreflectors. Then t together with the natural transformation is a functorial extension operator. By composition one defines a transfinite sequence of functorial extension operators:

$$t_0 = t,$$
$$t_{\alpha+1} = t \circ t_\alpha \text{ for ordinal numbers } \alpha,$$
$$t_\alpha = \lim_{\overrightarrow{\beta < \alpha}} t_\beta \text{ for limit ordinals } \alpha.$$

As in the proof of Proposition 8.8, the direct limit $T = \lim_{\overrightarrow{\alpha}} t_\alpha$ exists together with a natural transformation $\tau \colon Id_{\mathbf{C}} \to T$. It is clear that (T, τ) is an idempotent functorial extension operator, hence F is the composition of a monoreflector $u \colon \mathbf{C} \to \mathbf{F}$ with the inclusion functor $\mathbf{F} \to \mathbf{C}$ and τ is the unit of the corresponding adjunction (cf. section 2). From the construction it is clear that $\mathbf{F} = \mathbf{D} \cap \mathbf{E}$ and $u = r \vee s$.

Eventually it will be shown that Proposition 9A.1 is applicable to all pairs of monoreflectors $r \colon \mathbf{POR/N} \to \mathbf{D}$ and $s \colon \mathbf{POR/N} \to \mathbf{E}$ that have occured so far. At the moment we consider just one example:

Example 9A.3 *The supremum of* $r \colon \mathbf{POR/N} \to (\mathbf{A,P})\mathbf{POR/N}$ *and*
                an arbitrary monoreflector $s \colon \mathbf{POR/N} \to \mathbf{E}$

Suppose that (B,Q) belongs to the subcategory $(\mathbf{A,P})\mathbf{POR/N}$ and that
$f \colon (A,P) \to (B,Q)$ is the unique morphism giving (B,Q) the structure
of an (A,P)–algebra. The diagram

$$
\begin{array}{ccc}
(A,P) & \xrightarrow{\ f\ } & (B,Q) \\[4pt]
{\scriptstyle s_{(A,P)}}\Big\downarrow & & \Big\downarrow{\scriptstyle s_{(B,Q)}} \\[4pt]
s(A,P) & \xrightarrow{\ s(f)\ } & s(B,Q)
\end{array}
$$

shows that $s(B,Q)$ is an (A,P)–algebra as well. Thus, the hypothesis of
Proposition 9A.1 is satisfied. If $(B,Q) \in ob(\mathbf{POR/N})$ then

$$
(B,Q) \xrightarrow{\ r_{(B,Q)}\ } r(B,Q) \xrightarrow{\ s_{r(B,Q)}\ } s(r(B,Q))
$$

is the reflection morphism $(s \vee r)_{(B,Q)}$.   □

B The infimum of two reflectors

Suppose that $r \colon \mathbf{POR/N} \to \mathbf{D}$ and $s \colon \mathbf{POR/N} \to \mathbf{E}$ are monoreflectors.
It follows from Theorem 8.3 that there is a smallest epireflective subcat-
egory $\mathbf{G} \subseteq \mathbf{POR/N}$ that contains both $\mathbf{D}$ and $\mathbf{E}$. The corresponding
reflector is the infimum $r \wedge s$ with respect to the weaker–stronger rela-
tion. Other than in the case of $r \vee s$, in general there exists no simple
characterization of the $\mathbf{G}$–objects. This increases the importance of de-
termining $r \wedge s$ on the level of reflections. We shall prove

Proposition 9B.1   Let $r \colon \mathbf{POR/N} \to \mathbf{D}$ and $s \colon \mathbf{POR/N} \to \mathbf{E}$ be
monoreflectors. For the reduced poring (A,P), let $u_{(A,P)} \colon (A,P) \to$
$r(A,P) \cap s(A,P)$ be the morphism $\sigma_{(A,P)}$ with restricted codomain. (Recall that both $r(A,P)$ and $s(A,P)$ can be considered as subrings of
$\sigma(A,P)$, cf. Corollary 8.2, Theorem 8.10.) Then the $u_{(A,P)}$ are the reflections belonging to $r \wedge s$ if and only if $u_{(A,P)}$ is an epimorphism for every

(A, P).

The proposition will turn out to be a corollary of the following more general discussion. As in subsection A, let $\mathbf{C}$ be a wellpowered, co–wellpowered, complete and cocomplete category; let $r \colon \mathbf{C} \to \mathbf{D}$ and $s \colon \mathbf{C} \to \mathbf{E}$ be epireflectors. Let $t \colon \mathbf{C} \to \mathbf{F} = \mathbf{D} \cap \mathbf{E}$ be the supremum of r and s (subsection A). For any $C \in ob(\mathbf{C})$, let $\varphi_C \colon r(C) \to t(C)$ and $\psi_C \colon s(C) \to t(C)$ be the unique morphisms of Proposition 8.1 (c). The cartesian square

$$
\begin{array}{ccc}
u(C) & \xrightarrow{\ \pi_{r(C)}\ } & r(C) \\
{\scriptstyle \pi_{s(C)}} \downarrow & & \downarrow {\scriptstyle \varphi_C} \\
s(C) & \xrightarrow{\ \psi_C\ } & t(C)
\end{array}
$$

exists since $\mathbf{C}$ is complete. Let $u_C \colon C \to u(C)$ be the unique morphism with $r_C = \pi_{r(C)} u_C$ and $s_C = \pi_{s(C)} u_C$. For any morphism $f \colon C \to C'$ there is a unique morphism $u(f) \colon u(C) \to u(C')$ such that $u_{C'} f = u(f) u_C$. Therefore $u \colon \mathbf{C} \to \mathbf{C}$ is a functor and the u_C are the components of a natural transformation $Id_{\mathbf{C}} \to u$. If $v = r \wedge s$ then there is a unique morphism $\alpha_C \colon v(C) \to u(C)$ such that $\pi_{r(C)} \alpha_C \colon v(C) \to r(C)$ and $\pi_{s(C)} \alpha_C \colon v(C) \to s(C)$ are the morphisms of Proposition 8.1 (c). The nicest conceivable situation is that α_C is an isomorphism, i.e., that $u = r \wedge s$. The following result gives a criterion for when this is the case.

Proposition 9B.2 The fibre product functor u coincides with $r \wedge s$ if and only if u_C is an epimorphism for every $\mathbf{C}$–object C.

Proof If $u = r \wedge s$ then u is an epireflector, hence the reflection u_C is always epimorphic. Conversely, assume that every u_C is an epimorphism. Let $\mathbf{G} \subseteq \mathbf{C}$ be the full subcategory with objects G such that

$$
\begin{array}{ccc}
G & \xrightarrow{\ r_G\ } & r(G) \\
{\scriptstyle s_G} \downarrow & & \downarrow \\
s(G) & \longrightarrow & t(G)
\end{array}
$$

is cartesian. We shall show that every $u(C)$ belongs to $\mathbf{G}$. The commutative diagram

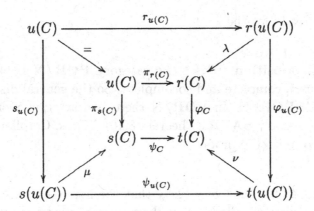

(where λ is the unique morphism $r(u(C)) \to r(C)$ with $\lambda r_{u(C)} = \pi_{r(C)}$, and μ and ν are defined similarly) shows that $id_{u(C)}$ factors as $f u_{u(C)}$ for some $f : u(u(C)) \to u(C)$. Thus, $u_{u(C)}$ is an initial epimorphic factor of an isomorphism, hence $u_{u(C)}$ is an isomorphism. This means that u can be considered as a functor $\mathbf{C} \to \mathbf{G}$. It is clear that every u_C has the universal property (2.1), hence u is a reflector, even an epireflector since the u_C are epimorphic. Proposition 8.1 implies that $r \wedge s \le u$ and that $u \le r, s$, i.e., $u = r \wedge s$. $\qquad \Box$

Even if the equivalent conditions of the proposition are satisfied, in general the description of $r \wedge s$ is less than explicit since it involves a fibre product (which may be difficult to determine) and the join $t = r \vee s$ (which may be known only as a direct limit). But there are important cases in which Proposition 9B.2 can be improved. Suppose that there is a monoreflector $w \colon \mathbf{C} \to \mathbf{H}$ which is stronger than t. Then all the reflectors occurring in the preceding discussion are monoreflectors; if x is one of them, then $w_{x(C)} \colon x(C) \to w(C)$ is a subobject (Corollary 8.2). In particular, since $t(C) \to w(C)$ is a subobject, the fibre products of $r(C)$ and $s(C)$ over $t(C)$ or over $w(C)$ are canonically isomorphic. So, in this situation it is not necessary to work with, and have a lot of information about, the reflector t. Moreover, $u(C)$ may be identified with the intersection of $r(C)$ and $s(C)$ in $w(C)$. Altogether we obtain the following improvement of Proposition 9B.2:

Corollary 9B.3 Suppose that there is a monoreflector $w : \mathbf{C} \to \mathbf{H}$ that is stronger than both r and s. Then, for every $C \in ob(\mathbf{C})$, $(r \wedge s)(C) = r(C) \cap s(C)$ (intersection inside $w(C)$) if and only if every canonical monomorphism $C \to r(C) \cap s(C)$ is an epimorphism. $\qquad \Box$

Now we are ready for the

Proof of Proposition 9B.1   The category **POR/N** is wellpowered, co–wellpowered, complete and cocomplete. So the general discussion applies with **C=POR/N**. In **POR/N** there is a smallest monoreflective subcategory, namely **SAFR** (Theorem 8.10). Thus, Corollary 9B.3 can be applied to finish the proof.   □

Remark 9B.4   It seems very likely that the hypotheses of Proposition 9B.1 are not always satisfied, although we do not have an example proving this. If the hypotheses fail then we still have a description of the infimum of r and s, however it is less explicit: If $u_{(A,P)} \colon (A,P) \to u(A,P)$ is the extension of Proposition 9B.1, let $v(A,P) \subseteq u(A,P)$ be the largest intermediate object of (A,P) and $u(A,P)$ such that $v_{(A,P)} \colon (A,P) \to v(A,P)$ (this is $u_{(A,P)}$ with restricted codomain) is epimorphic (Remark 8.6). We claim that $v_{(A,P)} = (r \wedge s)_{(A,P)}$: Let $\mathbf{G} \subseteq \mathbf{POR/N}$ be the reflective subcategory belonging to $r \wedge s$. Then $u(A,P) \in ob(\mathbf{G})$ since **G** is complete, and $v(A,P) \in ob(\mathbf{G})$ since **G** is closed under extremal subobjects (Theorem 8.3). Now let $\mathbf{H} \subseteq \mathbf{POR/N}$ be the full subcategory of objects for which $v_{(A,P)}$ is an isomorphism. It is easy to check that $\mathbf{D} \subseteq \mathbf{H}$ and $\mathbf{E} \subseteq \mathbf{H}$. We shall show that $\mathbf{H} \subseteq \mathbf{G}$ and that **H** is reflective. Then **G=H** since **G** is the smallest reflective subcategory containing both **D** and **E**.

The first claim is almost immediate: If $v_{(A,P)}$ is an isomorphism then (A,P) is an extremal subobject of $u(A,P) \in ob(\mathbf{G})$. It was noted above that this implies $(A,P) \in ob(\mathbf{G})$.

As a first step towards showing reflectivity of **H** we note that the construction of $v(A,P)$ is idempotent. For, one shows easily that $r(v_{(A,P)})$ and $s(v_{(A,P)})$ are isomorphisms. Therefore $u(A,P)$ can be identified canonically with $u(v(A,P))$. Since $v(A,P) \subseteq u(A,P)$ is extremal it follows that $v_{v(A,P)}$ is an isomorphism. The next and final step is to show that $v_{(A,P)}$ has the universal property (2.1). So, let $f \colon (A,P) \to (B,Q)$ be a morphism into some **H**–object. If there exists some $\overline{f} \colon v(A,P) \to (B,Q)$ with $\overline{f} v_{(A,P)} = f$ then $\overline{f}$ is unique since $v_{(A,P)}$ is an epimorphism. It remains to discuss the existence of $\overline{f}$. General properties of cartesian squares yield a homomorphism $u(f) \colon u(A,P) \to u(B,Q)$ with $u_{(B,Q)} f = u(f) u_{(A,P)}$. We claim that $u(f)$ maps $v(A,P)$ into $v(B,Q) = (B,Q)$. Recall that $v(A,P)$ is the colimit of all (A_i,P_i) such that there are an epimorphic extension $f_i \colon (A,P) \to (A_i,P_i)$ and a monomorphism $g_i \colon (A_i, P_i) \to u(A, P)$ with $u_{(A,P)} = g_i f_i$. For each i consider the diagram

$$
\begin{array}{ccccc}
u_{(A,P)} \colon (A,P) & \xrightarrow{\ f_i\ } & (A_i, P_i) & \xrightarrow{\ g_i\ } & u(A,P) \\[2pt]
{\scriptstyle f}\Big\downarrow & & {\scriptstyle \overline{f}_i}\Big\downarrow & & \Big\downarrow {\scriptstyle u(f)} \\[2pt]
u_{(B,Q)} \colon (B,Q) & \xrightarrow{\ f_i'\ } & (B_i, Q_i) & \xrightarrow{\ g_i'\ } & u(B,Q)
\end{array}
$$

in which the left hand square is cocartesian (note that **POR/N** is cocomplete). Since f_i is an epimorphism, so is f_i'. But $u_{(B,Q)}$ is an extremal subobject, hence f_i' is an isomorphism. Therefore we regard $\overline{f}_i$ as a morphism into (B, Q). It is obvious that the $\overline{f}_i$ fit together to define a morphism $\overline{f} \colon v(A, P) \to (B, Q)$ with $\overline{f} v_{(A,P)} = f$. $\qquad\square$

We finish this subsection with an example.

Example 9B.5 *The infimum of the reflectors Rep:* **POR/N** $\to$ **REPPOR** *and* $r\colon$ **POR/N** $\to$ **BIPOR/N** *is given by intersections*

None of the reflectors *Rep* and r is weaker than the other. For, the following totally ordered domain belongs to **REPPOR**, but not to **BIPOR/N**: Consider the polynomial ring $\mathbb{R}[T]$ with the total order defined by

$$
\sum_{i=0}^{n} a_i T^i > 0 \ \text{(with } a_n \neq 0\text{) if } a_n > 0.
$$

Conversely, the following partially ordered domain is in **BIPOR/N**, but not in **REPPOR**: As the underlying ring take $\mathbb{R}[[T]]$. The partial order is defined by

$$
\sum_{i \geq n} a_i T^i > 0 \ \text{(with } a_n \neq 0\text{) if } n \neq 1 \text{ and } a_n > 0.
$$

These examples show that $Rep \wedge r$ does not coincide with either of Rep and r. To check whether Proposition 9B.1 can be applied recall that the underlying ring of $Rep(A, P)$ is A. Thus, the underlying ring of $Rep(A, P) \cap r(A, P)$ is A as well. Let Q be the intersection of the partial orders on A given by $Rep(A,P)$ and $r(A,P)$. Thus, $id: (A,P) \to (A,Q)$ is a **POR/N**–morphism and $\sigma_{(A,P)}$ factors in the following way:

$$(A,P) \xrightarrow{id} (A,Q) \xrightarrow{\subseteq} Rep(A,P) \xrightarrow{\subseteq} \sigma(A,P).$$

Since $Sper(id): Sper(A,Q) \to Sper(A,P)$ is trivially injective we conclude that id is an epimorphism (Proposition 7.16). Thus, the conditions of Proposition 9B.1 are satisfied, and $id: (A,P) \to (A,Q)$ is the $Rep \wedge r$–reflection of (A,P).

Observe that $Rep \wedge r \neq id$. For, a partial order P is defined on the polynomial ring $\mathbb{R}[T]$ by setting

$$\sum_{i=0}^{n} a_i T^i > 0 \text{ (with } a_n \neq 0) \text{ if } n \neq 1 \text{ and } a_n > 0.$$

This poring does not belong to either of the categories. The partial orders of $Rep(\mathbb{R}[T], P)$ and $r(\mathbb{R}[T], P)$ are both total. Therefore $Rep(\mathbb{R}[T], P) \subseteq r(\mathbb{R}[T], P)$ is a sub–poring, hence $Rep(\mathbb{R}[T], P)$ is the reflection of $(\mathbb{R}[T], P)$ under $Rep \wedge r$. $\qquad \square$

C The intersection of a reflector with a subfunctor of Π

In subsection B it was shown that sometimes two reflectors $r : \mathbf{POR/N} \to \mathbf{D}$ and $s : \mathbf{POR/N} \to \mathbf{E}$ yield a new one by forming the intersection $r(A,P) \cap s(A,P)$ inside $\sigma(A,P)$. This approach can be generalized by replacing the reflector s with a subfunctor ([50], 0.1.7.1) of the functor Π. Let $F : \mathbf{POR/N} \to \mathbf{POR/N}$ be a subfunctor of Π such that $im(\Delta_{(A,P)}) \subseteq F(A,P)$ for all $(A,P) \in ob(\mathbf{POR/N})$. Then, restricting the codomains of the $\Delta_{(A,P)}$'s we obtain a family of homomorphisms $F_{(A,P)}: (A,P) \to F(A,P)$. For example, $F(A,P) = \prod_{\alpha \in Sper(A,P)} A/\alpha$ or $F(A,P) = \prod_{\alpha \in Sper(A,P)} \kappa_{(A,P)}(\alpha)$ are such subfunctors. Given such a subfunctor, for every (A,P) we form the intersection $s(A,P) = r(A,P) \cap F(A,P)$ and consider the morphism $s_{(A,P)}: (A,P) \to s(A,P)$ obtained from $\Delta_{(A,P)}$ by restriction of the codomain. We are looking for conditions ensuring that the morphisms $s_{(A,P)}$ are a family of reflection morphisms.

To start with, note that every **POR/N**–morphism $f\colon (A,P) \to$ (B,Q) yields morphisms $r(f)\colon r(A,P) \to r(B,Q)$ and $F(f)\colon F(A,P) \to$ $F(B,Q)$. These are both restrictions of $\Pi(f)\colon \Pi(A,P) \to \Pi(B,Q)$, hence they restrict to a homomorphism $s(f)\colon s(A,P) \to s(B,Q)$. Thus, s is a functor **POR/N** $\to$ **POR/N**. It is clear that the $s_{(A,P)}$'s are the components of a natural transformation $id_{\mathbf{POR/N}} \to s$.

Proposition 9C.1 Suppose that every $s_{(A,P)}$ is an epimorphism and that every $F(s_{(A,P)})$ is an isomorphism. Then s is a monoreflector.

Proof If s is a reflector at all then, obviously, it is a monoreflector. Let $\mathbf{E} \subseteq \mathbf{POR/N}$ be the full subcategory whose objects are the reduced porings (A,P) for which $s_{(A,P)}$ is an isomorphism. Suppose that $f\colon$ $(A,P) \to (B,Q)$ is a **POR/N**–morphism with $(B,Q) \in ob(\mathbf{E})$. Then there exists a morphism $\overline{f}\colon s(A,P) \to (B,Q)$ such that $f = \overline{f}s_{(A,P)}$, namely $\overline{f} = s_{(B,Q)}^{-1}s(f)$. Because $s_{(A,P)}$ is an epimorphism, $\overline{f}$ is the only such morphism. To complete the proof we only need to show that every $s(A,P)$ belongs to $\mathbf{E}$. For, then s is a functor **POR/N** $\to \mathbf{E}$ and the universal property (2.1) is satisfied. Because $s_{(A,P)}$ is an epimorphism and $\sigma_{(A,P)}$ factors as

$$(A,P) \xrightarrow{\;s_{(A,P)}\;} s(A,P) \subseteq r(A,P) \subseteq \sigma(A,P)$$

it follows from Proposition 7.16 that $\sigma(A,P) \cong \sigma(s(A,P))$. Therefore we may identify $\Pi(A,P)$ and $\Pi(s(A,P))$. We also identify $F(s(A,P))$ with $F(A,P)$ via the isomorphism $F(s_{(A,P)})$. Finally we show that also $r(s(A,P)) = r(A,P)$; for, then the idempotency of s is immediate. Consider the commutative diagram

$$
\begin{array}{ccccc}
r_{(A,P)}\colon (A,P) & \xrightarrow{\;s_{(A,P)}\;} & s(A,P) & \xrightarrow{\;\;f\;\;} & r(A,P) \\
\Big\downarrow{\scriptstyle r_{(A,P)}} & & \Big\downarrow{\scriptstyle r_{s(A,P)}} & & \Big\Vert{\scriptstyle =} \\
id_{r(A,P)}\colon r(A,P) & \xrightarrow{\;r(s_{(A,P)})\;} & r(s(A,P)) & \xrightarrow{\;\;\overline{f}\;\;} & r(A,P).
\end{array}
$$

Because $s_{(A,P)}$ is epimorphic, Proposition 2.4 implies that $r(s_{(A,P)})$ is an epimorphism. Since $\overline{f}r(s_{(A,P)}) = id_{r(A,P)}$ it follows that $r(s_{(A,P)})$ is an isomorphism. $\qquad\square$

We finish the subsection with two examples. The first one shows that the proposition is not always applicable, the second one uses the proposition to construct a reflector which is of great importance and will accompany us throughout the book.

Example 9C.2 *An intersection of a reflector of* **POR/N** *and a subfunctor of* Π *which is not a reflector*

Let F the subfunctor of Π defined by $F(A,P) = \prod\limits_{\alpha \in Sper(A,P)} A/\alpha$. We set $s(A,P) = \sigma(A,P) \cap F(A,P)$ and define $s_{(A,P)} \colon (A,P) \to s(A,P)$ to be $\sigma_{(A,P)}$ with restricted codomain. We claim that the hypotheses of Proposition 9C. 1 are not satisfied. To construct a counterexample, let $R = \mathbb{R}((\mathbb{Q}))$ be the formal power series field over $\mathbb{R}$ with exponents in $\mathbb{Q}$ ([101], Kapitel II, §5), let $A \subseteq R$ be the natural valuation ring ([101], p. 38). The valuation ring has value group $\mathbb{Q}$ and residue field $\mathbb{R}$. It carries a unique total order which we omit from the notation. The real spectrum has two points α, β with $\beta \in \overline{\{\alpha\}}$ and with residue fields $\rho(\alpha) = R$, $\rho(\beta) = \mathbb{R}$. Thus, $\Pi(A) = \sigma(A) = R \times \mathbb{R}$. Therefore $s(A) = F(A) = A \times \mathbb{R}$, and $s_A \colon A \to A \times \mathbb{R}$ is the identity in the first component and the canonical place in the second component. Since $|Sper(s(A))| = 3$ and $|Sper(A)| = 2$ we conclude that $Sper(s_A)$ is not an epimorphism (Theorem 5.2), hence Proposition 9C.1 does not apply. $\square$

Example 9C.3 *von Neumann regular f–rings*

We apply the intersection construction with the functor $\sigma :$ **POR/N** $\to$ **SAFR** and the subfunctor F of Π which is defined on the objects by $F(A,P) = \prod\limits_{\alpha} \kappa_{(A,P)}(\alpha)$. Then

$$s(A,P) = \{a \in \sigma(A,P); \ \forall\, \alpha : a(\alpha) \in \kappa_{(A,P)}(\alpha)\},$$

i.e., $s(A,P)$ is the ring of all semi–algebraic functions that take their values in $\kappa_{(A,P)}(\alpha)$ at each $\alpha \in Sper(A,P)$. Since $s(A,P)$ is the intersection of two von Neumann regular subrings of the von Neumann regular ring $\Pi(A,P)$ it is also von Neumann regular. For the same reason it is an f–ring. The real spectra $Sper(\sigma(A,P))$ and $Sper(s(A,P))$ both are homeomorphic to the prime spectra via the support functions. The functorial map $Spec(\sigma(A,P)) \to Spec(s(A,P))$ is surjective since every (minimal) prime ideal of $s(A,P)$ extends to a (minimal) prime ideal of $\sigma(A,P)$. But then $Sper(s_{(A,P)})$ is bijective, and Proposition 7.16 implies that $s_{(A,P)}$ is an epimorphism. Given $\alpha \in Sper(A, P)$, let $\beta \in Sper(s(A, P))$ be the unique prime cone extending α. Since β is determined by the homomorphism

$$s(A, P) \subseteq F(A, P) \xrightarrow{\ p_\alpha\ } \kappa_{(A,P)}(\alpha)$$

(where p_α is projection onto the α–th component) we see that $\kappa_{(A,P)}(\alpha) \to \kappa_{s(A,P)}(\beta)$ is an isomorphism. Thus, $F(s_{(A,P)})$ is an isomorphism, and Proposition 9C.1 shows that the functor s is a monoreflector.

The monoreflective subcategory belonging to s has the following explicit description. We have shown that every object of the subcategory is a von Neumann regular f–ring. On the other hand, suppose that (A, P) is a von Neumann regular f–ring. The support function is a homeomorphism onto the Brumfiel spectrum. The Brumfiel spectrum contains every minimal prime ideal. Since all prime ideals are minimal, the support function is a homeomorphism onto the prime spectrum. The canonical homomorphism $s_{(A,P)}$ induces a bijection between the prime spectra, hence a homeomorphism. The induced maps between the residue fields are isomorphisms. Since the rings (A, P) and $s(A, P)$ are both von Neumann regular, (7.13) applies and $s_{(A,P)}$ is an isomorphism. Therefore (A, P) belongs to the reflective subcategory belonging to the intersection of σ with the functor F.

We have shown that the reflective subcategory belonging to the intersection of σ with the functor F is the category of von Neumann regular f–rings. It is denoted by **VNRFR**; for the reflector we shall always use the notation ν. $\square$

von Neumann regular f–rings will occur throughout in many examples and results. They are the main topic of section 17, which is devoted to an investigation of the monoreflectors of **POR/N** that are stronger than ν.

D Intermediate reflectors between two reflectors r and s with $r \leq s$

If r: **POR/N** $\to$ **D** and s: **POR/N** $\to$ **E** are monoreflectors with **E** $\subset$ **D** then, using auxiliary subcategories of **POR/N**, it is frequently possible to construct a large number of intermediate reflectors between r and s.

We shall use this technique to show that **POR/N** has a *proper class of monoreflectors.*

The basic construction does not have anything to do with the category **POR/N**. Therefore we work with any category **C** that is wellpowered, co–wellpowered, complete and cocomplete and with any pair of reflectors $r: \mathbf{C} \to \mathbf{D}$ and $s: \mathbf{C} \to \mathbf{E}$ such that $\mathbf{E} \subseteq \mathbf{D}$. We shall use a full subcategory $\mathbf{C}' \subseteq \mathbf{C}$ having the following property:

(9D.1)   If $f: A \to B$ is a **C**–morphism and if $A \in ob(\mathbf{C}')$ then also $B \in ob(\mathbf{C}')$.

First, a new functor $t : \mathbf{C} \to \mathbf{C}$ will be defined together with a natural transformation $\mathrm{Id}_{\mathbf{C}} \to t$. For an object C let $t(C) = r(C)$ if $r(C) \notin ob(\mathbf{C}')$, $t(C) = s(C)$ if $r(C) \in ob(\mathbf{C}')$. In the first case define $t_C : C \to t(C)$ to be r_C, in the second case define it to be s_C. To define the functor t on morphisms, pick a **C**–morphism $f: A \to B$. If $r(B) \notin ob(\mathbf{C}')$ we set $t(f) = r(f)$; if $r(A) \in ob(\mathbf{C}')$ then $t(f) = s(f)$. Finally suppose that $r(A) \notin ob(\mathbf{C}')$ and $r(B) \in ob(\mathbf{C}')$. In this case we define $t(f)$ to be

$$r(A) \xrightarrow{\ r(f)\ } r(B) \xrightarrow{\ \varphi_B\ } s(B)$$

where φ_B is the unique morphism $\psi : r(B) \to s(B)$ with $s_B = \psi r_B$ (Proposition 8.1(c)). First we must check that t is actually a functor. The only questionable issue is whether $t(gf) = t(g)t(f)$ for $f : A \to B$, $g : B \to C$. This is clear in the two cases that $r(A) \in ob(\mathbf{C}')$ or $r(B) \notin ob(\mathbf{C}')$. So the only remaining case is that $r(A) \notin ob(\mathbf{C}')$ and $r(B) \in ob(\mathbf{C}')$. Then the two morphisms are

$$t(gf) \ : \ r(A) \xrightarrow{\ r(gf)\ } r(C) \xrightarrow{\ \varphi_C\ } s(C)$$

$$t(g)t(f) \ : \ r(A) \xrightarrow{\ r(f)\ } r(B) \xrightarrow{\ \varphi_B\ } s(B) \xrightarrow{\ s(g)\ } s(C).$$

We show that $\varphi_C r(g) = s(g)\varphi_B$ which then implies $t(gf) = t(g)t(f)$. First note that $s(g)s_B = s_C g$ (since s_B and s_C are reflections belonging to s), hence $s(g)\varphi_B r_B = s_C g = \varphi_C r(g) r_B$. By the universal property (2.1) there is only one morphism $\psi : r(B) \to s(C)$ such that $\psi r_B = s_C g$. Therefore $s(g)\varphi_B = \varphi_C r(g)$, and t is indeed a functor. Trivially, the t_C's are the components of a natural transformation.

Theorem 9D.2 Let $\mathbf{F} \subseteq \mathbf{C}$ be the full subcategory $(\mathbf{D}\backslash\mathbf{C}') \cup (\mathbf{E} \cap \mathbf{C}')$. Then $t(C) \in ob(\mathbf{F})$ for every $C \in ob(\mathbf{C})$. Considered as a functor from $\mathbf{C}$ to $\mathbf{F}$, t is a reflector. It is stronger than r and weaker than s.

Proof From the definition of $\mathbf{F}$ it is clear that $\mathbf{E} \subseteq \mathbf{F} \subseteq \mathbf{D}$. Thus, if t is a reflector at all then $r \leq t \leq s$. Next we pick any $C \in ob(\mathbf{C})$ and show that $t(C)$ belongs to $\mathbf{F}$: If $r(C) \notin ob(\mathbf{C}')$ then $t(C) = r(C) \in ob(\mathbf{D}\backslash\mathbf{C}')$; if $r(C) \in ob(\mathbf{C}')$ then $t(C) = s(C) \in ob(\mathbf{E} \cap \mathbf{C}')$. Thus, t is indeed a functor $\mathbf{C} \to \mathbf{F}$. It remains to check that the morphisms $t_C \colon C \to t(C)$ have the universal property (2.1) of reflections: Let $f \colon C \to F$ be any $\mathbf{C}$–morphism into an $\mathbf{F}$–object. Since $\mathbf{F} \subseteq \mathbf{D}$ there is a unique morphism $\overline{f} \colon r(C) \to F$ such that $f = \overline{f}r_C$. If $r(C) \notin ob(\mathbf{C}')$, then $r_C = t_C$, and we are finished. If $r(C) \in ob(\mathbf{C}')$ then (by (9D.1)) also $F \in ob(\mathbf{C}')$, hence $F \in ob(\mathbf{E})$. By the universal property of s there is a unique morphism $\overline{f} \colon s(C) = t(C) \to F$ such that $f = \overline{f}s_C = \overline{f}t_C$. $\square$

Observe that the reflector t of the theorem is an epireflector if both r and s are epireflectors. It is a monoreflector if s (and hence also r, cf. Corollary 8.2) is a monoreflector.

Theorem 9D.3 The monoreflectors of $\mathbf{POR/N}$ form a proper class.

Proof We use Theorem 9D.2 with $\mathbf{C} = \mathbf{POR/N}$, $\mathbf{D} = \mathbf{POR/N}$ and $\mathbf{E} = \mathbf{SAFR}$. For any ordinal number α pick some real closed field R_α of cardinality $\aleph_\alpha$. We define $\mathbf{C}_\alpha \subseteq \mathbf{POR/N}$ to be the full subcategory of R_α–algebras. Then (9D.1) is satisfied, hence each $\mathbf{C}_\alpha$ yields a monoreflector $t_\alpha \colon \mathbf{POR/N} \to \mathbf{F}_\alpha$ with $\sigma > t_\alpha > id_{\mathbf{POR/N}}$. If $\alpha \neq \beta$ it is clear that $t_\alpha \neq t_\beta$. Thus, the t_α's form a proper class. $\square$

Obviously, there are many variations of the theorem. For example, the real closed fields R_α in the proof can be chosen such that $R_\alpha \subset R_\beta$ if $\alpha < \beta$. Then it follows that $id_{\mathbf{POR/N}} < t_\beta < t_\alpha < \sigma$, i.e., there is even a chain which is a proper class. The same technique shows that there is a proper class of monoreflectors in between ν and σ or in between φ and ν, just to name a few examples.

10 *H*–closed epireflectors

In this section we study H–closed monoreflectors of **POR/N**. The notion
of H–closedness can be defined in much more general situations than the
one we are actually interested in: Let **C** be a wellpowered, co–wellpowered
and complete category that is concrete over the category **SETS** with forgetful functor $U: \mathbf{C} \to \mathbf{SETS}$. Suppose that **C** has free objects over
SETS, i.e., U has a left adjoint functor F. A **C**–morphism $f: C \to C'$ is
called *surjective*, if $U(f)$ is epimorphic (i.e., surjective) in **SETS**. Then
a class X of **C**–objects is said to be *H*–*closed* if, for all surjective **C**–
morphisms $f: C \to C'$, $C \in X$ implies that also $C' \in X$. A full and iso-
morphism closed subcategory $\mathbf{D} \subseteq \mathbf{C}$ is *H*–*closed* if $ob(\mathbf{D})$ is H–closed.
If $r: \mathbf{C} \to \mathbf{D}$ is a reflector into an H–closed subcategory then we also
say that r is *H*–*closed*. We show that the H–closed monoreflectors of
POR/N have a particularly simple description: They are determined
completely by the reflection of a single object, namely the free object
$F(\mathbb{N})$, or, equvalently, by the countably many reflections of the free objects $F(n), n \in \mathbb{N}$. It follows that the H–closed monoreflectors of **POR/N**
form a set of cardinality $2^{\aleph_0}$.

Examples in this and succeding sections show that many of the most
important reflectors of **POR/N** are H–closed. This fact is one explanation for our interest in H–closed reflectors. Another reason is that
H–closed reflectors arise directly from geometric situations. We adopt
the point of view that the reflection of the free object $F(n)$ in a monoreflective subcategory $\mathbf{D} \subseteq \mathbf{POR/N}$ represents the affine n–dimensional
space in **D**. Thus, H–closed reflectors are exactly those reflectors that
are completely determined by their affine n–dimensional spaces.

Remark 10.1 The notion of H–closedness for monoreflective subcategories of **POR/N** is closely related to *factor porings* modulo convex radical ideals. Given a reduced poring (A, P) and a convex radical
ideal I, let $\pi: A \to \overline{A} = A/I$ be the canonical homomorphism and define $\overline{P} = \pi(P)$. Then $(\overline{A}, \overline{P})$ is a reduced poring. Now suppose that
$r: \mathbf{POR/N} \to \mathbf{D}$ is an H–closed monoreflector. If (A, P) belongs to
D then H–closedness implies that $(\overline{A}, \overline{P}) \in ob(\mathbf{D})$ as well. We express this fact by saying that *factor porings exist in* **D**. One may ask
whether the converse of this statement is also true, i.e., supposing that

$r\colon \mathbf{POR/N} \to \mathbf{D}$ is a monoreflector and that factor porings exist in $\mathbf{D}$, is r H–closed? The answer to this question is no, see Example 10.12 and Example 10.13. But there is an additional condition satisfied by H–closed monoreflectors. We say that a monoreflective subcategory is *closed under strengthening of the partial order* if for any object (A, P) of $\mathbf{D}$ and for any partial order $Q \subseteq A$ that contains P, (A, Q) also belongs to $\mathbf{D}$. If $\mathbf{D}$ is H–closed then it is clear that $\mathbf{D}$ is closed under strengthening of the partial order. In fact, it is easy to prove that closedness under formation of factor porings and under strengthening of the partial order together are equivalent to H–closedness.

Now suppose that the monoreflective subcategory $\mathbf{D}$ of $\mathbf{POR/N}$ is contained in $\mathbf{FR/N}$. If $(A, P) \in ob(\mathbf{D})$ then there is no proper strengthening of the partial order P (by Proposition 1.11). Therefore $\mathbf{D}$ is H–closed if and only if factor porings exist in $\mathbf{D}$. □

Similar to the Characterization Theorems of section 8, there are criteria to decide whether a given subcategory is H–closed and epireflective. The following result is Theorem 2.2 in [53]. For the sake of completeness we include the proof.

Theorem 10.2 For a subcategory $\mathbf{D} \subseteq \mathbf{C}$ the following conditions are equivalent:

(a) $\mathbf{D}$ is H–closed and epireflective.

(b) The class $ob(\mathbf{D})$ is H–closed and is closed under the formation of products and extremal subobjects.

(c) $ob(\mathbf{D}) = Inj(\mathcal{E})$ for some class $\mathcal{E}$ of epimorphisms with free domain.

Proof The equivalence (a) $\Leftrightarrow$ (b) is immediate from the First Characterization Theorem (Theorem 8.3).

(a) $\Rightarrow$ (c) We know from the Second Characterization Theorem (Theorem 8.7) that $ob(\mathbf{D}) = Inj(\mathcal{E})$ where $\mathcal{E}$ is the class of all reflection morphisms. Let $\mathcal{E}_0 \subseteq \mathcal{E}$ be the subclass of reflection morphisms with free domain. Then it is trivial that $Inj(\mathcal{E}) \subseteq Inj(\mathcal{E}_0)$. We claim that the two classes coincide.

For the proof, pick $D \in Inj(\mathcal{E}_0)$ and recall that

$$U(D) \xrightarrow{\eta_{U(D)}} UFU(D) \xrightarrow{U(\mathcal{E}_D)} U(D)$$

is the identity, where η is the unit and ε is the counit of the adjunction between U and F ([83], Chapter IV, Section 1, Theorem 1). Thus, ε_D factors as

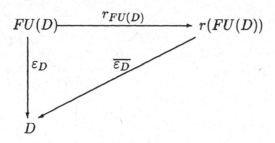

The surjectivity of ε_D implies that $\overline{\varepsilon_D}$ is surjective as well. Since **D** is *H*-closed we conclude that $D \in ob(\mathbf{D})$, and this proves the claim.

Finally we deal with the implication (**c**) $\Rightarrow$ (**a**): The Second Characterization Theorem shows immediately that **D** is epireflective. As usual, the reflector is denoted by r. Only *H*-closedness needs to be checked: Let $f \colon D \to C$ be a surjective morphism, $D \in ob(\mathbf{D})$, $C \in ob(\mathbf{C})$. It will be shown that $C \in Inj(\mathcal{E})$. Let $\mathcal{E}'$ be the class of reflection morphisms of all free objects that are the domain of some member of $\mathcal{E}$. We claim that $Inj(\mathcal{E}) = Inj(\mathcal{E}')$.

Because of $Inj(\mathcal{E}) = ob(\mathbf{D})$ it is clear that $Inj(\mathcal{E}) \subseteq Inj(\mathcal{E}')$. Now pick any $A \in Inj(\mathcal{E}')$ and suppose that $e \colon F(X) \to B$ belongs to $\mathcal{E}$. Since $r(F(X)) \in ob(\mathbf{D}) = Inj(\mathcal{E})$ there exits an extension $b \colon B \to r(F(X))$ of $r_{F(X)}$, i.e., $r_{F(X)} = be$. As $r_{F(X)} \in \mathcal{E}'$ and $A \in Inj(\mathcal{E}')$, every $a \colon F(X) \to A$ factors as $a = \overline{a}r_{F(X)}$. But then $\overline{a}b$ is the desired extension of a to a morphism with domain B, i.e., $A \in Inj(\mathcal{E})$.

Because of $Inj(\mathcal{E}) = Inj(\mathcal{E}')$ we may and shall assume that $\mathcal{E}$ consists of reflection morphisms of free objects. Now, pick any morphism $r_{F(X)} \colon F(X) \to r(F(X))$ belonging to $\mathcal{E}$ and any morphism $g \colon F(X) \to C$. The surjective map $U(f)$ has a section $s \colon U(C) \to U(D)$. Thus, we have the following diagram in **SETS**:

$$X \xrightarrow{\eta_X} UF(X) \xrightarrow{U(g)} U(C) \xrightarrow{=} U(C)$$

$$U(D)$$

with s and $U(f)_0$ to $U(D)$.

This diagram together with the reflector $r \colon \mathbf{C} \to \mathbf{D}$ yields the following diagram in **C**:

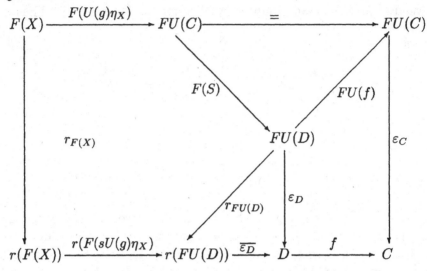

Note that $g = \varepsilon_C F(U(g)\eta_X)$ hence

$$g = (f\overline{\varepsilon_D}r(F(sU(g)\eta_X)))r_{F(X)}$$

is the factoriztation of g needed to finish the proof.   □

Just as the First Characterization Theorem showed that the class of epire-flective subcategories of a given category is a complete lattice, the same is true about the class of H–closed epireflective subcategories (due to con-dition (b) of the theorem). However, this does not mean that the class of H–closed epireflective subcategories is a sublattice of the class of all epireflective subcategories. The meet operations clearly agree (being just intersection), but the join operations are different, in general. Note that if **E** $\subseteq$ **D** $\subseteq$ **C** are epireflective subcategories then **E** is H–closed in **D** if it is H-closed in **C**; if **E** $\subseteq$ **D** and **D** $\subseteq$ **C** are H–closed then **E** $\subseteq$ **C** is H–closed.

The third condition of the theorem is of particular interest to us. If the equivalent conditions hold then it is clearly possible to choose as $\mathcal{E}$ the class of all reflections $r_{F(X)}\colon F(X) \to r(F(X))$ where r is the reflector belonging to **D** and X varies in the category **SETS**. However, usually this class $\mathcal{E}$ is extravagantly large, cf. the comments in [53], §5. One of our main concerns in this section and in section 13 is to find out how small $\mathcal{E}$ can be chosen.

Our main tool are functional formulas. Therefore we start by an-alyzing the relationship between reflectors of **POR/N** and functional formulas. Until further notice, $r: \mathbf{POR/N} \to \mathbf{D}$ is a fixed monoreflector. Every reflection $r(A, P)$ will be identified with its canonical image in $\sigma(A, P)$. For each $n \in \mathbb{N}$, we consider two sets of functional formulas:

- $\mathcal{F}_r(n)$ is the set of functional formulas $\Theta = \Theta(X_1, \dots, X_n, Y)$ such that the associated semi–algebraic function $\omega_\Theta(T_1, \dots, T_n)$ belongs to $r(F(n))$; and
- $\mathcal{F}_r^+(n)$ is the set of functional formulas $\Theta \in \mathcal{F}_r(n)$ for which the associated semi–algebraic function belongs to $r^+(F(n))$.

Also we define

$$\mathcal{F}_r(\mathbb{N}) = \bigcup_{n \in \mathbb{N}} \mathcal{F}_r(n), \ \mathcal{F}_r^+(\mathbb{N}) = \bigcup_{n \in \mathbb{N}} \mathcal{F}_r^+(n).$$

For every reduced poring (A, P) these sets of functional formulas determine the following subsets of $\sigma(A, P)$:

$$
\begin{aligned}
s(A, P) \ = \ & \{c \in \sigma(A, P); \ \exists \, \Theta = \Theta(X_1, \dots, X_n, Y) \in \mathcal{F}_r(\mathbb{N}) \\
& \exists \, a_1, \dots, a_n \in A : c = \omega_\Theta(a_1, \dots, a_n)\}, \\
s'(A, P) \ = \ & \{c \in s(A, P); \ \exists \, \Theta = \Theta(X_1, \dots, X_n, Y) \in \mathcal{F}_r^+(\mathbb{N}) \\
& \exists \, a_1, \dots, a_n \in A : c = \omega_\Theta(a_1, \dots, a_n)\}, \\
s^+(A, P) \ = \ & \{c \in s(A, P); \ \exists \, c_1, \dots, c_k \in s'(A, P) \\
& \exists \, a_1, \dots, a_k \in P : c = \sum_{i=1}^{k} \sigma_{(A,P)}(a_i) c_i\}.
\end{aligned}
$$

Lemma 10.3 The set $s(A, P)$ is a subring of $r(A, P)$. It is partially ordered with positive cone $s^+(A, P) \subseteq r^+(A, P)$. The reflection $r_{(A,P)}$ factors through $(s(A, P), s^+(A, P))$.

Proof First we observe that $r_{(A,P)}(A) = \sigma_{(A,P)}(A) \subseteq s(A, P)$: Because $\Theta \equiv (Y = X_1) \in \mathcal{F}_r(\mathbb{N})$ it follows that

$$r_{(A,P)}(a) = \omega_\Theta(a) \in s(A, P)$$

for every $a \in A$. The functional formula $\Theta \equiv (Y = 1)$ belongs to $\mathcal{F}_r^+(\mathbb{N})$, hence $1 \in s'(A, P)$ and, given $a \in P$,

$$r_{(A,P)}(a) = \sigma_{(A,P)}(a) \cdot 1 \in s^+(A, P).$$

Thus, $r_{(A,P)}(P) \subseteq s^+(A,P)$. The next step is to show that $s(A,P)$ is closed under addition and multiplication. We do this for addition; the proof for multiplication is practically identical. Suppose that $c = \omega_\Theta(a_1,\dots,a_m)$, $d = \omega_H(b_1,\dots,b_n) \in s(A,P)$. The formula

$$Z \equiv \exists\, U,V\, [\Theta(X_1,\dots,X_m,U) \,\&\, H(X_{m+1},\dots,X_{m+n},V) \,\&\, Y = U + V]$$

is functional and belongs to $\mathcal{F}_r(m+n)$ (since

$$\omega_Z(T_1,\dots,T_{m+n}) = \omega_\Theta(T_1,\dots,T_m) + \omega_H(T_{m+1},\dots,T_{m+n})$$

in $r(F(m+n))$). But then

$$c + d = \omega_Z(T_1,\dots,T_{m+n}) \in s(A,P).$$

Now we turn to the partial order. The set $s^+(A,P)$ is additively closed by definition. To show that $s^+(A,P)$ is multiplicatively closed one needs to show that $s'(A,P)$ is closed under multiplication. This can be done with the same method as in the proof of additive closedness of $s(A,P)$; we omit the details. If $c = \omega_\Theta(a_1,\dots,a_n) \in s(A,P)$ then the functional formula

$$H \equiv \exists\, Y\, [\Theta(X_1,\dots,X_n,Y) \,\&\, Z = Y^2]$$

belongs to $\mathcal{F}_r^+(n)$. Thus,

$$c^2 = \omega_H(a_1,\dots,a_n) \in s'(A,P),$$

hence $s^+(A,P)$ contains the squares.

We finish the proof by showing that $s(A,P) \subseteq r(A,P)$ and that $s^+(A,P) \subseteq r^+(A,P)$: Suppose that $c = \omega_\Theta(a_1,\dots,a_n) \in s(A,P)$. We define a homomorphism $f\colon F(n) \to (A,P)$ by $T_i \to a_i$. Then

$$c = r(f)(\omega_\Theta(T_1,\dots,T_n)) \in r(A,P).$$

If $c \in s'(A,P)$ then we choose $\Theta \in \mathcal{F}_r^+(n)$, and, because $r(f)$ is order-preserving, it follows that

$$c = r(f)(\omega_\Theta(T_1,\dots,T_n)) \in r^+(A,P).$$

It is trivial now that also $s^+(A,P) \subseteq r^+(A,P)$. In particular,

$$s^+(A,P) \cap -s^+(A,P) \subseteq r^+(A,P) \cap -r^+(A,P) = \{0\},$$

and $s^+(A,P)$ is indeed a partial order. $\square$

The partially ordered ring $(s(A,P), s^+(A,P))$ will be denoted just by $s(A,P)$. By restriction of the codomain, $r_{(A,P)}$ (or $\sigma_{(A,P)}$) yields a morphism $(A,P) \to s(A,P)$ which we denote by $s_{(A,P)}$. It is immediate from the definitions that any homomorphism $\varphi \colon F(n) \to (A,P)$, $n \in \mathbb{N}$, has a unique extension $\overline{\varphi} \colon r(F(n)) \to s(A,P)$, i.e., $\overline{\varphi} r_{F(n)} = s_{(A,P)}\varphi$. We shall show that the $s_{(A,P)}$'s are the reflection morphisms belonging to an *H*–closed reflector of **POR/N**. Before proving this we need a few more preparations.

Lemma 10.4 For every set S, the partially ordered subrings $s(F(S))$ and $r(F(S))$ of $\sigma(F(S))$ coincide, as well as the homomorphisms $s_{F(S)}$ and $r_{F(S)}$.

Proof It was shown in Lemma 10.3 that $s(F(S)) \subseteq r(F(S))$ and $s^+(F(S)) \subseteq r^+(F(S))$. Moreover, $s_{F(S)}$ is the same map as $r_{F(S)}$, only the codomain is potentially smaller. Therefore it suffices to prove that $r(F(S)) \subseteq s(F(S))$ and that $r^+(F(S)) \subseteq s^+(F(S))$. So, suppose that $c \in r(F(S))$ (or $\in r^+(F(S))$). Being semi–algebraic, $c = \omega_\Theta(c_1, \dots, c_k)$ for some functional formula $\Theta = \Theta(X_1, \dots, X_k, Y)$ and some $c_1, \dots, c_k \in F(S)$. The underlying ring of $F(S)$ is the polynomial ring $\mathbb{Z}[T_s; s \in S]$, its partial order is the sums of squares. Each c_κ is a polynomial containing finitely many variables, hence there is a finite subset $S' \subseteq S$ such that $c_1, \dots, c_k \in \mathbb{Z}[T_s; s \in S']$. We enumerate S' as $\{s_1, \dots, s_n\}$ and consider the terms $c_\kappa(X_1, \dots, X_n)$ obtained from the c_κ's by replacing T_{s_i} with X_i. The formula

$$
\begin{aligned}
H \;&=\; H(X_1, \dots, X_n, Y) \\
&\equiv\; \exists\, Z_1, \dots, Z_k\, [\Theta(Z_1, \dots, Z_k, Y) \;\&\; Z_1 = c_1(X_1, \dots, X_n) \\
&\qquad\&\, \dots\, \&\; Z_k = c_k(X_1, \dots, X_n)]
\end{aligned}
$$

is functional and $c = \omega_H(T_{s_1}, \dots, T_{s_n}) \in r(F(S))$. Let $f \colon F(S) \to F(n)$ be any homomorphism with $f(T_{s_i}) = T_i$. The extension $r(f) \colon r(F(S)) \to r(F(n))$ yields

$$
\omega_H(T_1, \dots, T_n) = r(f)(\omega_H(T_{s_1}, \dots, T_{s_n})) = r(f)(c) \in r(F(n))
$$

(or $\in r^+(F(n))$ if $c \in r^+(F(S))$). It follows that $H \in \mathcal{F}_r(\mathbb{N})$ (or $\in \mathcal{F}_r^+(\mathbb{N})$). Now the definition of $s(F(n))$ implies $c \in s(F(S))$, or $\in s^+(F(S))$, and the proof is finished. $\square$

Lemma 10.5 Let S be any set. Then the reduced poring (A, P) belongs to the injectivity class of $r_{F(S)}$ if and only if, for all $n \in \mathbb{N}$, $n \leq |S|$, for all $\Theta = \Theta(X_1, \ldots, X_n, Y) \in \mathcal{F}_r(n)$ (or $\in \mathcal{F}_r^+(n)$), and for all $a_1, \ldots, a_n \in A$, the semi-algebraic function $\omega_\Theta(a_1, \ldots, a_n)$ belongs to the image of A (or P) in $r(A, P)$.

Proof First suppose that $(A, P) \in Inj(\{r_{F(S)}\})$, let $n \leq |S|$, let $\Theta \in \mathcal{F}_r(n)$ (or $\in \mathcal{F}_r^+(n)$) and let $a_1, \ldots, a_n \in A$. Pick any subset $\{s_1, \ldots, s_n\} \subseteq S$ of cardinality n and consider $c = \omega_\Theta(T_{s_1}, \ldots, T_{s_n}) \in r(F(S))$ (or $\in r^+(F(S))$). There is a homomorphism $f \colon F(S) \to (A, P)$ with $f(T_{s_i}) = a_i$. By hypothesis it extends uniquely to $\overline{f} \colon r(F(S)) \to (A, P)$, and $r_{(A,P)}\overline{f} = r(f)$. But then

$$
\begin{aligned}
\omega_\Theta(a_1, \ldots, a_n) &= r(f)(\omega_\Theta(T_{s_1}, \ldots, T_{s_n})) \\
&= r_{(A,P)}(\overline{f}(\omega_\Theta(T_{s_1}, \ldots, T_{s_n}))) \\
&\in r_{(A,P)}(A) \quad (\text{or } \in r_{(A,P)}(P)).
\end{aligned}
$$

Conversely, we assume that the condition about the $\omega_\Theta(a_1, \ldots, a_n)$ holds. Let $f \colon F(S) \to (A, P)$ be any homomorphism. We want to show that $r(f) \colon r(F(S)) \to r(A, P)$ factors through (A, P), i.e., that $r(f)(c) \in r_{(A,P)}(A)$, or $\in r_{(A,P)}^+(P)$, whenever $c \in r(F(S))$, or $\in r^+(F(S))$. So, as in the proof of Lemma 10.4, for a given c, there is a formula $\Theta = \Theta(X_1, \ldots, X_n, Y) \in \mathcal{F}_r(n)$ (or $\in \mathcal{F}_r^+(n)$) with $n \leq |S|$ and $c = \omega_\Theta(T_{s_1}, \ldots, T_{s_n})$. The hypothesis shows that

$$
r(f)(c) = \omega_\Theta(f(T_{s_1}), \ldots, f(T_{s_n})) \in r_{(A,P)}(A) \ (\text{or } \in r_{(A,P)}(P)),
$$

as claimed. $\square$

Now the preparations are finished for the proof of the main result about H-closed monoreflectors of **POR/N**:

Theorem 10.6 Let $\mathbf{E} \subseteq \mathbf{POR/N}$ be the subcategory of those reduced porings (A, P) for which $s_{(A,P)}$ is an isomorphism. Then $\mathbf{E}$ is the injectivity class of any one of the following classes of epimorphisms:

- $\{r_{F(S)}; S \in ob(\mathbf{SETS})\}$;
- $\{r_{F(\mathbb{N})}\}$;
- $\{r_{F(n)}; n \in \mathbb{N}\}$.

In particular, **E** is an *H*–closed monoreflective subcategory of **POR/N**. The reflection of $(A, P) \in ob(\textbf{POR/N})$ is $s_{(A,P)} \colon (A, P) \to s(A, P)$. The category **D** is contained in **E**; they are equal if and only if r is *H*–closed. In particular, if r is *H*–closed then $ob(\textbf{D})$ is the injectivity class of $\{r_{F(\mathbb{N})}\}$ and $\{r_{F(n)}; n \in \mathbb{N}\}$.

Proof It is clear from Lemma 10.5 that the injectivity classes of the three classes of epimorphisms coincide. If (A, P) belongs to this injectivity class then Lemma 10.5 shows that $s_{(A,P)} \colon (A, P) \to s(A, P)$ is an isomorphism, i.e., $(A, P) \in ob(\textbf{E})$. Conversely, suppose that $s_{(A,P)} \colon (A, P) \to s(A, P)$ is an isomorphism. We have to prove that (A, P) is an element of the injectivity class: For any homomorphism $f \colon F(n) \to (A, P)$ we have the diagram

$$
\begin{array}{ccc}
F(n) & \xrightarrow{\ f\ } & (A, P) \\
{\scriptstyle r_{F(n)}} \downarrow & & \downarrow {\scriptstyle r_{(A,P)}} \\
r(F(n)) & \xrightarrow{\ r(f)\ } & r(A, P)
\end{array}
$$

Via $r_{(A,P)}$ we consider (A, P) as a subring of $r(A, P)$. It is claimed that, for $c \in r(F(n))$ (resp., $c \in r^+(F(n))$), $r(f)(c) \in A$ (resp., $r(f)(c) \in P$). So, pick a functional formula $\Theta = \Theta(X_1, \ldots, X_n, Y) \in \mathcal{F}_r(n)$ (or $\in \mathcal{F}_r^+(n)$) such that $c = \omega_\Theta(T_1, \ldots, T_n)$. Then $r(f)(c) = \omega_\Theta(f(T_1), \ldots, f(T_n))$ belongs to $s(A, P) = A$ (or to $s^+(A, P) = P$), and this shows that (A, P) is in the injectivity class of $r_{F(n)}$. We conclude that $ob(\textbf{E}) = Inj(\{r_{F(\mathbb{N})}\})$.

From Theorem 10.2 we know that **E** is an *H*–closed monoreflective subcategory of **POR/N**. It will be shown that $s_{(A,P)} \colon (A, P) \to s(A, P)$ is the reflection of (A, P) in **E**. First we check that $s(A, P) \in ob(\textbf{E}) = Inj(\{r_{F(n)}; n \in \mathbb{N}\})$: Suppose that $\varphi \colon F(k) \to s(A, P)$ is any **POR/N**-morphism. We have to show that φ can be extended to a **POR/N**-morphism $\overline{\varphi} \colon r(F(k)) \to s(A, P)$. Let $T_1, \ldots, T_k$ be the variables of $F(k)$ and set $c_\kappa = \varphi(T_\kappa) \in s(A, P)$. By definition of $s(A, P)$ there are formulas $\Theta_\kappa = \Theta_\kappa(X_{\kappa 1}, \ldots, X_{\kappa l_\kappa}, Y_\kappa) \in \mathcal{F}_r(l_\kappa)$ such that $c_\kappa = \omega_{\Theta_\kappa}(a_{\kappa 1}, \ldots, a_{\kappa l_\kappa})$ for suitable elements $a_{\kappa \lambda} \in A$. For all κ, λ we set

$$
X_{l_1 + \ldots l_{\kappa-1} + \lambda} = X_{\kappa \lambda}, \quad a_{l_1 + \ldots l_{\kappa-1} + \lambda} = a_{\kappa \lambda}.
$$

With $l = l_1 + \ldots + l_k$ we define a homomorphism

$$
\psi \colon F(l) \to (A, P) \colon X_i \to a_i.
$$

By the remark preceding Lemma 10.4 there is an extension $\overline{\psi}\colon r(F(l)) \to$ $s(A, P)$ of ψ, i.e., $\overline{\psi}r_{F(l)} = s_{(A,P)}\psi$. If we define $b_\kappa = \omega_{\Theta_\kappa}(X_{\kappa 1}, \dots, X_{\kappa l_\kappa})$ in $r(F(l))$ then $\overline{\psi}(b_\kappa) = c_\kappa$. Thus, the homomorphism φ factors as $\varphi = \overline{\psi}\vartheta$ where $\vartheta\colon F(k) \to r(F(l))$ is defined by $\vartheta(T_\kappa) = b_\kappa$. Since r is a reflector there exists a unique extension $\overline{\vartheta}\colon r(F(k)) \to r(F(l))$ of ϑ. But then $\overline{\psi}\circ\overline{\vartheta}$ is the desired extension of $\varphi = \overline{\psi}\vartheta$.

Finally, let $f\colon (A, P) \to (B, Q)$ be any **POR/N**–morphism with $(B, Q) \in ob(\mathbf{E})$. Via $r_{(B,Q)}$ we consider (B, Q) as a partially ordered subring of $r(B, Q)$. We show that $r(f)\colon r(A, P) \to r(B, Q)$ maps $s(A, P)$ into B and $s^+(A, P)$ into Q: Suppose that $c = \omega_\Theta(a_1, \dots, a_n) \in s(A, P)$ with $\Theta \in \mathcal{F}_r(\mathbb{N})$, or $\in \mathcal{F}_r^+(\mathbb{N})$. Then

$$r(f)(c) = \omega_\Theta(f(a_1), \dots, f(a_n)) \in s(B, Q) = B$$

(by definition of $s(B, Q)$) and $r(f)(c) \in s'(B, Q) \subseteq Q$ if $\Theta \in \mathcal{F}_r^+(\mathbb{N})$ (by definition $s'(B, Q)$). But then $f(P) \subseteq Q$ implies that $r(f)(s^+(A, P)) \subseteq$ $s^+(B, Q) = Q$. Thus, the universal property (2.1) is satisfied by $s_{(A,P)}$, i.e., $s_{(A,P)}$ is the reflection morphism of (A, P).

According to Lemma 10.4 the r–reflections and the s–reflections of free objects coincide. Every object of $\mathbf{D}$ belongs to the injectivity class of the r–reflections of the free objects in **POR/N**. Since this injectivity class is exactly $ob(\mathbf{E})$ we conclude that $\mathbf{D} \subseteq \mathbf{E}$. Moreover, $\mathbf{D}$ is H–closed if and only if $ob(\mathbf{D}) = Inj(\{r_{F(S)}; S \in ob(\mathbf{SETS})\})$ (Theorem 10.2), hence $\mathbf{D}$ is H–closed if and only if $\mathbf{D} = \mathbf{E}$. $\square$

It is a consequence of Theorem 10.2 that the class of H–closed monoreflective subcategories of **POR/N** is a complete lattice. In the next result we use Theorem 10.6 to obtain some more information about this lattice:

Theorem 10.7 The class of H–closed monoreflective subcategories of **POR/N** is a set of cardinality $2^{\aleph_0}$. It is a complete lattice with largest element **POR/N** and smallest element **SAFR**. Every chain in the lattice is at most countable.

Proof For the cardinality, note that any H–closed monoreflector r of **POR/N** is completely determined by the intermediate poring $r(F(\mathbb{N}))$ of $F(\mathbb{N})$ and $\sigma(F(\mathbb{N}))$ (Theorem 10.6). The underlying ring of $F(\mathbb{N})$ is the polynomial ring over $\mathbb{Z}$ in countably many variables, hence it is countable. The language of porings is countable, hence there are countably

many operations. There are countably many different substitutions for
the variables of each of these countably many operations. Altogether it
follows that $\sigma(F(\mathbb{N}))$ is countable. There are at most $2^{\aleph_0}$ different intermediate rings of $F(\mathbb{N})$ and $\sigma(F(\mathbb{N}))$, each of them admitting at most
$2^{\aleph_0}$ different partial orders. Thus, the class of *H*–closed monoreflective
subcategories of **POR/N** is a set of cardinality at most $2^{\aleph_0}$. We show
that the cardinality is exactly $2^{\aleph_0}$ by exhibiting a sufficiently large set of
H–closed monoreflectors. Note that any subcategory of **POR/N** having
property (9D.1) is *H*–closed. In particular, for every totally ordered ring
(A, P) between $\mathbb{Z}$ and R_0 the subcategory $(\mathbf{A,P})\mathbf{POR/N}$ has this property. Thus, it suffices to show that there are $2^{\aleph_0}$ different intermediate
rings between $\mathbb{Z}$ and R_0: Let $\mathbb{P}$ be the set of all prime numbers, let $\mathbb{P}'$
be any subset. Then we define $\mathbb{Q}_{\mathbb{P}'}$ to be the field $\mathbb{Q}(\sqrt{p}; p \in \mathbb{P}')$; it is
totally ordered by the restriction of the total order of R_0. If $\mathbb{P}' \neq \mathbb{P}''$
then $\mathbb{Q}_{\mathbb{P}'} \neq \mathbb{Q}_{\mathbb{P}''}$, hence $\mathbb{Q}_{\mathbb{P}'}\mathbf{POR/N} \neq \mathbb{Q}_{\mathbb{P}''}\mathbf{POR/N}$. Thus, there are at
least as many different *H*–closed monoreflectors as there are subsets of
$\mathbb{P}$, hence the cardinality is $2^{\aleph_0}$.

Let $(r_i)_{i \in I}$, I some totally ordered set, be a proper chain of *H*-closed
monoreflectors of **POR/N**. This chain corresponds to the proper chain
$(r_i(F(\mathbb{N}))_{i \in I}$ of intermediate porings of $F(\mathbb{N})$ and $\sigma(F(\mathbb{N}))$. Since these
porings are both countable the chain must also be at most countable.

It remains to prove that **SAFR** is *H*–closed. For, since **SAFR** is
the smallest of *all* monoreflective subcategories of **POR/N**, this implies
that **SAFR** is also the smallest *H*–closed* monoreflective subcategory. To
prove the claim we use the set $\mathcal{F}_\sigma(\mathbb{N})$ of functional formulas associated
with the reflector σ. It is clear from the definition that $\mathcal{F}_\sigma(\mathbb{N})$ is the
set of all functional formulas. Suppose that $(A, P) \in Inj(\{\sigma_{F(\mathbb{N})}\})$ and
consider the reflection $\sigma_{(A,P)} \colon (A, P) \to \sigma(A, P)$. It suffices to show that
$\sigma_{(A,P)}$ is surjective (since then the rings are isomorphic, even as porings
since the ring $\sigma(A, P)$ admits only one partial order, cf. Proposition 7.7).
But surjectivity is clear since, if $c = \omega_\Theta(a_1, \ldots, a_n)$ is any semi–algebraic
function, then $\Theta \in \mathcal{F}_\sigma(\mathbb{N})$, hence $c \in A$ by Lemma 10.5. $\square$

Our next result relates *H*–closedness of a reflector r to the preservation
of surjectivity of morphisms by the functor r.

Proposition 10.8 (a) Let $r \colon \mathbf{POR} \to \mathbf{D}$ be an *H*–closed monoreflector.
Then surjectivity of a **POR/N**–morphism $f \colon (A, P) \to (B, Q)$ implies
surjectivity of $r(f) \colon r(A, P) \to r(B, Q)$.

(b) Let $r: \mathbf{FR/N} \to \mathbf{D}$ be a monoreflector. The following conditions are equivalent:

(i) r is H–closed.
(ii) If $f: (A, P) \to (B, Q)$ is a surjective $\mathbf{FR/N}$–morphism then $r(f)$ is surjective as well.

Proof (a) Let $B' = im(r(f)) \subseteq r(B, Q)$, let Q' be the restriction of the partial order of $r(B, Q)$ to a partial order of B'. Then $(B', Q') \in ob(\mathbf{D})$ (since r is H-closed), and we have a diagram

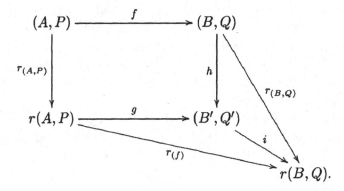

The universal property (2.1) of the reflection $r_{(B,Q)}$ yields a unique homomorphism $\overline{h}: r(B, Q) \to (B', Q')$ with $h = \overline{h} r_{(B,Q)}$. Then $r_{(B,Q)} = ih = i\overline{h} r_{(B,Q)}$ implies that $i\overline{h} = id_{r(B,Q)}$ (once again using the universal property). But then i is surjective, and the proof of (a) is finished.
(b) The implication (i) $\Rightarrow$ (ii) is clear because of (a). So, we only need to prove (ii) $\Rightarrow$ (i). Suppose that $f: (A, P) \to (B, Q)$ is an $\mathbf{FR/N}$–morphism with (A, P) in $\mathbf{D}$. Again, let (B', Q') be the image of f, let $f': (A, P) \to (B', Q')$ be f with restricted codomain. Since f' is surjective condition (ii) implies that $r(f'): (A, P) \cong r(A, P) \to r(B', Q')$ is surjective as well. Then $r_{(B',Q')}$ is both injective and surjective. Since we are dealing with reduced f–rings this means that $r_{(B',Q')}$ is an isomorphism, i.e., $(B', Q') \in ob(\mathbf{D})$. $\square$

Before turning to explicit examples we briefly discuss the behavior of H–closed monoreflectors under the constructions of section 9. If r and s are H–closed then so is their supremum (by Theorem 10.2). Concerning the infimum, there are examples showing that $r \wedge s$ can be H–closed. But we do not know any general rule governing the behavior of $r \wedge s$ with respect to H–closedness. The same is true for the construction of section 9C, i.e., forming the intersection of an H–closed monoreflector with a subfunctor of the complete ring of functions functor Π. Finally, for the construction of section 9D we have a negative result: Suppose that $r: \mathbf{POR/N} \to \mathbf{D}$ is H–closed and that $s: \mathbf{POR/N} \to \mathbf{E}$ is a monoreflector which is properly stronger than r. Suppose that $\mathbf{C} \subseteq \mathbf{POR/N}$ is a full subcategory having property (9D.1). Let $t: \mathbf{POR/N} \to \mathbf{E}$ be the monoreflector of section 9D, i.e., $t_{(A,P)} = r_{(A,P)}$ if $r(A,P) \notin ob(\mathbf{C})$, $t_{(A,P)} = s_{(A,P)}$ if $r(A,P) \in ob(\mathbf{C})$. We show that t is not H–closed if $t \neq r,s$. Assume by way of contradiction that t is H–closed. Then r and t are both determined by the reflections of the free objects $F(X)$, X varying in $\mathbf{SETS}$ (Theorem 10.1). If $r(F(X)) \in ob(\mathbf{C})$ for some set X then $\mathbf{D} \subseteq \mathbf{C}$. For, let (B,Q) be any member of $\mathbf{D}$. There is a morphism $f: F(X) \to (B,Q)$ which can be extended uniquely to $\overline{f}: r(F(X)) \to (B,Q)$ (by the universal property (2.1)). But then it follows that $(B,Q) \in ob(\mathbf{C})$. Since $\mathbf{D} \subseteq \mathbf{C}$ the definition of t implies that $t = s$. This contradicts our assumptions about t, hence $r(F(X)) \notin ob(\mathbf{C})$ for all sets X. Now the construction of t shows that $t_{F(X)} = r_{F(X)}$ for all sets X. Since r and t are H–closed this means that $r = t$, a contradiction. Therefore t cannot be H–closed.

In the rest of the section a large number of examples will be discussed showing that many, but not all, important reflectors are H–closed.

Example 10.9 POR and POR/N are not H–closed in PREOR
Suppose that (A,P) is any reduced poring. Then (A,A) is preordered, but is not a poring unless A is the zero ring. Since the identity map id_A, considered as a map $(A,P) \to (A,A)$ is a surjective $\mathbf{PREOR}$–morphism it follows that both $\mathbf{POR/N}$ and $\mathbf{POR}$ are not H–closed in $\mathbf{PREOR}$.□

Example 10.10 POR/N is not H–closed in POR
Let (A,P) be the formal power series ring $\mathbb{R}[[T]]$ with the total order determined by $0 < T$. The ideal (T^2) is convex, but not radical. Therefore the totally ordered factor ring $\mathbb{R}[[T]]/(T^2)$ is not in $\mathbf{POR/N}$, but the canonical homomorphism $\mathbb{R}[[T]] \to \mathbb{R}[[T]]/(T^2)$ is surjective.   □

Example 10.11 (A,P)POR/N is H–closed in POR/N
If $\mathbf{D} \subseteq \mathbf{POR/N}$ is a reflective subcategory of $\mathbf{POR/N}$ having property (9D.1) then $\mathbf{D}$ is clearly H–closed. If (A,P) is a totally ordered ring between $\mathbb{Z}$ and R_0 then the category $\mathbf{(A,P)POR/N}$ satisfies (9D.1),

hence it is H–closed. $\square$

Example 10.12 WBIPOR/N *is H–closed in* **POR/N**
Suppose that (A, P) is a reduced poring belonging to **WBIPOR/N**, i.e.,
$1 + \sum A^2 \subseteq A^\times$ (Example 2.10). If $f: (A, P) \to (B, Q)$ is a surjective
POR/N–morphism then $f(1 + \sum A^2) = 1 + \sum B^2$. Therefore $(B, Q) \in$
$ob(\textbf{WBIPOR/N})$. (The same argument applies to the categories **POR**
and **PREOR** instead of **POR/N**.) $\square$

Example 10.13 BIPOR/N *is not H–closed in* **POR/N**
Let (A, P) be the ring $\mathbb{R}[T]$ with its weak partial order. The reflection
of (A, P) in **BIPOR/N** is (A_{1+P}, P_{1+P}) (Example 2.11). Note that
$T \notin A_{1+P}^\times$. If Q is the total order of A for which T is positive and
infinitely large then the identity map of A is a morphism $(A, P) \to (A, Q)$
and induces the surjective morphism $(A_{1+P}, P_{1+P}) \to (A_{1+P}, Q_{1+P})$. But
(A_{1+P}, Q_{1+P}) is not a bounded inversion ring since $1 < T$ but $T \notin A_{1+P}^\times$.
Note however that factor porings exist in **BIPOR/N** (Remark 10.1).
This can be shown using the argument of Example 10.12. (The same
example shows that **BIPOR** is not H–closed in **POR**.) $\square$

Example 10.14 REPPOR *is not H–closed in* **POR/N**
As in the preceding example, let (A, P) be the ring $\mathbb{R}[T]$ with its weak
partial order. Since P consists of the positive semidefinite polynomials
([61], p. 344), (A, P) is representable. Let Q be the partial order of
A defined in Example 6.4, i.e., $\sum_{i=m}^{n} a_i T^i \in A$ is positive if and only if
$a_m > 0$ and $m = 0$ or $m \geq 2$. It is clear that $P \subseteq Q$, hence the identity
map of A is a surjective morphism $(A, P) \to (A, Q)$. But (A, Q) is not
representable. $\square$

Example 10.15 FR/N *is H–closed in* **POR/N**
Let $f: (A, P) \to (B, Q)$ be a surjective morphism of reduced porings
where (A, P) is a reduced f–ring. Composing with $\Delta_{(B,Q)}: (B, Q) \to$
$\Pi(B, Q)$ we obtain the morphism $\Delta_{(B,Q)} f: (A, P) \to \Pi(B, Q)$ of reduced
f–rings. Since it is an l–homomorphism ([44], Lemma 2.2), $ker(f) =$
$ker(\Delta_{(B,Q)} f)$ is an l–ideal of (A, P), hence $(\overline{A}, \overline{P}) = (A/ker(f), P +$
$ker(f)/ker(f))$ is an f–ring as well. Because the morphism $\overline{f}: (\overline{A}, \overline{P}) \to$

(B,Q) induced by f is bijective it follows from Proposition 1.11 that $\overline{f}$ is an isomorphism in **POR/N**, hence $(B,Q) \in ob(\textbf{FR/N})$. $\qquad\qquad\square$

Example 10.16 BIFR/N *is H–closed in* **FR/N** *and* **POR/N**
According to Example 10.13 the subcategory **BIPOR/N** $\subseteq$ **POR/N** is not *H*–closed. Therefore it is not immediately clear whether **BIFR/N** is or is not *H*–closed in **POR/N**. However, it was noted in Example 10.13 that factor porings exist in **BIPOR/N**. Since **FR/N** is *H*–closed they also exist in **FR/N** (Example 10.15, Remark 10.1). But then factor porings also exist in **BIFR/N**. According to Remark 10.1, for subcategories of **FR/N** the existence of factor porings is equivalent to *H*–closedness. Thus, **BIFR/N** is *H*–closed in **FR/N** and in **POR/N**. $\qquad\square$

Example 10.17 VNRFR *is H–closed in* **POR/N**
Suppose that (A,P) is a von Neumann regular f–ring and that $f\colon (A,P) \to (B,Q)$ is a surjective **POR/N**–morphism. It follows from Example 10.15 that (B,Q) is an f–ring as well. Moreover, homomorphic images of von Neumann regular rings are always von Neumann regular. Together this implies that $(B,Q) \in ob(\textbf{VNRFR})$. $\qquad\square$

According to Theorem 10.6 every *H*–closed monoreflector $r\colon \textbf{POR/N} \to \textbf{D}$ is completely determined by the reflections $r_{F(n)}, n \in \mathbb{N}$. Using Corollary 8.2 and Theorem 8.10 one may consider $r(F(n))$ as an intermediate poring between $F(n)$ and $\sigma(F(n))$, i.e., $r(F(n))$ is a ring of semi-algebraic functions $R_0^n \to R_0$. These subrings of $\sigma(F(n))$ put the reflector r into a rather concrete geometric context. Now suppose that, for each $n \in \mathbb{N}$, an intermediate poring (A_n, P_n), between $F(n)$ and $\sigma(F(n))$ is given (with canonical homomorphisms $f_n\colon F(n) \to (A_n, P_n)$, $g_n\colon (A_n, P_n) \to \sigma(F(n))$). It is clear that the f_n are the reflections $r_{F(n)}$ for some *H*–closed monoreflector r if and only if each f_n is an epimorphism and each (A_n, P_n) belongs to the injectivity class of the set $\{f_n; n \in \mathbb{N}\}$. The latter condition means that the functions belonging to $\bigcup_{n\in\mathbb{N}} A_n$ and $\bigcup_{n\in\mathbb{N}} P_n$ are stable with respect to composition. More precisely, suppose that $a \in A_n$ (or $\in P_n$) and that $a_1, \dots, a_n \in A_m$ for some m. Then

$$R_0^m \to R_0\colon x \to a(a_1(x), \dots, a_n(x))$$

must belong to A_m (or P_m). Here are a few examples of families satisfying the stability condition:

Example 10.18 *Continuous semi–algebraic functions*
For each $n \in \mathbb{N}$, let $A_n \subseteq \sigma(F(n))$ be the ring of continuous semi–algebraic functions ([39], section 6; [41], Chapter I, §1), let $P_n = A_n \cap \sigma^+(F(n))$. Since compositions of continuous functions are continuous these rings have the stability property. In fact, the rings of continuous semi–algebraic functions define a reflector. Section 12 is devoted entirely to this reflector. $\qquad\qquad\square$

Example 10.19 *Nash functions*
For each $n \in \mathbb{N}$, let $A_n \subseteq \sigma(F(n))$ be the ring of Nash functions on R_0^n ([19], Definition 8.1.8) and let $P_n = A_n \cap \sigma^+(F(n))$. The Nash functions are clearly stable under composition. We show that the canonical homomorphisms $f_n \colon F(n) \to (A_n, P_n)$ are epimorphisms: The subcategory $\mathbf{R_0POR/N} \subseteq \mathbf{POR/N}$ is H–closed and contains every (A_n, P_n); let $r \colon \mathbf{POR/N} \to \mathbf{R_0POR/N}$ be the corresponding reflector. Then f_n factors uniquely as $\overline{f}_n r_{F(n)}$ where $\overline{f}_n \colon r(F(n)) \to (A_n, P_n)$ is the canonical homomorphism $R_0[T_1, \ldots, T_n] \to (A_n, P_n)$. According to [19], Proposition 8.8.1, $Sper(\overline{f}_n)$ is a homeomorphism. Now Proposition 7.16 implies that $\overline{f}_n$ and f_n are epimorphisms. Therefore the set $\{f_n; n \in \mathbb{N}\}$ defines a reflector which will be called the *Nash reflector*.

There exists a completely different approach to Nash functions on real spectra, originating with [105] (see also [2]; [37]; [102]; [103]). With every ring A it associates a sheaf $\mathcal{N}$ of Nash functions on the real spectrum $Sper(A)$. The global ring of sections, N_A, is called the *ring of Nash of functions* over A. There is a canonical homomorphism $\varphi_A \colon A \to N_A$. This construction of rings of Nash functions is functorial and the homomorphisms φ_A are the components of a natural transformation. But it is an open problem whether the construction is a reflector, in particular, whether it is idempotent ([102]). We shall return to this presentation of Nash functions at the end of section 14. $\qquad\qquad\square$

Example 10.20 *Differentiable semi–algebraic functions*
Suppose that $A_n \subseteq \sigma(F(n))$ is the ring of r times continuously differentiable semi–algebraic functions $R_0^n \to R_0$; let $P_n = A_n \cap \sigma^+(F(n))$. Once again it is clear that these functions are stable under compo-

sition.   It will be shown in Example 14.11 that the canonical functions $f_n\colon F(n) \to (A_n, P_n)$ are epimorphisms, hence that the family $(A_n, P_n), n \in \mathbb{N}$, defines a reflector.   □

11 Quotient–closed reflectors

According to Remark 10.1, one important property of an H–closed monoreflective subcategory of **POR/N** is that it contains every reduced factor poring of any of its objects. The present section is devoted to the equally important construction of porings of quotients (cf. section 1): If (A, P) is a reduced poring and if $S \subseteq A$ is a multiplicative subset then the ring of quotients A_S together with the set

$$P_S = \{\frac{a}{s} \in A_S;\ \exists\, t \in S\colon ast^2 \in P\}$$

is a reduced poring and the canonical homomorphism $i_S\colon A \to A_S$ is a morphism of reduced porings. Given a monoreflector $r\colon \mathbf{POR/N} \to \mathbf{D}$ we study the question whether (A_S, P_S) belongs to $\mathbf{D}$ for all $(A, P) \in ob(\mathbf{D})$ and all multiplicative subsets $S \subseteq A$. A monoreflective subcategory having this property and the corresponding reflector are called *quotient–closed*. We show that the quotient–closed monoreflectors form a complete lattice with smallest element **SAFR**. We do not know whether this lattice is a proper class and, if it is a set, what its cardinality is. Examples show that many, but not all, of the most important monoreflectors are quotient–closed. None of the two notions of quotient–closedness and H–closedness implies the other one. However, for subcategories of **VNRFR** they are equivalent.

We shall see below that it is essential and powerful information to know that a reflector is quotient–closed. Also note that the notion of quotient–closedness arises very naturally in connection with sheaf theoretic considerations (which are never very far beyond the horizon in real algebraic geometry and semi–algebraic geometry). For, if (A, P) belongs to a quotient–closed reflective subcategory $\mathbf{D} \subseteq \mathbf{POR/N}$, then the rings of sections of the affine scheme $Spec(A)$ can be studied as $\mathbf{D}$–objects.

Our first result shows that for a quotient–closed reflector of **POR/N** the formation of porings of quotients and the formation of reflections are commuting operations.

Remark 11.1 Suppose that $\mathbf{C}$ is a category and that $\mathbf{E}$ is a reflective subcategory. Then $\mathbf{E}$ is also a reflective subcategory of every intermediate category $\mathbf{D}$ between $\mathbf{E}$ and $\mathbf{C}$. Let P be any property which reflective subcategories may or may not have. We call P an *absolute* property if

either **E** has the property P as a reflective subcategory of every intermediate category **D** of **E** and **C**, or **E** does not have the property with
respect to any intermediate category **D**. Otherwise, P is a *relative property*. It is clear from the definition that quotient–closedness is an absolute
property. By contrast, being H–closed is a relative property. For example, **BIPOR/N** is not H–closed in **POR/N** (Example 10.13), but it is
trivially H–closed in itself. □

Proposition 11.2   Suppose that $r: \textbf{POR/N} \to \textbf{D}$ is a quotient–
closed epireflector. Then the canonical morphism $f_S : r(A, P)_{r_{(A,P)}(S)} \to$
$r(A_S, P_S)$ is an isomorphism.

Proof   Various universal properties yield the unique existence of the
morphism f_S with $f_S i_{r_{(A,P)}(S)} r_{(A,P)} = r_{(A_S, P_S)} i_S$. On the other hand,
quotient–closedness of **D** implies that $r(A, P)_{r_{(A,P)}(S)} \in ob(\textbf{D})$. Hence
there is a unique morphism $g_S : r(A_S, P_S) \to r(A, P)_{r_{(A,P)}(S)}$ with
$g_S r_{(A_S,P_S)} i_S = i_{r_{(A,P)}(S)} r_{(A,P)}$. Since i_S and $i_{r_{(A,P)}(S)}$ are epimorphisms
of rings and r is an epireflector one concludes that f_S and g_S are inverse
to each other. □

The intersection of any class of quotient–closed monoreflective subcategories of **POR/N** is obviously quotient–closed. Therefore the quotient–
closed monoreflectors form a complete lattice. As with H–closed monoreflectors, the supremum in this lattice coincides with the supremum formed
in the lattice of all monoreflectors. We conjecture that the corresponding
statement about infima is false, but we do not know any examples to this
effect.

Concerning the constructions discussed in section 9, we know already
that $r \vee s$ is quotient–closed if r and s are quotient–closed. Frequently
quotient–closedness of r and s implies that $r \wedge s$ is quotient–closed, but
we do not know any general criterion to decide when this is the case. The
same is true for reflections obtained as intersections of reflections with
subfunctors of Π (cf. section 9C). For the final construction (section 9D)
we can settle the question completely:

Proposition 11.3   Suppose that $r: \textbf{POR/N} \to \textbf{D}$ and $s: \textbf{POR/N} \to$
E are quotient–closed monoreflectors with $r \leq s$. Let **C** $\subseteq$ **POR/N**

be a subcategory satisfying condition (9D.1). Then the monoreflector $t\colon \mathbf{POR}/\mathbf{N} \to \mathbf{F} = (\mathbf{D}\backslash\mathbf{C}) \cup (\mathbf{E} \cap \mathbf{C})$ given by $t(A,P) = r(A,P)$ if $r(A,P) \notin ob(\mathbf{C})$, $t(A,P) = s(A,P)$ if $r(A,P) \in ob(\mathbf{C})$, is not quotient–closed whenever $r \neq t \neq s$.

Proof Assume by way of contradition that t is quotient–closed. Since $r \neq t \neq s$ there exist (A,P) and (B,Q) such that $t(A,P) = r(A,P) \neq s(A,P)$ and $r(B,Q) \neq t(B,Q) = s(B,Q)$. Thus, $r(A,P)$ belongs to $\mathbf{D}\backslash\mathbf{C}$ and $r(B,Q)$ belongs to $\mathbf{D} \cap \mathbf{C}$. Because $r(A,P)$ is an algebra over the direct product $r(A,P) \times r(B,Q)$ via the first projection it follows that $r(A,P) \times r(B,Q) \notin ob(\mathbf{C})$. From Proposition 2.6 we know that $r(A,P)\times r(B,Q) \in ob(\mathbf{D})$, hence $r(A,P)\times r(B,Q) \in ob(\mathbf{F})$. Now we form the poring of quotients with respect to the multiplicative set $S = \{e\}$, $e = (0,1) \in r(A,P) \times r(B,Q)$:

$$(r(A,P) \times r(B,Q))_S \cong r(B,Q).$$

According to Proposition 11.2 the poring $t(r(B,Q)) = s(B,Q)$ must be isomorphic to

$$(t(r(A,P) \times r(B,Q)))_{t_{r(A,P)\times r(B,Q)}(S)} \cong r(B,Q).$$

This contradicts the choice of (B,Q) and the proof is finished. $\qquad\square$

Proposition 11.4 Suppose that $r\colon \mathbf{POR}/\mathbf{N} \to \mathbf{D}$ is a monoreflector with $\mathbf{D} \subseteq \mathbf{VNRFR}$. Then r is quotient–closed if and only if r is H–closed.

Proof Note that, for $\mathbf{D} \subseteq \mathbf{FR}/\mathbf{N}$, being H–closed is equivalent to the existence of factor porings in $\mathbf{D}$ (Remark 10.1). Now suppose that $(A,P) \in ob(\mathbf{D})$. If $S \subseteq A$ is multiplicative then $i_S\colon A \to A_S$ is surjective and $i_S(P) = P_S$. Therefore, if $I = ker(i_S) \subseteq A$ then $(A/I, P + I/I) \cong (A_S, P_S)$. This shows that $\mathbf{D}$ is quotient–closed if it is H–closed. Now let $I \subseteq A$ be some convex ideal and let $\pi\colon (A,P) \to (\overline{A},\overline{P})$ be the canonical homomorphism onto the factor poring. The subset $S = \pi^{-1}(\overline{A}^{\times}) \subseteq A$ is multiplicative. Then π factors uniquely as $\pi = \overline{\pi}i_S$ with $\overline{\pi}\colon (A_S, P_S) \to (\overline{A},\overline{P})$. Because $ker(i_S) = I$ it follows that $\overline{\pi}$ is an isomorphism. Thus, quotient–closedness implies H–closedness. $\qquad\square$

Corollary 11.5 **SAFR** and **VNRFR** are quotient–closed monoreflective subcategories of **POR/N**.

Proof This follows from Proposition 11.4 in view of Theorem 10.7 and Example 10.17. □

Corollary 11.6 The class of quotient–closed monoreflective subcategories of **POR/N** is a complete lattice with largest element **POR/N** and smallest element **SAFR**. □

The notion of quotient–closedness can be used to determine another sufficient condition for the supremum of two reflectors to be their composition.

Proposition 11.7 Let $r\colon \mathbf{POR/N} \to \mathbf{D}$, $s\colon \mathbf{POR/N} \to \mathbf{E}$ be monoreflectors. Suppose that r is quotient–closed and that $s(A,P)$ is of the form (A_S, P_S) for a suitable multiplicative subset of A. Then $(r \vee s)(A,P) = s(r(A,P))$ for all $(A,P) \in ob(\mathbf{POR/N})$.

Proof This is an immediate consequence of Proposition 9A.1. □

Two of the monoreflective subcategories of **POR/N** discussed so far satisfy the condition about the reflector s in the proposition, namely **WBIPOR/N** (Example 2.10) and **BIPOR/N** (Example 2.11). Thus, if $\mathbf{D} \subseteq \mathbf{POR/N}$ is quotient–closed then the reflections of a reduced poring (A,P) in **WBIPOR/N** or in **BIPOR/N** are obtained easily from the reflection of (A,P) in **D**. A large number of quotient–closed reflectors will be exhibited below.

An H–closed monoreflector $r\colon \mathbf{POR/N} \to \mathbf{D}$ is completely determined by the reflections $r_{F(n)}, n \in \mathbb{N}$ (Theorem 10.6). Therefore these reflections must contain enough information about r to decide whether the reflector is quotient–closed. However, it is not clear how this information can be retrieved. Although the following result does not solve this problem it may be considered at least as an approximation of an answer.

Proposition 11.8 Let $r\colon \mathbf{POR/N} \to \mathbf{D}$ be an H–closed monoreflector. Then r is quotient–closed if and only if the poring of quotients $r(F(n +$

$1))_{T_{n+1}}$ belongs to $\mathbf{D}$ for all $n \in \mathbb{N}$ (where $T_1, \ldots, T_{n+1}$ are the variables of $F(n+1)$).

Proof If r is quotient–closed then $r(F(n+1))_{T_{n+1}} \in ob(\mathbf{D})$ by the definition of quotient–closedness. Conversely, assume that the condition holds. To prove that r is quotient–closed, let $(A, P) \in ob(\mathbf{D})$ and suppose that $S \subseteq A$ is multiplicative. It is claimed that (A_S, P_S) belongs to the injectivity class of $\{r_{F(n)}; n \in \mathbb{N}\}$. So, let $f \colon F(n) \to (A_S, P_S)$ be any homomorphism and consider the functorial map $r(f) \colon r(F(n)) \to r(A_S, P_S)$. We have to prove that $im(r(f)) \subseteq im(r_{(A_S, P_S)})$. The images of the variables $T_1, \ldots, T_n$ can be written in the form $f(T_i) = \frac{a_i}{s_i}$, $a_i \in A$, $s_i \in S$. We (may) assume that $s_1 = \ldots = s_m = s$. The homomorphism f is the composition of the two homomorphisms $g \colon F(n) \to F(n+1)_{T_{n+1}}$, $h \colon F(n+1)_{T_{n+1}} \to (A_S, P_S)$ defined by $g(T_i) = \frac{T_i}{T_{n+1}}$ for $i = 1, \ldots, n$, and $h(T_i) = a_i$ for $i = 1, \ldots, n$, $h(T_{n+1}) = s$. The restriction $h' \colon F(n+1) \to (A, P)$ of h extends uniquely to $\overline{h'} \colon r(F(n+1)) \to (A, P))$ (because (A, P) belongs to $\mathbf{D}$). The universal property of porings of quotients yields the morphisms k and l in the following commutative diagram

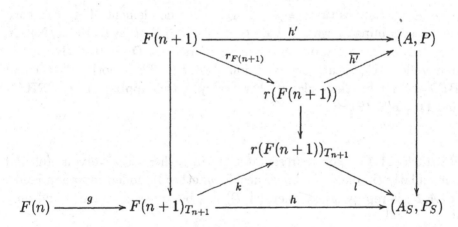

By hypothesis $r(F(n+1))_{T_{n+1}}$ belongs to $\mathbf{D}$; therefore there is a unique morphism $m \colon r(F(n)) \to r(F(n+1))_{T_{n+1}}$ with $kg = mr_{F(n)}$. The universal property of $r_{F(n)}$ together with

$$r(f)r_{F(n)} = r_{(A_S, P_S)}f = r_{(A_S, P_S)}lmr_{F(n)}$$

implies that $r(f) = r_{(A_S, P_S)}lm$, hence the desired containment $im(r(f)) \subseteq im(r_{(A_S, P_S)})$ holds. $\square$

Example 11.9 *The categories* **(A,P)POR/N** *are quotient–closed*
Let (A, P) be an intermediate totally ordered ring between $\mathbb{Z}$ and R_0.
If (B, Q) is an (A, P)–algebra and $S \subseteq B$ is a multiplicative set then
(B_S, Q_S) is an (A, P)–algebra as well. $\square$

As a consequence of this example we note that the class of quotient–closed
monoreflective subcategories of **POR/N** has cardinality at least $2^{\aleph_0}$ (cf.
the proof of Theorem 10.7).

Example 11.10 *The categories* **WBIPOR/N** *and* **BIPOR/N** *are not*
                        quotient–closed
Let (A, P) be the local ring $\mathbb{R}[X, Y]_{(X,Y)}$ with the total order defined by:
$0 < X \ll 1$ and $0 < Y < X^n$ for all $n \in \mathbb{N}$. The maximal ideal of A is
convex, hence (A, P) has bounded inversion. In particular, (A, P) also has
weak bounded inversion. We show now that (A_S, P_S) with $S = \{Y^n; n \in \mathbb{N}\}$ does not have weak bounded inversion, hence bounded inversion also
fails. Assume that (A_S, P_S) has weak bounded inversion. Then $1 + \left(\frac{X}{Y}\right)^2 \in A_S^\times$ implies that $\frac{1}{X^2+Y^2} \in A_S$. Since no element of A_S can have
such a denominator we conclude that $(A_S, P_S) \notin ob(\textbf{WBIPOR/N})$. In
fact, this counterexample shows much more: Let $\mathbf{D} \subseteq \textbf{POR/N}$ be any
monoreflective subcategory containing (A, P). Then both **WBID** and
BID are not quotient–closed. For example, this applies to $\mathbf{D} = \textbf{FR/N}$
and $\mathbf{D} = \textbf{REPPOR}$. $\square$

Remark 11.11 For representable porings there is a very useful and
particularly simple criterion to decide whether bounded inversion holds.
If (A, P) is representable then (A, P) has bounded inversion if and only
if
$$A^\times = \{a \in A; \ \forall \alpha \in Sper(A, P) \colon a(\alpha) \neq 0\}.$$

One direction is trivial; for the other one, suppose that (A, P) is a
bounded inversion poring. If $a \in A^\times$ then $a(\alpha) \neq 0$ for $\alpha \in Sper(A, P)$
is trivial. Now let $a(\alpha) \neq 0$ for every α. It suffices to show that $a^2 \in A^\times$,
hence we may assume that $a(\alpha) > 0$ for every α. Since $a(\alpha) \in A/\alpha$ cannot belong to any proper convex prime ideal there is some $b_\alpha \in A$ such
that $1 \leq a(\alpha)b_\alpha(\alpha)$ in A/α. The sets
$$C_\alpha = \{\beta \in Sper(A, P); \ 1 \leq a(\beta)b_\alpha(\beta)\}$$

are closed and constructible. Because they form a cover of $Sper(A, P)$ there exist $\alpha_1, \ldots, \alpha_r$ with $Sper(A, P) = C_\alpha \cup \ldots \cup C_{\alpha_r}$. If we define $b = \sum_i (1 + b_{\alpha_i})^2$ then $1 \leq (ab)(\alpha)$ for every α. Because (A, P) is representable we conclude that $ab \geq 1$ in (A, P). From bounded inversion it follows that $ab \in A^\times$, hence $a \in A^\times$. □

Example 11.12 REPPOR *is quotient–closed*

Let (A, P) be a representable poring and suppose that $S \subseteq A$ is multiplicative. Then $Sper(A_S, P_S)$ is identified canonically with the generically closed proconstructible subspace $\{\alpha \in Sper(A, P); S \cap supp(\alpha) = \emptyset\}$ of $Sper(A, P)$. We claim that $\frac{a}{s} \in P_S$ if $\frac{a(\alpha)}{s(\alpha)} \geq 0$ for all $\alpha \in Sper(A_S, P_S)$. For, given such a quotient, the set

$$C = \{\alpha \in Sper(A, P); as(\alpha) \geq 0\}$$

is closed constructible and contains $Sper(A_S, P_S)$. If $\beta \notin C$ then there is some $s_\beta \in S$ such that $s_\beta(\beta) = 0$. Replacing s_β by its square we may assume that $0 \leq s_\beta$. For each β the set

$$C_\beta = \{\alpha \in Sper(A, P); s_\beta(\alpha) = 0\}$$

is closed constructible and $C_\beta \cap Sper(A_S, P_S) = \emptyset$. These sets form a constructible cover of $Sper(A, P) \backslash C$, hence there is a finite subcover by sets, say, $C_1, \ldots, C_k$. With $t = s_{\beta_1} \cdot \ldots \cdot s_{\beta_k} \in S$ one has $t(\alpha) \geq 0$ for all $\alpha \in Sper(A, P)$ and $t(\alpha) = 0$ for $\alpha \notin C$. It follows that $ast^2(\alpha) \geq 0$ for every α. Representability of (A, P) implies that $ast^2 \in P$. Now $\frac{a}{s} \in P_S$ follows immediately from the definition of P_S. □

Example 11.13 FR/N *is quotient–closed*

Given a reduced f–ring (A, P) and a multiplicative subset $S \subseteq A$ the preceding example shows that (A_S, P_S) is a subporing of $\Pi(A_S, P_S)$. The claim of the present example is that it is even a sublattice. It suffices to show that $\frac{a}{s} \vee \frac{b}{t}$ belongs to A_S for any two elements $\frac{a}{s}, \frac{b}{t} \in A_S$ (where the supremum is formed in the f–ring $\Pi(A_S, P_S)$). But this follows from

$$\frac{a}{s} \vee \frac{b}{t} = \frac{ast^2}{s^2t^2} \vee \frac{bts^2}{s^2t^2} = \frac{ast^2 \vee bts^2}{s^2t^2}$$

since (A, P) is a sub–f–ring of $\Pi(A, P)$. □

The examples show that there are many monoreflectors of **POR/N** that are both H–closed and quotient–closed (Proposition 11.4, Example 11.9, Example 11.13). The subcategory **WBIPOR/N** is H–closed (Example 10.12), but not quotient–closed (Example 11.10), whereas **REPPOR** is not H–closed (Example 10.14), but is quotient–closed (Example 11.12). Finally **BIPOR/N** is neither H–closed (Example 10.13) nor quotient–closed (Example 11.10).

The reflectors that have both properties are particularly well–behaved. Any such reflective subcategory **D** $\subseteq$ **POR/N** is closely related to its intersection with **TOF**. This is due to the following observation: If $(A, P) \in ob(\mathbf{D})$ and if $\alpha \in Sper(A, P)$ then A/α belongs to **D** (by H–closedness) and $\kappa_{(A,P)}(\alpha)$ belongs to **D** (by quotient–closedness). This connection between H–closed and quotient–closed reflectors and totally ordered fields will be exploited in later sections.

Remark 11.14 The notion of quotient–closed reflective subcategories can be extended from **POR/N** to **POR** and **PREOR**. Then **POR/N** and **POR** are quotient–closed subcategories of **PREOR**. Recall that they are not H–closed (Example 10.9, Example 10.10). □

12 The real closure reflector

The notion of real closed rings was first introduced in [108] (see also [110]; [111]) to extend the theory of semi–algebraic spaces as developed by Delfs and Knebusch ([39]; [41]) to real spectra. (For a simultaneous and independent attempt with a different motivation, cf. [28].) Generalizing the approach of [108] in [110] and [111], a real closed ring was associated with any ring A and any proconstructible subset $K \subseteq Sper(A)$. If (A, P) is a poring then $Sper(A, P) \subseteq Sper(A)$ is proconstructible, hence the general construction can be applied to associate a real closed ring with (A, P), namely the real closed ring belonging to A and $Sper(A, P)$. This real closed ring is called the *real closure* of (A, P). From [108], [110] and [111] it is clear that the construction of the real closure is a reflector (although the word "reflector" was not used). The real closure reflector is of great importance for the topological investigation of real spectra. The reason is the following extremal property of this reflector: If r: **PREOR** $\rightarrow$ **D** is any reflector such that every map $Sper(r_{(A,P)})$ is a homeomorphism then **RCR** $\subseteq$ **D** (Theorem 12.12). We consider this as the Main Theorem about real closed rings.

In this section we propose a new construction of the real closure of a reduced poring. It is based on the determination of an H–closed monoreflector of **POR/N** by the reflection morphisms $r_{F(n)}, n \in \mathbb{N}$. The idea is to emulate on the level of real spectra the relationship between a polynomial ring $R[X_1, \dots, X_n]$ (R a real closed field) and the ring of continuous semi–algebraic functions on R^n by associating something like a "ring of continuous semi–algebraic functions on $Sper(A, P)$" with every reduced poring (A, P). One way to execute this idea is evident from section 10: We start with $\sigma_{(A,P)}(A) \subseteq \sigma(A, P)$ as a basic supply of continuous semi–algebraic functions on $Sper(A, P)$. For the reduced poring $F(n)$ we already know the ring we want to study, namely: we define $\rho(F(n)) \subseteq \sigma(F(n))$ to be the subring of continuous semi–algebraic functions. The functions belonging to the rings $\rho(F(n))$ are used as generic continuous semi–algebraic functions, i.e., if $a \in \rho(F(n))$ and if $a_1, \dots, a_n \in A$ then we substitute the basic continuous semi–algebraic functions $a_1, \dots, a_n$ for the variables of a to obtain a continuous semi–algebraic function in $\sigma(A, P)$. Technically, for the substitution we use the n–ary operation ω belonging to a; thus, $\omega(a_1, \dots, a_n) \in \sigma(A, P)$ is a typical continuous semi–algebraic function over (A, P).

We shall prove that this procedure leads to an H–closed reflector of **POR/N** whose reflections of the $F(n)$'s are the subrings $\rho(F(n)) \subseteq \sigma(F(n))$ of continuous semi–algebraic functions. It was noted in Example 10.18 that the continuous semi–algebraic functions are stable under composition. Let $\rho_{F(n)} \colon F(n) \to \rho(F(n))$ be the canonical homomorphism. In order to prove the existence of an H–closed monoreflector whose reflection morphisms belonging to the $F(n)$'s are the maps $\rho_{F(n)}$ we only need to prove that the $\rho_{F(n)}$'s are epimorphisms. Before doing this we recall a few elementary properties of the rings $\rho(F(n))$.

The ring $\rho(F(n))$ is a reduced poring with the partial order induced by $\sigma(F(n))$. If $a, b \in \rho(F(n))$ then the functions $a \wedge b$, $a \vee b \in \sigma(F(n))$ are continuous. Thus, $\rho(F(n))$ is a sub–f–ring of $\sigma(F(n))$. Suppose that $0 \leq a \in \rho(F(n))$. The unique function $0 \leq b \in \sigma(F(n))$ with $b^2 = a$ (Proposition 7.7) obviously belongs to $\rho(F(n))$. Therefore, the positive functions in $\rho(F(n))$ are exactly the squares. Finally note that every function $a \in \rho(F(n))$ with $a(x) \neq 0$ for all $x \in R_0^n$ has a continuous inverse. In particular, $\rho(F(n))$ has bounded inversion.

Proposition 12.1 For each $n \in \mathbb{N}$, the canonical homomorphism $\rho_{F(n)} \colon F(n) \to \rho(F(n))$ is an epimorphism.

Proof Let $i_n \colon \rho(F(n)) \to \sigma(F(n))$ be the inclusion. By Proposition 7.16 it suffices to show that $Sper(i_n)$ is surjective. If this is false then

$$im(Sper(i_n)) \subset Sper(\rho(F(n)))$$

is a proper proconstructible subspace, hence there is a nonempty locally closed constructible subset

$$C = \{\beta \in Sper(\rho(F(n))); \, a(\beta) = 0, a_1(\beta) > 0, \dots, a_k(\beta) > 0\}$$

of $Sper(\rho(F(n)))$ such that $C \cap im(Sper(i_n)) = \emptyset$. Since $a, a_1, \dots, a_k$ are semi–algebraic functions the set

$$B = Sper(\rho_{F(n)})(C) \subseteq Sper(F(n))$$

is nonempty and constructible (cf. Lemma 7.6 (a)). Remember that $\mathbf{R_0 POR/N} \subseteq \mathbf{POR/N}$ is monoreflective and that the reflection of $F(n)$ is the polynomial ring $R_0[T_1, \dots, T_n]$ with the weak partial order. Therefore the real spectra of the two porings are homeomorphic with respect to the constructible topology. (They are even homeomorphic using the spectral topology; but this is not needed here.) Therefore R_0^n can be considered as a dense subspace of $Sper(\sigma(F(n)))$ ([33], Théorème 5.1; [19], Théorème 7.2.3). Since $Sper(F(n))$ with the constructible topology is homeomorphic to $Sper(\sigma(F(n)))$ we consider B as a closed and open subset of $Sper(\sigma(F(n)))$. Because $B \neq \emptyset$ there exists some point $x \in R_0^n \cap B$. The set $Sper(\rho_{F(n)})^{-1}(\{x\})$ contains the element $Sper(i_n)(x)$. In fact, this is its only element since every function of $\rho(F(n))$ is given by a functional formula. By definition of B, $Sper(i_n)(x)$ must belong to C, i.e., $C \cap im(Sper(i_n)) \neq \emptyset$. This contradiction finishes the proof. $\qquad \square$

Because of Proposition 12.1 we can now make the following definition:

Definition 12.2 The H–closed monoreflector of **POR/N** determined by the epimorphisms $\rho_{F(n)}$, $n \in \mathbb{N}$, is called the *real closure reflector* and is denoted by ρ. The corresponding subcategory **RCR** is the category of *real closed rings*. The reflection of any reduced poring in **RCR** is its *real closure*. The elements of $\rho(A, P)$ are called *continuous semi-algebraic function (abstract case)* over (A, P). $\qquad \square$

Since **POR/N** $\subseteq$ **POR** and **POR** $\subseteq$ **PREOR** are reflective subcategories, **RCR** is also reflective in **POR** and in **PREOR**. The corresponding reflectors of **POR** and **PREOR** are also denoted by ρ and are called *real closure reflectors*.

To start with, ρ is just one reflector among many. But we shall show that ρ has truly exceptional properties making it as important as the reflector σ. In fact, for studying the topology of real spectra, ρ is far more important than σ. The next couple of results establish a few basic properties of real closed rings and real closures.

Proposition 12.3

(a) For every $n \in \mathbb{N}$, the functorial map $Sper(\rho_{F(n)})$ is a homeomorphism.

(b) For every $n \in \mathbb{N}$, the support map $supp_{\rho(F(n))}$ is a homeomorphism.

(c) For every $n \in \mathbb{N}$ and every $\alpha \in Sper(\rho(F(n)))$, the canonical homomorphism $\kappa(\alpha) = \kappa_{\rho(F(n))}(\alpha) \to \rho(\alpha) = \rho_{\rho(F(n))}(\alpha)$ is an isomorphism.

Proof (a) It is known from Proposition 7.9 that $Sper(\sigma_{F(n)})$ is bijective. Because $\sigma_{F(n)}$ factors through $\rho_{F(n)}$ this implies that $Sper(\rho_{F(n)})$ is surjective. Injectivity holds since $\rho_{F(n)}$ is an epimorphism (Proposition 12.1). It remains to show that $Sper(\rho_{F(n)})$ is open. So, let

$$U = \{\beta \in Sper(\rho(F(n))); \; a(\beta) > 0\}$$

for some $a \in \rho(F(n))$. Then $Sper(\rho_{F(n)})(U) \subseteq Sper(F(n))$ is the constructible subset corresponding to the open semi–algebraic subset

$$V = \{x \in R_0^n; \; a(x) > 0\}$$

of R_0^n. According to [33], Théorème 5.1, or [19], Théorème 7.2.3, the set $Sper(\rho_{F(n)})(U)$ is open.

(b) Since $\rho(F(n))$ is an f–ring the support map is injective and is a homeomorphism onto the image. Only surjectivity needs to be checked. So, let $p \subseteq \rho(F(n))$ be a prime ideal and let $a, b \in \rho^+(F(n))$ with $a+b \in p$. In R_0^n, let $Z(a), Z(b)$ and $Z(a + b)$ be the zero sets of the functions a, b and $a + b$. It is clear that $Z(a + b) \subseteq Z(a), Z(b)$. Therefore

$$c: R_0^n \;\rightarrow\; R_0: x \longrightarrow \begin{cases} 0 & \text{if } x \in Z(a) \\ \dfrac{a(x)^2}{a(x) + b(x)} & \text{if } x \notin Z(a), \end{cases}$$

$$d: R_0^n \;\rightarrow\; R_0: x \longrightarrow \begin{cases} 0 & \text{if } x \in Z(b) \\ \dfrac{b(x)^2}{a(x) + b(x)} & \text{if } x \notin Z(b) \end{cases}$$

are continuous semi–algebraic functions. Since $a^2 = c(a + b), b^2 = d(a + b) \in p$ and since p is a prime ideal it follows that $a, b \in p$. Thus, p is convex and belongs to the Brumfiel spectrum (= the image of the support map), cf. Proposition 4.3.

(c) It is claimed that $\kappa(\alpha)$ is real closed. First we show that positive elements in $\kappa(\alpha)$ are squares. So, consider some element $0 \leq \frac{a(\alpha)}{b(\alpha)} \in \kappa(\alpha)$ with $a, b \in \rho(F(n))$, $b(\alpha) \neq 0$. This is a square if $a(\alpha)b(\alpha) \in \rho(F(n))/\alpha$ is a square. Thus, it suffices to deal with some $0 \leq a(\alpha) \in \rho(F(n))/\alpha$. Since the residue homomorphism $\rho(F(n)) \rightarrow \rho(F(n))/\alpha$ is order preserving it follows that $(a \vee 0)(\alpha) = a(\alpha) \vee 0 = a(\alpha)$. Therefore we may assume that $0 \leq a \in \rho(F(n))$. But then there exists a unique positive square root $b \in \rho(F(n))$ of a, hence $a(\alpha) = b(\alpha)^2$.

It remains to show that any odd degree polynomial in $\kappa(\alpha)[X]$ has a root in $\kappa(\alpha)$. So, suppose that

$$F_0 = X^{2n+1} + \frac{a_{2n}(\alpha)}{b(\alpha)}X^{2n} + \ldots + \frac{a_0(\alpha)}{b(\alpha)} \in \kappa(\alpha)[X]$$

with $a_i, b \in \rho(F(n))$, $b(\alpha) \neq 0$, is irreducible. Then

$$F_1 = X^{2n+1} + a_{2n}(\alpha)X^{2n} + \ldots + (b^{2n}a_0)(\alpha) \in \rho(F(n))/\alpha[X]$$

is irreducible as well. The polynomial

$$G_1 = X^{2n+1} + a_{2n}X^{2n} + \ldots + b^{2n}a_0 \in \rho(F(n))[X]$$

is a representative of F_1. If $\beta \in Sper(\rho(F(n)))$ then $G_1(\beta)$ denotes the image of G_1 in $\rho(F(n))/\beta[X]$. By [19], Lemma 1.2.9, the absolute value of every root of $G_1(\beta)$ in $\rho(\beta)$ is bounded by $M(\beta)$ where

$$M = 1 + |a_{2n}| + \ldots + |b^{2n}a| \in \rho(F(n)).$$

By bounded inversion it follows that $M \in \rho(F(n))^\times$. Therefore

$$G_2 = X^{2n+1} + \frac{a_{2n}}{M}X^{2n} + \ldots + \frac{b^{2n}}{M^{2n+1}} \in \rho(F(n))[X],$$

and an element $t(\beta) \in \rho(\beta)$ is a root of $G_2(\beta)$ if and only if $t(\beta)M(\beta)$ is a root of $G_1(\beta)$. In particular, if $t \in \sigma(F(n))$ is a root of G_2 then t is a bounded function. Note that such a function t exists by Proposition 7.10. More explicitly, suppose that $t(\beta) \in \rho(\beta)$ is the largest root of $G_2(\beta)$ for all $\beta \in Sper(\rho(F(n)))$. Let $d = d(G_2)$ be the discriminant of G_2 and define

$$U = \{\beta \in Sper(\rho(F(n))); \ d(\beta) \neq 0\}.$$

It is clear that $\alpha \in U$. Now the function $c = dt$ is continuous and c is a root of

$$G_3 = X^{2n+1} + \frac{da_{2n}}{M}X^{2n} + \ldots + \frac{d^{2n+1}b^{2n}}{M^{2n+1}} \in \rho(F(n))[X].$$

One checks that $\frac{c(\alpha)M(\alpha)}{d(\alpha)b(\alpha)}$ is a root of F_0. Because each one of the functions b, c, d and M is continuous this root belongs to $\kappa(\alpha)$.   □

Proposition 12.4 Let (A, P) be a real closed ring.

(a) (A, P) is a reduced f–ring.
(b) (A, P) has bounded inversion.
(c) If $a \in P$ then there is some $b \in P$ with $b^2 = a$.
(d) The support map $supp_{(A,P)} : Sper(A, P) \to Spec(A)$ is a homeomorphism.
(e) For every $\alpha \in Sper(A, P)$, the canonical homomorphism $\kappa(\alpha) = \kappa_{(A,P)}(\alpha) \to \rho(\alpha) = \rho_{(A,P)}(\alpha)$ is an isomorphism.

Proof (a),(b) The rings $\rho(F(n))$ belong to **BIFR/N**, hence they belong to the injectivity class of $\{r_{F(n)}; n \in \mathbb{N}\}$, where r is the reflector of **BIFR/N**. It follows immediately that every real closed ring also belongs to this injectivity class. (Recall that **BIFR/N** is H-closed, Example 10.16.)

(c) In $\rho(F(1))$ there is a unique function c such that $0 \leq c$ and $c^2 = T \vee 0$ (where T is the variable of $F(1)$). Given $a \in P$, define $f \colon F(1) \to (A, P)$ by $T \to a$. Let $\overline{f} \colon \rho(F(1)) \to (A, P)$ be the unique extension of f. Then it follows from $0 \leq \overline{f}(c)$ and $\overline{f}(c)^2 = \overline{f}(c^2) = \overline{f}(T \vee 0) = a$ that $\overline{f}(c)$ is the desired square root of a.

(d) The support map is injective since (A, P) is a reduced f-ring. For surjectivity, let $p \subseteq (A, P)$ be a prime ideal. It is claimed that p is convex. So, pick $a, b \in P$ with $a + b \in p$. There are operations ω_a and ω_b (belonging to continuous semi-algebraic functions) with $a = \omega_a(a_1, \dots, a_k)$, $b = \omega_b(b_1, \dots, b_l)$. We define

$$f \colon F(k + l) \to (A, P) \colon T_i \to \begin{cases} a_i & \text{if} \quad i = 1, \dots, k \\ b_{i-k} & \text{if} \quad i = k + 1, \dots, k + l \end{cases}$$

and extend f uniquely to $\overline{f} \colon \rho(F(k + l)) \to (A, P)$. Setting $u = \omega_a(T_1, \dots, T_k)$, $v = \omega_b(T_{k+1}, \dots, T_l)$ one sees that $a = \overline{f}(u)$ and $b = \overline{f}(v)$. Moreover, $\overline{f}$ is a homomorphism of reduced f-rings, hence also $\overline{f}(u \vee 0) = \overline{f}(u) \vee 0 = a$, $\overline{f}(v \vee 0) = \overline{f}(v) \vee 0 = b$. Thus, $u \vee 0$, $v \vee 0 \in \rho(F(k + l))$ and $u \vee 0 + v \vee 0 \in \overline{f}^{-1}(p)$. Proposition 12.3(b) implies that $u \vee 0 \in \overline{f}^{-1}(p)$ and $v \vee 0 \in \overline{f}^{-1}(p)$. But then $a = \overline{f}(u \vee 0), b = \overline{f}(v \vee 0) \in p$, as claimed. This finishes the proof that the support map is bijective. Since (A, P) is an f-ring the support is a homeomorphism onto the image.

(e) If $0 \leq \frac{a(\alpha)}{b(\alpha)} \in \kappa(\alpha)$ then $\frac{a(\alpha)}{b(\alpha)}$ is a square if and only if $(ab)(\alpha)$ is a square. Therefore it suffices to prove that every $0 \leq a(\alpha) \in A/\alpha$ is a square. There is some function $0 \leq c \in \rho(F(n))$ with corresponding operation ω such that $a = \omega(a_1, \dots, a_n)$ for suitable $a_1, \dots, a_n \in A$. The homomorphism $f \colon F(n) \to (A, P) \colon T_i \to a_i$ extends uniquely to $\overline{f} \colon \rho(F(n)) \to (A, P)$ and $\overline{f}(c) = a$. Because there exists some $d \in \rho(F(n))$ with $c = d^2$ it follows that $\overline{f}(d)^2 = a$, hence also $f(d)(\alpha)^2 = a(\alpha)$.

Finally it will be shown that polynomials of odd degree in $\kappa(\alpha)[X]$ have a root in $\kappa(\alpha)$. It suffices to deal with polynomials

$$F = X^{2n+1} + a_{2n}(\alpha)X^{2n} + \dots + a_0(\alpha) \in A/\alpha[X],$$

i.e., $F = G(\alpha)$ with

$$G = X^{2n+1} + a_{2n}X^{2n} + \dots + a_0 \in A[X].$$

For $i = 0, \ldots, 2n$, let $c_i \in \rho(F(n_i))$ be a function with corresponding operation ω_i such that $a_i = \omega_i(a_{i1}, \ldots, a_{in_i})$. Now $f \colon F(\sum n_i) \to (A, P)$ is defined by $T_{ij} \to a_{ij}$; let $\overline{f} \colon \rho(F(\sum n_i)) \to (A, P)$ be the unique extension. Then $\overline{f}(c_i) = a_i$, hence $\overline{f}(H) = G$ where

$$H = X^{2n+1} + c_{2n}X^{2n} + \ldots + c_0 \in \rho(F(\textstyle\sum n_i))[X].$$

Let $\gamma = \overline{f}^{-1}(\alpha) \in Sper(\rho(F(\sum n_i)))$. According to Proposition 12.3(c) there is some $u \in \kappa(\gamma)$ with $H(\gamma)(u) = 0$. But then

$$F(\kappa_{\overline{f}/\alpha}(u)) = \kappa_{\overline{f}/\alpha}(H(\gamma)(u)) = 0,$$

i.e., $\kappa_{\overline{f}/\alpha}(u)$ is the desired root of F. $\qquad\qquad\square$

Proposition 12.5 For every reduced poring (A, P) the map $Sper(\rho_{(A,P)})$ is a homeomorphism.

Proof Since $\rho_{(A,P)}$ is an epimorphism the map is injective. Surjectivity follows from the fact that $\sigma_{(A,P)}$ factors through $\rho_{(A,P)}$ and $Sper(\sigma_{(A,P)})$ is bijective (Proposition 7.9). Thus, with respect to the constructible topology $Sper(\rho_{(A,P)})$ is a homeomorphism. To finish the proof we must show that for any $\gamma, \delta \in Sper(\rho(A, P))$ with $\rho_{(A,P)}^{-1}(\gamma) \subseteq \rho_{(A,P)}^{-1}(\delta)$ we also have $\gamma \subseteq \delta$. So, suppose that $\gamma \not\subseteq \delta$. Then there is some $a \in \rho(A, P)$ with $a(\delta) > 0$, and $a(\gamma) \leq 0$. Pick a function $c \in \rho(F(n))$ with corresponding operation ω such that $a = \omega(a_1, \ldots, a_n)$ for suitable $a_1, \ldots, a_n \in A$. The homomorphism $f \colon F(n) \to (A, P) \colon T_i \to a_i$ extends uniquely to $\overline{f} \colon \rho(F(n)) \to \rho(A, P)$. From the definition of f it is clear that $\overline{f}(c) = a$. Let $\alpha = \overline{f}^{-1}(\gamma)$ and $\beta = \overline{f}^{-1}(\delta)$. Then $c(\alpha) \leq 0$ and $c(\beta) > 0$, hence $\alpha \not\subseteq \beta$. Proposition 12.3(a) implies that $\rho_{F(n)}^{-1}(\alpha) \not\subseteq \rho_{F(n)}^{-1}(\beta)$. But then

$$f^{-1}(\rho_{(A,P)}^{-1}(\gamma)) = \rho_{F(n)}^{-1}(\alpha), \; f^{-1}(\rho_{(A,P)}^{-1}(\delta)) = \rho_{F(n)}^{-1}(\beta)$$

shows that $\rho_{(A,P)}^{-1}(\gamma) \not\subseteq \rho_{(A,P)}^{-1}(\delta)$, finishing the proof. $\qquad\square$

The real closure reflector on **POR/N** is H–closed by its very definition. We shall show now that it is also quotient–closed. The tool is Proposition 11.9:

Proposition 12.6 The real closure reflector $\rho\colon \mathbf{POR/N} \to \mathbf{RCR}$ is quotient–closed.

Proof It suffices to show that $\rho(F(n+1))_{T_{n+1}}$ is real closed for every $n \in \mathbb{N}$. Suppose that $f\colon F(k) \to \rho(F(n+1))_{T_{n+1}}$ is any homomorphism. If $a \in \rho(F(k))$ then

$$\rho(f)(a) = a(f(T_1), \dots, f(T_k)) \in \rho(\rho(F(n+1))_{T_{n+1}}).$$

Because $f(T_1), \dots, f(T_k)$ are continuous semi–algebraic functions on the semi–algebraic set $M = \{x \in R_0^{n+1}; x_{n+1} \neq 0\}$ the same is true for the composition $\rho(f)(a)$. We are done if we can show that $\rho(F(n+1))_{T_{n+1}}$ contains every continuous semi–algebraic function on M. So, let $a\colon M \to R_0$ be any continuous semi–algebraic function. For every $r \in R_0$, $r \geq 2$, let

$$M_r = \{x \in M;\ \frac{1}{r} \leq |x_{n+1}| \leq 1, |x_1|, \dots, |x_n| \leq r\}.$$

The sets M_r are closed and bounded, hence $b(r) = sup\{|a(x)|; x \in M_r\}$ exists. The map $b\colon [2, \infty) \to R_0$ defined in this way is a continuous nonnegative monotonically increasing semi–algebraic function. It follows from [19], Proposition 2.6.2, that $b(r) \leq r^N$ for some $N \in \mathbb{N}$ and all $r \in [2, \infty)$. Therefore the continuous semi–algebraic function

$$c\colon M \to R_0\colon x \to a(x)x_{n+1}^N$$

is bounded in a neighborhood of the coordinate plane $R_0^{n+1} \backslash M$. This implies that the semi–algebraic function

$$d\colon R_0^{n+1} \to R_0\colon x \to \begin{cases} c(x)x_{n+1} & \text{if} \quad x \in M \\ 0 & \text{if} \quad x \notin M \end{cases}$$

is continuous, hence belongs to $\rho(F(n+1))$. But then

$$a = \frac{d}{T_{n+1}^{N+1}} \in \rho(F(n+1))_{T_{n+1}},$$

and the proof is finished. $\square$

We pointed out at the beginning of this section that (without using the term "reflector") the real closure reflector was first studied in [108], [110] and [111]. It was introduced as a construction applied to rings. But,

properly interpreted, the results also showed that the real closure is an
H–closed monoreflector of **POR/N**. The reflections of the free object
$F(n)$ were identified as the rings of continuous semi–algebraic functions
on R_0^n. Therefore we know from Theorem 10.6 that the real closure
as defined in this section coincides with the original one defined in the
literature. We shall make extensive use of both descriptions of the real
closure. To keep this work reasonably self–contained we include a proof
of the equivalence of the definitions.

To start with we recall the definition given in [111], Chapter I, §2: Let
(A, P) be any reduced poring and pick $\alpha, \beta \in Sper(A, P)$ with $\alpha \subseteq \beta$.
In the real closed field $\rho(\alpha)$ there is a *largest* convex subring $C_{\beta\alpha}$ with
maximal ideal $M_{\beta\alpha}$ such that $\rho_\alpha^{-1}(M_{\beta\alpha}) = supp(\beta)$. Let $\lambda_{\beta\alpha} \colon C_{\beta\alpha} \to \rho_{\beta\alpha}$
be the residue map. Then $\lambda_{\beta\alpha}\rho_\alpha$ factors through A/β and the homo-
morphism $A/\beta \to \rho_{\beta\alpha}$ extends uniquely to $\rho(\beta) \to \rho_{\beta\alpha}$. We consider
$\rho(\beta)$ as a subfield of $\rho_{\beta\alpha}$. A semi–algebraic function $a \in \sigma(A, P)$ is
called *compatible* if, for all $\alpha, \beta \in Sper(A, P)$ with $\alpha \subseteq \beta$, $a(\alpha) \in C_{\beta\alpha}$
and $\lambda_{\beta\alpha}(a(\alpha)) = a(\beta)$. If $C \subseteq Sper(A, P)$ and if $a \in \sigma(A, P)$ then
$a|_C$ is said to be *compatible* if the compatibility condition is satisfied
for all $\alpha, \beta \in C$, $\alpha \subseteq \beta$. Trivially, the compatible semi–algebraic functions are a subring of $\sigma(A, P)$ that contains $im(\sigma_{(A,P)})$; we denote the
subring by $\overline{\rho}(A, P)$. A partial order is defined for $\overline{\rho}(A, P)$ by setting
$\overline{\rho}^+(A, P) = \overline{\rho}(A, P) \cap \sigma^+(A, P)$. The homomorphism $(A, P) \to \overline{\rho}(A, P)$
obtained from $\sigma_{(A,P)}$ by restriction of the codomain is denoted by $\overline{\rho}_{(A,P)}$.

The definition of compatibility for semi–algebraic functions contains
the condition that $a(\alpha) \in C_{\beta\alpha}$ for all $\alpha, \beta \in Sper(A, P)$ with $\alpha \subseteq \beta$. This
condition can be strengthened as follows:

Lemma 12.7 Suppose that (A, P) is a reduced poring and that $a \in$
$\sigma(A, P)$ is compatible. Then $a(\alpha) \in \overline{V}(\alpha)$ for all $\alpha \in Sper(A, P)$.

Proof Let β be the unique closed point of $\overline{\{\alpha\}}$. By compatibility, $a(\alpha) \in$
$C_{\beta\alpha}$ and $\lambda_{\beta\alpha}(a(\alpha)) \in \rho(\beta) \subseteq \rho_{\beta\alpha}$. Since β is a closed point of $Sper(A, P)$
it follows that $\rho_\beta(A)$ is cofinal in $\rho(\beta)$, i.e, $\overline{V}(\beta) = \rho(\beta)$. If $V \subseteq \rho_{\beta\alpha}$ is the
convex hull of $\overline{V}(\beta) = \rho(\beta)$ then $\overline{V}(\alpha) = \lambda_{\beta\alpha}^{-1}(V)$, hence $a(\alpha) \in \overline{V}(\alpha)$. $\square$

The next result is useful for comparing compatibility of a semi–algebraic
function with respect to different subrings of a given ring of semi–algebraic
functions.

Lemma 12.8 Let $f\colon (A, P) \to (B, Q)$ be a homomorphism of reduced porings such that $\sigma(f)\colon \sigma(A, P) \to \sigma(B, Q)$ is an isomorphism and $Sper(f)$ is a homeomorphism. Then a semi–algebraic function $a \in \sigma(A, P)$ is compatible if and only if $\sigma(f)(a) \in \sigma(B, Q)$ is compatible.

Proof Consider $\alpha, \beta \in Sper(A, P)$ with $\alpha \subseteq \beta$ and let $\gamma, \delta \in Sper(B, Q)$ be the corresponding points. Then $\gamma \subseteq \delta$ because $Sper(f)$ is a homeomorphism. From

$$\rho_\alpha^{-1}(\rho_{f/\gamma}^{-1}(M_{\delta\gamma})) = f^{-1}(\rho_\gamma^{-1}(M_{\delta\gamma})) = supp(\beta)$$

it follows that $\rho_{f/\gamma}^{-1}(C_{\delta\gamma}) \subseteq C_{\beta\alpha}$. Let $\varepsilon \in \overline{\{\alpha\}}$ and $\zeta \in \overline{\{\gamma\}}$ be the closed points. By hypothesis, $\rho_{f/\zeta}\colon \rho(\varepsilon) \to \rho(\zeta)$ is an isomorphism, hence $\rho_{f/\gamma}(\rho_\alpha(A))$ is cofinal in $\rho_\gamma(B)$. This implies that $\rho_{f/\gamma}(\overline{V}(\alpha)) = \overline{V}(\gamma)$. We assume by way of contradiction that $\rho_{f/\gamma}^{-1}(C_{\delta\gamma}) \subset C_{\beta\alpha}$. Then $\rho_{f/\gamma}(M_{\beta\alpha}) \subset M_{\delta\gamma} \subseteq \overline{V}(\gamma)$ are distinct convex prime ideals. (Recall that $\rho_{f/\gamma}$ is an isomorphism.) They determine prime cones $\gamma_1, \gamma_2 \in Sper(\overline{V}(\gamma))$. The definition of $C_{\delta\gamma}$ implies that $\rho_\gamma^{-1}(\gamma_1) \subset \gamma = \rho_\gamma^{-1}(\gamma_2)$. On the other hand, $f^{-1}(\rho_\gamma^{-1}(\gamma_1)) = \beta = f^{-1}(\gamma)$, which is impossible since $Sper(f)$ is bijective. This contradiction proves that $\rho_{f/\gamma}^{-1}(C_{\delta\gamma}) = C_{\beta\alpha}$. Thus, $a(\alpha) \in C_{\beta\alpha}$ if and only if $\sigma(f)(a)(\gamma) \in C_{\delta\gamma}$. Now the claim follows from the commutativity of the following diagram:

$$
\begin{array}{ccc}
\rho(\alpha) & \overset{\cong}{\longrightarrow} & \rho(\gamma) \\
\subseteq \uparrow & & \uparrow \subseteq \\
C_{\beta\alpha} & \overset{\cong}{\longrightarrow} & C_{\delta\gamma} \\
\lambda_{\beta\alpha} \downarrow & & \downarrow \lambda_{\gamma\delta} \\
\rho_{\beta\alpha} & \overset{\cong}{\longrightarrow} & \rho_{\delta\gamma} \\
\subseteq \uparrow & & \uparrow \subseteq \\
\rho(\beta) & \overset{\cong}{\longrightarrow} & \rho(\delta)
\end{array}
$$

$\square$

For compatible semi–algebraic functions we obtain the following strengthening of Lemma 7.6(a):

Lemma 12.9 If $a, b \in \sigma(A, P)$ are compatible semi–algebraic functions then the set
$$C = \{\alpha \in Sper(A, P); \ a(\alpha) = b(\alpha)\}$$
is closed and constructible.

Proof Constructibility was shown in Lemma 7.6 (a). To prove closedness we show that $\overline{\{\alpha\}} \subseteq C$ for every $\alpha \in C$. Pick any prime cone $\beta \in \overline{\{\alpha\}}$ and consider the homomorphism $\lambda_{\beta\alpha}: C_{\beta\alpha} \to \rho_{\beta\alpha}$. Then
$$a(\beta) = \lambda_{\beta\alpha}(a(\alpha)) = \lambda_{\beta\alpha}(b(\alpha)) = b(\beta)$$
proves the claim. $\qquad\qquad\qquad\qquad\qquad\qquad\qquad\qquad\qquad\qquad\qquad$ $\square$

Theorem 12.10 For any reduced poring (A, P) the subrings $\rho(A, P)$ and $\overline{\rho}(A, P)$ of $\sigma(A, P)$ coincide.

Proof We break the proof up into several steps.

Step 1 Continuous semi–algebraic functions are compatible.
Proof Let $\alpha, \beta \in Sper(A, P)$ be prime cones with $\alpha \subseteq \beta$. The corresponding prime cones of $\rho(A, P)$ are also denoted by α and β. Given $a \in \rho(A, P)$ we first check that $a(\alpha) \in \overline{V}(\alpha)$. Let $a = \omega(a_1, \ldots, a_n)$ and let $f: R_0^n \to R_0$ be the continuous semi–algebraic function corresponding to the operation ω. Then there are natural numbers c and N such that $|f| \leq c(1 + T_1^2 + \ldots + T_n^2)^N$ ([19], Proposition 2.6.1). Substituting a_i for T_i one gets
$$|a| \leq c(1 + a_1^2 + \ldots + a_n^2)^N \in A,$$
hence $a(\alpha) \in \rho(\alpha)$ is bounded by some element of A/α. This means that $a(\alpha) \in \overline{V}(\alpha)$.

From $a(\alpha) \in \overline{V}(\alpha)$ it follows that $\lambda_{\beta\alpha}(a(\alpha))$ is a well–defined element of $\rho_{\beta\alpha}$. To see that it coincides with $a(\beta)$, consider the commutative diagram

$$
\begin{array}{ccccc}
(A, P) & \longrightarrow & \rho(A, P) & & \\
\downarrow & & \downarrow & & \\
A/\alpha & \longrightarrow & \rho(A, P)/\alpha & \longrightarrow & C_{\beta\alpha} \subseteq \rho(\alpha) \\
\downarrow & & \downarrow & & \downarrow \\
A/\beta & \longrightarrow & \rho(A, P)/\beta & \longrightarrow & \rho(\beta) \subseteq \rho_{\beta\alpha}
\end{array}
$$

Proposition 7.3 yields

$$
\begin{aligned}
\lambda_{\beta\alpha}(a(\alpha)) &= \lambda_{\beta\alpha}(\omega(a_1(\alpha),\dots,a_n(\alpha))) \\
&= \omega(\lambda_{\beta\alpha}(a_1(\alpha)),\dots,\lambda_{\beta\alpha}(a_n(a))) \\
&= \omega(a_1(\beta),\dots,a_n(\beta)) \\
&= a(\beta)
\end{aligned}
$$

which finishes the proof of the claim.

Step 2 The map $Sper(\overline{\rho}_{(A,P)})$ is a homeomorphism.
Proof The map is continuous by definition and surjective since $\sigma_{(A,P)}$
factors through $\overline{\rho}(A,P)$ (note that $Sper(\sigma_{(A,P)})$ is surjective, Proposition
7.9). Let $\gamma, \delta \in Sper(\overline{\rho}(A,P))$ and $\alpha = \overline{\rho}^{-1}_{(A,P)}(\gamma)$, $\beta = \overline{\rho}^{-1}_{(A,P)}(\delta)$. Suppose that $\alpha \subseteq \beta$ and that $\gamma \not\subseteq \delta$. Then there is some semi–algebraic
function $a \in \gamma \backslash \delta$, say $a = \omega(a_1,\dots,a_n)$. By Proposition 7.3, the canonical homomorphisms $\rho(\alpha) \to \rho(\gamma)$ and $\rho(\beta) \to \rho(\delta)$ map $a(\alpha) \to a(\gamma) \geq 0$
and $a(\beta) \to a(\delta) < 0$, resp. This implies that $a(\alpha) \geq 0$ and $a(\beta) < 0$. But
then $a(\beta) = \lambda_{\beta\alpha}(a(\alpha))$ yields a contradiction. We conclude that $\gamma \subseteq \delta$.
This argument proves two things:

- $Sper(\overline{\rho}_{(A,P)})$ is injective (for, if $\alpha = \beta$ then it follows that $\gamma = \delta$);
- $Sper(\overline{\rho}_{(A,P)})$ is open, hence a homeomorphism.

Thus, the proof of Step 2 is finished.

Step 3 The canonical homomorphism $\overline{\rho}_{(A,P)}$ is an epimorphism.
Proof By Proposition 7.16 this is an immediate consequence of the injectivity of $Sper(\overline{\rho}_{(A,P)})$ (Step 2).

Step 4 The construction of $\overline{\rho}$ is idempotent.
Proof By injectivity of $Sper(\overline{\rho}_{(A,P)})$ it follows that $\sigma(f)\colon \sigma(A,P) \to$
$\sigma(\overline{\rho}(A,P))$ is an isomorphism (Proposition 7.16). Since $Sper(\overline{\rho}_{(A,P)})$
is even a homeomorphism (Step 2) the hypotheses of Lemma 12.8 are
satisfied. Thus, the isomorphism $\sigma(f)$ restricts to an isomorphism
$\overline{\rho}(A,P) \to \overline{\rho}(\overline{\rho}(A,P))$.

Step 5 Let $\mathbf{D} \subseteq \mathbf{POR/N}$ be the subcategory of those objects (A,P) for
which $\overline{\rho}_{(A,P)}$ is an isomorphism. Then $\overline{\rho}$ is a reflector $\mathbf{POR/N} \to \mathbf{D}$.
Proof The universal property (2.1) will be checked for the canonical
homomorphisms $\overline{\rho}_{(A,P)}$. Suppose that $f\colon (A,P) \to (B,Q)$ is a homomorphism into a member of $\mathbf{D}$. Then f extends to the homomorphism

$\sigma(f)\colon \sigma(A,P) \to \sigma(B,Q)$ between the rings of semi–algebraic functions. We consider (B,Q) as a subporing of $\sigma(B,Q)$. It suffices to show that $\sigma(f)\overline{p}(A,P) \subseteq B$. So, pick $a \in \overline{p}(A,P)$ and let $\gamma,\delta \in Sper(B,Q)$ with $\gamma \subseteq \delta$. If $\alpha = f^{-1}(\gamma)$ and $\beta = f^{-1}(\delta)$ then also $\alpha \subseteq \beta$. It follows from

$$f^{-1}\rho_\gamma^{-1}(M_{\delta\gamma}) = f^{-1}(supp(\delta)) = supp(\beta)$$

that

$$\rho_{f/\gamma}^{-1}(C_{\delta\gamma}) \subseteq C_{\beta\alpha} \subseteq \rho(\alpha).$$

Therefore $\rho_{f/\gamma}^{-1}(C_{\delta\gamma})/M_{\beta\alpha} \subseteq \rho_{\beta\alpha}$ is a convex subring with maximal ideal $\rho_{f/\gamma}^{-1}(M_{\delta\gamma})/M_{\beta\alpha}$. There is a canonical homomorphism $\lambda\colon \rho_{f/\gamma}^{-1}(C_{\delta\gamma})/M_{\beta\alpha} \to \rho_{\delta\gamma}$ making the following diagram commutative:

$$
\begin{array}{ccc}
\rho(\alpha) & \xrightarrow{\rho_{f/\gamma}} & \rho(\gamma) \\
\subseteq \uparrow & & \uparrow \subseteq \\
C_{\beta\alpha} \supseteq \rho_{f/\gamma}^{-1}(C_{\delta\gamma}) & \xrightarrow{\quad\quad} & C_{\delta\gamma} \\
\lambda_{\beta\alpha} \downarrow & & \downarrow \lambda_{\delta\gamma} \\
\rho_{\beta\alpha} \supseteq \rho_{f/\gamma}^{-1}(C_{\delta\gamma})/M_{\beta\alpha} & \xrightarrow{\lambda} & \rho_{\delta\gamma} \\
\subseteq \uparrow & & \uparrow \subseteq \\
\rho(\beta) & \xrightarrow{\rho_{f/\delta}} & .\rho(\delta)
\end{array}
$$

Note that $\overline{V}(\alpha) \subseteq \rho_{f/\beta}^{-1}(C_{\delta\gamma})$, hence $a(\alpha) \in \rho_{f/\gamma}^{-1}(C_{\delta\gamma})$. Now

$$
\begin{aligned}
\lambda_{\delta\gamma}(\sigma(f)(a)(\gamma)) &= \lambda_{\delta\gamma}(\rho_{f/\gamma}(a(\alpha))) \\
&= \lambda(\lambda_{\beta\alpha}(a(\alpha))) = \lambda(a(\beta)) \\
&= \rho_{f/\delta}(a(\beta)) = \sigma(f)(a)(\delta)
\end{aligned}
$$

shows that $\sigma(f)(a)$ is compatible, finishing the proof of Step 5.

Step 6 For every $n \in \mathbb{N}$, $\rho(F(n)) = \overline{p}(F(n))$.
Proof Every element $a \in \overline{p}(F(n))$ is a semi–algebraic function on R_0^n; it is claimed that a is also continuous. If $a^{-1}([c,d])$ is a *closed* semi–algebraic subset of R_0^n for all $c,d \in R_0$, $c \le d$, then this is the case. The semi–algebraic subset $a^{-1}([c,d])$ corresponds to the constructible subset

$$C = \{\alpha \in Sper(F(n)); \ c \le a(\alpha) \le d\}$$

([19], Théorème 7.1.3; [33], Théorème 5.1). Moreover, $a^{-1}([c,d])$ is closed if and only if so is C. Because $\bar{p}(F(n))$ is real closed (Step 1 and Step 5), the semi–algebraic function $b = (a \wedge d) \vee c$ belongs to $\bar{p}(F(n))$. But then

$$C = \{\alpha \in Sper(F(n)); \ a(\alpha) = b(\alpha)\}$$

is closed by Lemma 12.9.

Step 7 The reflector $\bar{p}$ has the following extension property: Suppose that $(A, P) \in ob(\mathbf{D})$. If $U \subseteq Sper(A, P)$ is open constructible then there is some $0 \leq u \in A$ such that

$$U = \{\alpha \in Sper(A, P); \ u(\alpha) > 0\}$$

(note that (A, P) is real closed by Step 1 and Step 4). Let $a \in \sigma(A, P)$ be a semi–algebraic function that satisfies the compatibility condition for all pairs $\alpha, \beta \in U$ with $\alpha \subseteq \beta$. If there are some $b \in A$ and some $1 \leq n \in \mathbb{N}$ with $b(\gamma) = 0$ for all $\gamma \notin U$ and $|a(\gamma)|^n \leq b(\gamma)$ for all $\gamma \in Sper(A, P)$ then $a \in A$.

Proof The compatibility condition must be checked for a. So pick $\alpha, \beta \in Sper(A, P)$, $\alpha \subseteq \beta$. If $\alpha, \beta \in U$ then the condition is satisfied by hypothesis. If $\alpha, \beta \notin U$ then there is nothing to do since $a(\alpha) = 0$, $a(\beta) = 0$ and, hence, $\lambda_{\beta\alpha}(a(\alpha)) = a(\beta)$. Finally, if $\alpha \in U$ and $\beta \notin U$ then $|\lambda_{\beta\alpha}(a(\alpha))|^n \leq \lambda_{\beta\alpha}(b(\alpha)) = b(\beta) = 0$ shows that $\lambda_{\beta\alpha}(a(\alpha)) = 0 = a(\beta)$.

Step 8 The reflector $\bar{p}$ is quotient–closed.
Proof Suppose that (A, P) belongs to $\mathbf{D}$ and that $S \subseteq A$ is multiplicative. Because $\mathbf{D}$ is cocomplete (Proposition 2.7(d)) and

$$(A_S, P_S) = \varinjlim_{s \in S}(A_s, P_s)$$

it suffices to show that each $(A_s, P_s), s \in S$, is a member of $\mathbf{D}$. Once again, (A, P) is real closed (by Step 1 and Step 5). In particular, (A, P) is an f–ring. Therefore

$$(A_s, P_s) = (A_{s^2}, P_{s^2}) = (A_{|s|^2}, P_{|s|^2}) = (A_{|s|}, P_{|s|}),$$

and it suffices to consider the case that $0 \leq s$. Note that $Sper(A_s, P_s)$, is canonically homeomorphic to

$$U = \{\alpha \in Sper(A, P); \ s(\alpha) \neq 0\}$$

and that real closed residue fields of corresponding points $\alpha \in U$ and $\gamma \in Sper(A_s, P_s)$ are isomorphic.

Now pick some element $a \in \bar{\rho}(A_s, P_s)$. For every $\gamma \in Sper(A_s, P_s)$ there are $a_\gamma \in A$ and $n_\gamma \in \mathbb{N}$ with $0 \le \frac{a_\gamma}{s^{n_\gamma}} \in A_s$ and $|a(\gamma)| \le \frac{a_\gamma(\gamma)}{s(\gamma)^{n_\gamma}}$ (because $a(\gamma) \in \bar{V}(\gamma)$). The sets

$$C_\gamma = \{\delta \in Sper(A_s, P_s); \ |a(\delta)| \le \frac{a_\gamma(\delta)}{s(\delta)^{n_\gamma}}\}$$

form a constructible cover of $Sper(A_s, P_s)$. By compactness there is a finite subcover by sets $C_{\gamma_i}, i = 1, \ldots, r$. We may assume that $N = n_{\gamma_1} = \ldots = n_{\gamma_r}$. The function $b = s(a_{\gamma_1} + \ldots + a_{\gamma_r})$ belongs to A, is nonnegative and has the property that $b(\alpha) = 0$ for all $\alpha \notin U$. A function $c \in \sigma(A, P)$ is defined by

$$c(\alpha) = \begin{cases} a(\alpha)s^{N+1}(\alpha) & \text{if} \quad \alpha \in U \\ 0 & \text{if} \quad \alpha \notin U \end{cases}$$

From the construction it is clear that $|c| \le b$. By Step 7, $c \in A$ if c satisfies the compatibility condition for all pairs $\alpha, \beta \in U$ with $\alpha \subseteq \beta$. Let $\gamma, \delta \in Sper(A_s, P_s)$ be the prime cones corresponding to α and β. We identify $\rho(\alpha) = \rho(\gamma)$ and $\rho(\beta) = \rho(\delta)$. Since $supp(\beta)$ is the kernel of

$$(A, P) \longrightarrow (A_s, P_s) \xrightarrow{\rho_\gamma} C_{\delta\gamma} \xrightarrow{\lambda_{\delta\gamma}} \rho_{\delta\gamma}$$

it follows that $C_{\delta\gamma} \subseteq C_{\beta\alpha}$. On the other hand, $\rho_\alpha : (A, P) \to C_{\beta\alpha}$ factors through (A_s, P_s) because $s(\alpha) \in C_{\beta\alpha}^\times$. This implies that $C_{\beta\alpha} \subseteq C_{\delta\gamma}$. Since the convex subrings $C_{\beta\alpha}$ and $C_{\delta\gamma}$ coincide, $\lambda_{\beta\alpha}(c(\alpha)) = c(\beta)$ if and only if $\lambda_{\delta\gamma}(c(\gamma)) = c(\delta)$. But the latter condition is satisfied because the functions a and s used to define c are compatible on $Sper(A_s, P_s)$.

We have shown that the canonical homomorphism $(A_s, P_s) \to \bar{\rho}(A_s, P_s)$ maps $\frac{c}{s^{N+1}} \to a$. Since it is clearly injective we conclude that it is an isomorphism. (Note that we do not have to worry about the partial orders since both rings are known to be real closed.)

Step 9 The reflector $\bar{\rho}$ is H–closed.

Proof This was proved in [114], Lemma 3.7. Or, it is a consequence of the more general lemma following immediately after this proof. The hypotheses of the lemma are either clear ($\kappa_{(A,P)}(\alpha) = \rho_{(A,P)}(\alpha)$ for all $\alpha \in Sper(A, P)$ and for all $(A, P) \in ob(\mathbf{D})$ implies that $\kappa_{(A,P)}(\alpha) \in \mathbf{SAFR} \subseteq \mathbf{D}$) or have been proved above (Step 7, Step 8).

Step 10 To conclude the proof of the theorem, note that ρ and $\overline{\rho}$ are both H–closed monoreflectors of **POR/N** and that $\rho_{F(n)}$ and $\overline{\rho}_{F(n)}$ coincide for all $n \in \mathbb{N}$. According to Theorem 10.6 this means that $\rho = \overline{\rho}$. $\square$

To make our proof of Theorem 12.10 truly self–contained it remains to present a proof of Step 9. For later applications we consider a more general situation:

Lemma 12.11 Let $\overline{\rho}$: **POR/N** $\rightarrow$ **D** be the reflector of the proof of Theorem 12.10. Let r: **POR/N** $\rightarrow$ **E** be another reflector satisfying the following four conditions:

(i) **D** $\subseteq$ **E** $\subseteq$ **BIFR/N**;

(ii) r is quotient–closed;

(iii) if $(A, P) \in ob(\mathbf{E})$ and $\alpha \in Sper(A, P)$ then $\kappa_{(A,P)}(\alpha) \in ob(\mathbf{E})$;

(iv) for all $(A, P) \in ob(\mathbf{E})$ and all $0 \leq u \in A$, a semi–algebraic function $a \in \sigma(A, P)$ belongs to (A, P) if the following conditions are satisfied:

  • the restriction $a|_U$, with $U = \{\alpha \in Sper(A, P); \ u(\alpha) > 0\}$, belongs to the quotient ring A_u; and

  • there are some $0 \leq b \in A$ and some $1 \leq n \in \mathbb{N}$ such that $|a|^n \leq \sigma_{(A,P)}(b)$ and $b(\alpha) = 0$ if $\alpha \notin U$.

Then the reflector r is H–closed.

Proof The proof will be similar to the one of [111], Theorem I 4.5, or [114], Lemma 3.7. Let (A, P) be an object of **E**, let $f : (A, P) \rightarrow (B, Q)$ be a **POR/N**–morphism. Then $\Delta_{(B,Q)}f : (A, P) \rightarrow \Pi(B, Q)$ is a **POR/N**–morphism as well. The codomain is a reduced f–ring, hence the kernel $I = ker(f) = ker(\Delta_{(B,Q)}f)$ is an l–ideal of (A, P). Thus $(\overline{A}, \overline{P}) = (A/I, P + I/I)$ is a reduced f–ring. By Proposition 1.11 its partial order cannot be strengthened. Therefore $(\overline{A}, \overline{P})$ is isomorphic to $f(A)$ with the restriction of the partial order of B. To prove H–closedness of r it suffices to show that $r_{(\overline{A},\overline{P})} : (\overline{A}, \overline{P}) \rightarrow r(\overline{A}, \overline{P})$ is an isomorphism. In fact, we only need to check that $r_{(\overline{A},\overline{P})}$ is surjective. For, then $r_{(\overline{A},\overline{P})}$ is an isomorphism of the underlying rings. Both rings are reduced f–rings, hence it is an isomorphism in **POR/N** (Proposition 1.11).

Note that the map $Sper(r_{(A,P)})$ is a homeomorphism for all reduced porings (A, P). This follows from the fact that the same is true for the reflector $\overline{\rho}$ (Step 2 in the proof of Theorem 12.10) and from the hypothesis that r is weaker than $\overline{\rho}$. The real spectra are identified via $Sper(r_{(A,P)})$. As usual, we consider $Sper(\overline{A}, \overline{P})$ as a closed subspace of $Sper(A, P)$. Thus, $\kappa_{(\overline{A},\overline{P})}(\alpha) = \kappa_{(A,P)}(\alpha)$ belongs to $\mathbf{E}$ for all $\alpha \in Sper(\overline{A}, \overline{P})$.

The proof comes in two steps. First we deal with the case that $(\overline{A}, \overline{P})$ is totally ordered; afterwards the general situation will be considered.

Suppose that $(\overline{A}, \overline{P})$ is totally ordered, i.e., $I = supp(\alpha)$ for a unique $\alpha \in Sper(A, P)$ (recall that (A, P) is a reduced f–ring, hence the support is injective). Then $(\overline{A}, \overline{P}) = A/\alpha$ and $Sper(\overline{A}, \overline{P}) = \overline{\{\alpha\}} \subseteq Sper(A, P)$. If $a \in r(\overline{A}, \overline{P}) \subseteq \prod_{\beta \in \overline{\{\alpha\}}} \kappa_{(A,P)}(\beta)$ then for each β there are $a_\beta, u_\beta \in A$ with $u_\beta \geq 0$, $u_\beta(\beta) \neq 0$ and $a(\beta) = \frac{a_\beta(\beta)}{u_\beta(\beta)}$. The sets

$$
\begin{aligned}
U_\beta &= \{\gamma \in Sper(\overline{A}, \overline{P}); \ u_\beta(\gamma) > 0\}, \\
C_\beta &= \{\gamma \in U_\beta; \ a(\gamma)u_\beta(\gamma) = a_\beta(\gamma)\}
\end{aligned}
$$

are constructible, U_β is open, C_β is locally closed. Because $\beta \in C_\beta$ for every $\beta \in \overline{\{\alpha\}}$, we have a constructible cover $\overline{\{\alpha\}} = \bigcup_{\beta \in \overline{\{\alpha\}}} C_\beta$ by locally closed subsets. By quasi–compactness of $\overline{\{\alpha\}}$ there is a finite subcover.

So far we have shown the following: There are locally closed constructible subsets $C_i \subseteq \overline{\{\alpha\}}$, $i = 1, \ldots, r$, there are $a_i, u_i \in A$, $0 \leq u_i$, and there are closed constructible subsets $F_i \subseteq \overline{\{\alpha\}}$ such that $C_i = U_i \cap F_i$ with

$$
U_i = \{\gamma \in \overline{\{\alpha\}}; \ u_i(\gamma) > 0\}
$$

and $a_i(\gamma) = a(\gamma)u_i(\gamma)$ for all $\gamma \in C_i$.

Among all these covers, choose one with minimal length r. The C_i's are intervals in the chain $\overline{\{\alpha\}}$. Let $\beta_i \in C_i$ be the closed point. The enumeration of the C_i's may be arranged so that

$$
\beta_r \to \beta_{r-1} \to \cdots \to \beta_1
$$

is the specialization chain of the β_i's. In particular, β_1 is the closed point of $\overline{\{\alpha\}}$. Since β_1 is also closed in $Sper(A, P)$ there is some $v \in A$, $0 \leq v$, such that $vu_1(\beta_1) > 1$. As (A, P) is an f–ring, $vu_1 \vee 1$ exists in A. Note that $vu_1|_{C_1} = (vu_1 \vee 1)|_{C_1}$. Replacing a_1 by va_1 and u_1 by $vu_1 \vee 1$ one may assume that $u_1 \geq 1$. Then $u_1 \in A^\times$ by bounded inversion, hence a_1 may be replaced by $a_1 u_1^{-1}$, and we may assume that $u_1 = 1$. We define

$$
V_i = \{\gamma \in Sper(A, P); \ u_i(\gamma) > 0\}.
$$

If $r = 1$ then $C_1 = \overline{\{\alpha\}}$. In this case a_1 is mapped to a by the canonical map

$$(A, P) \longrightarrow (\overline{A}, \overline{P}) \xrightarrow{r_{(\overline{A}, \overline{P})}} r(\overline{A}, \overline{P}),$$

hence $a \in im(r_{(\overline{A}, \overline{P})})$.

Next we suppose that $r > 1$. Our aim is to show that the cover of $\overline{\{\alpha\}}$ can be shortened. Let γ_1 be the generic point of $C_1 \backslash C_2$. With respect to specialization, γ_1 is the immediate successor of β_2. In $r(\overline{A}, \overline{P})/\beta_2$ there is a smallest convex prime ideal, namely the support of γ_1/β_2. As a convex prime ideal it is generated by $u_2(\beta_2)$. Since $a(\gamma_1) - a_1(\gamma_1) = 0$, $a(\beta_2) - a_1(\beta_2)$ is an element of this convex prime ideal. Thus, there exists some $n \in \mathbb{N}$ such that

$$\left| \frac{a_2(\beta_2)}{u_2(\beta_2)} - a_1(\beta_2) \right|^n = |a(\beta_2) - a_1(\beta_2)|^n < u_2(\beta_2).$$

The set

$$F = \{\gamma \in V_2; \ \left| \frac{a_2(\gamma)}{u_2(\gamma)} - a_1(\gamma) \right|^n \geq u_2(\gamma)\}$$

is closed in V_2 and is disjoint from $\overline{\{\alpha\}} \cap V_2$. The hypothesis of quotient–closedness implies that (A_{u_2}, P_{u_2}) belongs to $\mathbf{E}$. The real spectrum of (A_{u_2}, P_{u_2}) is identified canonically with V_2. At each $\alpha \in V_2$ the residue fields of (A, P) and (A_{u_2}, P_{u_2}) coincide. Since (A_{u_2}, P_{u_2}) is a reduced f–ring there are $w_1, w_2 \in A_{u_2}$, $0 \leq w_1, w_2$ such that $w_1|_F = 0$, $w_1|_{\overline{\{\alpha\}} \cap V_2} > 0$, $w_2|_F > 0$, $w_2|_{\overline{\{\alpha\}} \cap V_2} = 0$ and $w_1 + w_2 > 0$. By bounded inversion, $w = w_1(w_1 + w_2)^{-1} \in A_{u_2}$, $0 \leq w \leq 1$, $w|_F = 0$, $w|_{\overline{\{\alpha\}} \cap V_2} = 1$. If $\gamma \in V_2$ then $|w(\gamma)(\frac{a_2(\gamma)}{u_2(\gamma)} - a_1(\gamma))|^n < u_2(\gamma)$ by definition of w. By condition (iv), the semi–algebraic function $b \in \sigma(A, P)$ defined by $b|_{V_2} = w(\frac{a_2}{u_2} - a_1)$, $b|_{Sper(A,P) \backslash V_2} = 0$, belongs to A. Now we define $a_0 = a_1 + b$. If $\gamma \in C_1 \backslash C_2$ then $b(\gamma) = 0$, hence $a_0(\gamma) = a_1(\gamma) = a(\gamma)$. If $\gamma \in C_2$ then $b(\gamma) = \frac{a_2(\gamma)}{u_2(\gamma)} - a_1(\gamma)$, hence $a_0(\gamma) = \frac{a_2(\gamma)}{u_2(\gamma)} = a(\gamma)$. To wrap up this part of the proof we define $C_0 = C_1 \cup C_2$, $F_0 = F_1 \cup F_2$, $U_0 = U_1 \cup U_2$, $u_0 = 1$. The length of the cover $\overline{\{\alpha\}} = C_0 \cup \bigcup_{i=3}^{r} C_i$ is $r - 1$, and $a_i(\gamma) = a(\gamma) u_i(\gamma)$ for all $\gamma \in C_i$, $i = 0, 3, \ldots, r$. Thus, we have achieved the desired shortening of the cover. This contradicts the minimality of r, and the proof is complete for the case that $(\overline{A}, \overline{P})$ is totally ordered.

Now suppose that I is any radical l–ideal. Once again we pick any element $a \in r(\overline{A}, \overline{P})$ and try to find some $b \in A$ with image a under the canonical homomorphism

$$(A, P) \longrightarrow (\overline{A}, \overline{P}) \xrightarrow{r(\overline{A}, \overline{P})} r(\overline{A}, \overline{P}).$$

For each $\alpha \in Sper(\overline{A}, \overline{P}) = Sper(r(\overline{A}, \overline{P}))$ there is some $a_\alpha \in A$ which is mapped to $a(\alpha) \in r(\overline{A}, \overline{P})/\alpha = A/\alpha$. (Note that $A/\alpha \in ob(\mathbf{E})$ by the first step of the proof). The set

$$F_\alpha = \{\beta \in Sper(\overline{A}, \overline{P}); \ a_\alpha(\beta) = a(\beta)\}$$

is a closed constructible subset of $Sper(\overline{A}, \overline{P})$ containing α. The F_α's form a cover of $Sper(\overline{A}, \overline{P})$, hence there is a finite subcover.

So far it has been shown that there are finitely many closed constructible subsets $F_1, \ldots, F_r \subseteq Sper(\overline{A}, \overline{P})$ covering $Sper(\overline{A}, \overline{P})$ and $b_1, \ldots, b_r \in A$ such that $b_i|_{F_i} = a|_{F_i}$.

Among these covers we pick one with r as small as possible. If $r = 1$ then $F_1 = Sper(\overline{A}, \overline{P})$ and $a = b_1|_{F_1}$, i.e., $b_1 \in A$ is mapped onto $a \in r(\overline{A}, \overline{P})$. Finally, assume by way of contradiction that $r > 1$. Once again we shall shorten the cover. We define

$$\begin{aligned} F &= \{\alpha \in Sper(A, P); \ b_1(\alpha) = b_2(\alpha)\}, \\ U &= Sper(A, P) \backslash F. \end{aligned}$$

Since (A, P) is a reduced f–ring there is some $u \in A$, $0 \le u$, such that

$$U = \{\alpha \in Sper(A, P); \ u(\alpha) > 0\}.$$

Because $\mathbf{E}$ is assumed to be quotient–closed, $(A_u, P_u) \in ob(\mathbf{E})$. We identify $Sper(A_u, P_u) = U$ and $\kappa_{(A, P)}(\alpha) = \kappa_{(A_u, P_u)}(\alpha)$ for $\alpha \in U$. In U the sets

$$C_i = \{\alpha \in U; \ b_i(\alpha) = a(\alpha)\}, \ i = 1, 2,$$

are closed and disjoint. As in the first part of the proof, there is some $w \in A_u$, $0 \le w \le 1$, with $w|_{C_1} = 1$, $w|_{C_2} = 0$. Now consider $c = w\frac{b_1 - b_2}{1} \in A_u$. Since $b_1 - b_2|_F = 0$ there are some $d \in A^\times$ and some $n \in \mathbb{N}$ such that $|b_1 - b_2|^n < du$, hence $|c|^n \le (du)|_U$. Condition (iv) implies that the semi–algebraic function $e \in \sigma(A, P)$ defined by $e|_U = c$, $e|_F = 0$, belongs to A. Finally, we define $F_0 = F_1 \cup F_2$ and $b_0 = b_2 + e \in A$. One checks that $b_0(\alpha) = a(\alpha)$ for all $\alpha \in F_0 \cup F_1$. Thus, $F_0, F_3, \ldots, F_r$ and $b_0, b_3, \ldots, b_r \in A$ provide the shorter cover that is needed for the contradiction. $\qquad\square$

The introduction of the real closure reflector and the discussion of its most basic properties are now complete. We continue with a result which we consider to be the Main Theorem about real closed rings. It exhibits an extremal property of the category **RCR**. This property explains the importance of real closed rings in semi–algebraic geometry.

Theorem 12.12 Among all reflectors $r\colon$ **PREOR** $\to$ **D**, the real closure is the strongest one such that $Sper(r_{(A,P)})$ is a homeomorphism for every preordered ring (A, P).

Proof Given a preordered ring (A, P), let $\pi\colon (A, P) \to (\overline{A}, \overline{P})$ be the reflection in **POR/N**. Since the real spectra of (A, P) and $(\overline{A}, \overline{P})$ are homeomorphic and the real closures coincide, it follows from Proposition 12.5 that $Sper(\rho_{(A,P)})$ is a homeomorphism. Now suppose that r is any reflector with the property that every map $Sper(r_{(A,P)})$ is a homeomorphism and assume by way of contradiction that **RCR** $\not\subseteq$ **D**.

In a first step we show that **SAFR** $\subseteq$ **D**. By Theorem 8.12 it suffices to check that every real closed field belongs to **D**. So, suppose that R is a real closed field not belonging to **D**. Because $Sper(r_R)$ is a homeomorphism, $|Sper(r(R))| = 1$ and $\sigma(r(R))$ is a real closed field. Thus,

$$\sigma_{r(R)} r_R \colon R \to r(R) \to \sigma(r(R))$$

is a proper extension of real closed fields, hence it is not an epimorphism in the category of real closed fields. There is another real closed field S and there are homomorphisms $f, g\colon \sigma(r(R)) \to S$, $f \neq g$, such that $f\sigma_{r(R)} r_R = g\sigma_{r(R)} r_R$. The universal property of the reflection $\sigma_{r(R)}$ implies that $f\sigma_{r(R)} \neq g\sigma_{r(R)}$. The reflection $r_S\colon S \to r(S)$ is a monomorphism hence $r_S f\sigma_{r(R)} \neq r_S g\sigma_{r(R)}$. Because of the universal property of the reflection r_R we see that $r_S f\sigma_{r(R)} r_R \neq r_S g\sigma_{r(R)} r_R$. But this is impossible since $f\sigma_{r(R)} r_R = g\sigma_{r(R)} r_R$. This contradiction shows that every real closed field belongs to **D**, hence **SAFR** $\subseteq$ **D**.

There exists some real closed ring (A, P) that does not belong to **D** (by the assumption that **RCR** $\not\subseteq$ **D**). The universal property of the reflection $r_{(A,P)}$ provides a unique factorization $\sigma_{(A,P)} = \overline{\sigma_{(A,P)}} r_{(A,P)}$ with $\overline{\sigma_{(A,P)}} \colon r(A, P) \to \sigma(A, P)$. Using the universal property of $\rho_{r(A,P)}$ one writes $\overline{\sigma_{(A,P)}} = \widetilde{\sigma_{(A,P)}} \rho_{r(A,P)}$. The hypothesis that r preserves the real spectrum and the fact that the same is true for ρ (by Proposition 12.5) imply that $Sper(\rho(r(A, P)))$ and $Sper(A, P)$ are homeomorphic. Using

Proposition 7.16 one sees that the hypotheses of Lemma 12.8 are satisfied. Thus, if $a \in \sigma(A, P)$ then compatibility of a over the ring (A, P) coincides with compatibilty of a over the ring $\rho(r(A, P))$. We conclude that $\rho_{r(A,P)} r_{(A,P)}$ is an isomorphism. Being an initial epimorphic factor of an isomorphism, $r_{(A,P)}$ is an isomorphism as well. But this contradicts the choice of $(A, P) \notin ob(\mathbf{D})$, and the proof is finished. $\qquad \Box$

Theorem 12.12 deals with functorial constructions. Through a slight modification of the final part of the proof of the theorem we get the following result dealing with individual porings:

Theorem 12.13 Let (A, P) be a reduced poring, let (B, Q) be a poring between (A, P) and $\sigma(A, P)$ with inclusion homomorphisms $f: (A, P) \to (B, Q)$ and $i: (B, Q) \longrightarrow \sigma(A, P)$. If $Sper(f)$ is a homeomorphism then $(B, Q) \subseteq \rho(A, P)$.

Proof Every map between real spectra induced by any of the homomorphism in the following diagram is bijective:

$$
\begin{array}{ccccc}
(A, P) & \xrightarrow{f} & (B, Q) & \xrightarrow{i} & \sigma(A, P) \\
\rho_{(A,P)} \downarrow & & \rho_{(B,Q)} \downarrow & & \downarrow = \\
\rho(A, P) & \xrightarrow{\rho(f)} & \rho(B, Q) & \xrightarrow{\rho(i)} & \sigma(A, P).
\end{array}
$$

Therefore $\rho(f)$ and $\rho(i)$ are injective. It is claimed that $\rho(f)$ is an isomorphism. Since $Sper(f)$ is a homeomorphism, the same is true for $Sper(\rho(f))$. By Proposition 7.16, the homomorphism $\rho(f)$ satisfies the hypotheses of Lemma 12.8. Therefore, given any semi–algebraic function $a \in \sigma(A, P)$, compatibility of a over (A, P) is the same as compatibility over (B, Q). Since every element of $\rho(B, Q)$ is compatible relative (B, Q), it is also compatible with respect to (A, P), hence it belongs to $\rho(A, P)$ (Theorem 12.10). $\qquad \Box$

In the rest of this section we are concerned with examples of reflectors that are related to the real closure. In particular, the constructions of section 9B and section 9C will be used to produce new reflectors. It is clear from Proposition 9B.2 and Proposition 9C.1 that in order to apply the constructions one needs to recognize whether certain morphisms are epimorphisms. The entire section 14 will be concerned with this question. Right now we prove one result that suffices for our immediate needs:

Proposition 12.14 Suppose that (A, P) is a reduced poring and that $\rho_{(A,P)} = gf$ where $f : (A, P) \to (B, Q)$ is a monomorphism and $g \colon (B, Q) \to \rho(A, P)$ is a sub–f–ring. Then f is an epimorphism.

Proof Consider the diagram

$$
\begin{array}{ccccc}
(A, P) & \xrightarrow{\ f\ } & (B, Q) & \xrightarrow{\ g\ } & \rho(A, P) \\
\rho_{(A,P)} \downarrow & & \rho_{(B,Q)} \downarrow & & \downarrow = \\
\rho(A, P) & \xrightarrow{\rho(f)} & \rho(B, Q) & \xrightarrow{\rho(g)} & \rho(A, P)
\end{array}
$$

Because of $\rho(g)\rho(f) = id_{\rho(A,P)}$ the functorial map $Sper(\rho(g))$ is a homeomorphism onto a closed subspace of $Sper(\rho(B, Q))$, hence $Sper(g)$ is a homeomorphism onto a closed subspace of $Sper(B, Q)$. Since (B, Q) is an f–ring the sets $pos(b)$, $b \in B$, form a basis of the topology of $Sper(B, Q)$. So, if $im(Sper(g))$ is a proper closed subspace of $Sper(B, Q)$ then there is some $b \in B$ with $pos(b) \neq \emptyset$, but $pos(b) \cap im(Sper(g)) = \emptyset$. Replacing b by b^+ we may assume that $b \geq 0$. But then $b \neq 0$ since $pos(b) \neq \emptyset$, whereas $g(b) = 0$ since $pos(b) \cap im(Sper(g)) = \emptyset$. By injectivity of g this is impossible. We conclude that $Sper(g)$ is a homeomorphism. But then $Sper(f)$ is a homeomorphism as well. Let $\alpha \in Sper(A, P)$, $\beta \in Sper(B, Q)$, $\gamma \in Sper(\rho(A, P))$ be corresponding points under the homeomorphisms $Sper(f)$ and $Sper(g)$. Then

$$
\rho(\alpha) \xrightarrow{\rho_{f/\beta}} \rho(\beta) \xrightarrow{\rho_{g/\gamma}} \rho(\gamma)
$$

is an isomorphism. Altogether this shows that f satisfies the conditions of Theorem 5.2, i.e., f is an epimorphism. $\qquad\qquad\square$

Example 12.15 *The supremum of ν and ρ is σ*
The reflective subcategory belonging to $\rho \vee \nu$ is **RCR $\cap$ VNRFR**. Thus, it is claimed that **SAFR = RCR $\cap$ VNRFR**. The inclusion **SAFR $\subseteq$ RCR $\cap$ VNRFR** being clear, it remains to check the other inclusion. Suppose that (A, P) is a von Neumann regular real

closed ring. Then $supp : Sper(A, P) \to Spec(A)$ is a homeomorphism. Since A is von Neumann regular these spaces are Boolean. Thus, $Sper(\sigma_{(A,P)}) \colon Sper(\sigma(A, P)) \to Sper(A, P)$ is a homeomorphism. Proposition 7.10 shows that corresponding real closed residue fields are isomorphic. Since (A, P) is real closed it follows that $\kappa_{(A,P)}(\alpha) = \rho_{(A,P)}(\alpha)$ for each $\alpha \in Sper(A, P)$. Now (7.13) can be used to conclude that $\sigma_{(A,P)}$ is an isomorphism, i.e., $(A, P) \in ob(\mathbf{SAFR})$.

So far $\rho \vee \nu$ has been determined on the level of reflective subcategories. It will be shown now that $\rho \vee \nu$ is also easy to determine directly from the reflections, namely $\sigma = \nu\rho = \rho\nu$. For $\rho \vee \nu = \nu\rho$ we pick a real closed ring (A, P) and show that $\nu(A, P) \in ob(\mathbf{SAFR})$: We can consider $\nu(A, P)$ as a subring of $\sigma(A, P)$. All maps in the diagram

$$
\begin{array}{ccc}
Sper(\sigma(A, P)) & \longrightarrow & Sper(\nu(A, P)) \\
\downarrow & & \downarrow \\
Spec(\sigma(A, P)) & \longrightarrow & Spec(\nu(A, P))
\end{array}
$$

are known to be homeomorphisms. If $\alpha \in Sper(A, P)$ and the corresponding points of $Sper(\nu(A, P))$ and $Sper(\sigma(A, P))$ are denoted by β and γ, resp., then

$$
\kappa(\alpha) = \rho(\alpha) \to \kappa(\beta) \to \kappa(\gamma) = \rho(\gamma)
$$

is an isomorphism, hence so is $\kappa(\beta) \to \kappa(\gamma)$. Once again the claim follows from (7.13).

Finally, for $\nu \vee \rho = \rho\nu$ it suffices to show that $\rho(A, P)$ is von Neumann regular if (A, P) is a member of $\mathbf{VNRFR}$ (Proposition 9A.1). The support functions of (A, P) and $\rho(A, P)$ are both homeomorphisms. Since $Sper(\rho_{(A,P)})$ is a homeomorphism as well, so is the map $Spec(\rho_{(A,P)})$. Thus, $\rho(A, P)$ is von Neumann regular, and the proof is finished. $\qquad \square$

Example 12.16 *Continuous piecewise rational functions*
We are concerned with the infimum of the reflectors ρ and ν. The best conceivable result is that the reflection of (A, P) under $\rho \wedge \nu$ is the intersection $\rho(A, P) \cap \nu(A, P)$ (see the discussion in section 9B). According to Proposition 9B.1 this is the case if and only if the morphism

$$
u_{(A,P)} \colon (A, P) \to \rho(A, P) \cap \nu(A, P)
$$

(obtained from $\sigma_{(A,P)}$ by restriction of the codomain) is an epimorphism. But this follows immediately from Proposition 12.14: It suffices to note that the inclusion $i_{(A,P)}\colon \rho(A,P) \cap \nu(A,P) \to \rho(A,P)$ is an embedding of f–rings and that $\rho_{(A,P)} = i_{(A,P)}u_{(A,P)}$.

The category corresponding to $\rho \vee \nu$ is denoted by **CPWRFR** and is called the category of *rings of continuous piecewise rational functions*. Sections 21 and 22 are devoted to a closer examination of this category and its subcategories. $\square$

Example 12.17 *Continuous piecewise polynomial functions*
Suppose that (A, P) is a reduced poring.   We define a functor $F\colon \textbf{POR/N} \to \textbf{POR/N}$ by $F(A,P) = \displaystyle\prod_{\alpha \in Sper(A,P)} A/\alpha$. This is a sub-functor of Π, the canonical maps $F_{(A,P)}\colon (A,P) \to F(A,P)$ are the components of a natural transformation $Id_{\textbf{POR/N}} \to F$. Inside $\Pi(A,P)$ we form the intersection $r(A,P) = F(A,P) \cap \rho(A,P)$. It was shown in section 9C that this construction gives us a functor $r\colon \textbf{POR/N} \to \textbf{POR/N}$ together with a natural transformation. We wish to prove that r is a reflector. Proposition 9C.1 tells us what we have to do. First note that $r_{(A,P)}\colon (A,P) \to r(A,P)$ is an epimorphism by Proposition 12.14. Moreover, $Sper(r_{(A,P)})$ is a homeomorphism. If $\alpha \in Sper(A,P)$ and $\beta \in Sper(r(A,P))$ are corresponding points, then β is given by the homomorphism

$$r(A,P) \subseteq F(A,P) \xrightarrow{p_\alpha} A/\alpha$$

(where p_α is projection onto the α–th component). Therefore the canonical homomorphism $A/\alpha \to r(A,P)/\beta$ is an isomorphism for every α, hence $F(r_{(A,P)})$ is an isomorphism. Thus, the hypotheses of Proposition 9C.1 are satisfied and r is a reflector.

The monoreflective subcategory of **POR/N** belonging to r is denoted by **CPWPFR**, its objects are called *continuous piecewise polynomial function rings*. We claim that the reflection of (A,P) can also be described as the subring of $\rho(A,P)$ consisting of the elements a with the following property:

(12.18)
There is a closed constructible cover $Sper(A,P) = C_1 \cup \ldots \cup C_r$ and there are elements $a_1, \ldots, a_r \in A$ such that $a(\alpha) = a_i(\alpha)$ for all $\alpha \in C_i$.

First note that these elements obviously belong to $r(A,P)$. Conversely, suppose that $a \in r(A,P) \subseteq F(A,P)$. For every $\alpha \in Sper(A,P)$ there is some $a_\alpha \in A$ with $a_\alpha(\alpha) = a(\alpha)$. The sets

$$C_\alpha = \{\beta \in Sper(A, P); \ a_\alpha(\beta) = a(\beta)\}$$

are closed and constructible in $Sper(A, P)$ (since all the elements involved belong to $\rho(A, P)$, Lemma 12.9), and they cover $Sper(A, P)$. By compactness there is a finite subcover $Sper(A, P) = C_{\alpha_1} \cup \ldots \cup C_{\alpha_r}$. These closed constructible sets together with the a_{α_i} satisfy the condition (12.18).

This concrete description of the elements of $r(A, P)$ also explains the name of continuous piecewise polyomial functions. For, the reflection $r(Fn))$, $n \in \mathbb{N}$, consists exactly of the continuous functions $a \colon R_0^n \to R_0$ that agree with restrictions of polynomial functions on the sets belonging to a finite closed semi-algebraic cover of R_0^n.

A careful investigation of rings of continuous piecewise polynomial functions, in particular of their real and prime spectra, may be found in [115]. The category **CPWPFR** is obviously contained in **FR/N**. The precise relationship between these categories or the corresponding reflections is rather delicate. The notorious *Pierce-Birkhoff Conjecture* (cf. [85]; [89]; [91]) is concerned with this question. The conjecture is that for every polynomial ring $A_n = \mathbb{R}[T_1, \ldots, T_n]$ the rings $\varphi(A_n, P_w(A_n))$ and $r(A_n, P_w(A_n))$ coincide. Mahé was the first to note that **CPWPFR** is a proper subcategory of **FR/N** ([89], p. 985), other examples can be obtained from [91]. In Mahé's example the decisive idea is to find a closed semi-algebraic subset $X \subseteq \mathbb{R}^2$ and a continuous piecewise polynomial function $h \colon X \to \mathbb{R}$ that cannot be extended to a continuous piecewise polynomial function on the entire plane. Thus, the category **CPWPFR** is not H–closed.

We claim that **CPWPFR** is quotient–closed. So, we pick $(A, P) \in ob(\mathbf{CPWPFR})$, and consider a multiplicative subset $S \subseteq A$. First of all, note that the poring of quotients (A_S, P_S) is an f–ring (Example 11.13). The set

$$S' = \{t \in P; \ \exists\, a \in P : at \in S\}$$

is multiplicative as well. For every $s \in S$ it contains some multiple of s; to be precise: $s^2 = |s|^2 \in S \cap P \subseteq S'$. Conversely if $s' \in S'$ then there is some multiple $as' \in S$ by the definition of S'. Therefore the porings (A_S, P_S) and $(A_{S'}, P_{S'})$ are isomorphic, and we may assume that $S = S'$. To show that the canonical homomorphism $(A_S, P_S) \to (\ \cdot \prod\limits_{\alpha \in Sper(A_S, P_S)} A_S/\alpha) \cap$ $\rho(A_S, P_S)$ is an isomorphism we first note that injectivity is clear; to check surjectivity, pick some element c in the codomain. There is a cover $Sper(A_S, P_S) = C_1 \cup \ldots \cup C_r$ by closed constructible sets and there are $a_i \in$

A, $s_i \in S$, $i = 1, \ldots, r$ such that $\frac{a_i(\alpha)}{s_i(\alpha)} = c(\alpha)$ for $\alpha \in C_i$. Without loss of generality we may make the following assumptions about the quotients $\frac{a_i}{s_i}$:

- The denominators are all equal, say $s = s_1 = \ldots = s_r$.
- If $s(\alpha) = 0$ for some $\alpha \in Sper(A, P)$ then also $a_1(\alpha) = \ldots = a_r(\alpha) = 0$.

We identify $Sper(A_S, P_S)$ with the generically closed proconstructible subspace $\{\alpha \in Sper(A, P); \ \forall t \in S : t(\alpha) \neq 0\}$ of $Sper(A, P)$. There are closed constructible sets $D_i \subseteq Sper(A, P)$ with $D_i \cap Sper(A_S, P_S) = C_i$; let $D = \cup D_i$. The set

$$D' = \{\alpha \in D; \ \exists \, i, j : i \neq j \ \& \ \alpha \in D_i \cap D_j \ \& \ a_i(\alpha) \neq a_j(\alpha)\}$$

is a constructible subset of $Sper(A, P)$, and $D' \cap Sper(A_S, P_S) = \emptyset$. There exists some $t \in S$ such that $t(\alpha) = 0$ for every $\alpha \in D'$. Now we replace s by st, hence we have $s(\alpha) = 0$ for $\alpha \in D'$. But then $\alpha \in D'$ implies $a_1(\alpha) = \ldots = a_r(\alpha) = 0$, hence $D' = \emptyset$. Next we pick an element $t \in S$ such that

$$Sper(A_S, P_S) \subseteq \{\alpha \in Sper(A, P); \ t(\alpha) \neq 0\} \subseteq D$$

and replace s by st. The set

$$D_0 = \{\alpha \in Sper(A, P); \ s(\alpha) = 0\}$$

is closed and constructible and covers $Sper(A, P)$ together with $D_1, \ldots, D_r$. If we definie $a_0 = 0$ then a function $b \in \Pi(A, P)$ is well–defined by: $b(\alpha) = a_i(\alpha)$ if $\alpha \in D_i$. It is also clear that $b \in \prod_{\alpha \in Sper(A,P)} A/\alpha$. Using the technique of inverse real closed spaces ([114], §1) one sees that $b \in \rho(A, P)$. Altogether it has been shown that b belongs to (A, P). If $\alpha \in C_i$ then

$$\frac{b(\alpha)}{s(\alpha)} = \frac{a_i(\alpha)}{s(\alpha)} = c(\alpha),$$

i.e., c is the canonical image of $\frac{b}{s} \in (A_S, P_S)$, and (A_S, P_S) belongs to **CPWPFR**.

We shall return to the category **CPWPFR** repeatedly in examples. The H–closed reflective subcategories $\mathbf{D} \subseteq \mathbf{POR/N}$ that lie between **RCR** and **CPWPFR** will be examined in section 20. $\qquad \square$

It was shown in Example 9C.2 that in Example 12.17 the reflector ρ cannot be replaced by the reflector σ. For, we saw that the canonical homomorphism $(A, P) \to F(A, P) \cap \sigma(A, P)$ is not always an epimorphism.

Example 12.19 *Euclidean f–rings*

We call a reduced f–ring (A, P) *euclidean* if $P = A^2$. (This notion generalizes euclidean fields, cf. [13].) Examples of euclidean f–rings are the real closed rings (Proposition 12.4). The subcategory of **POR/N** consisting of the euclidean f–rings is denoted by **EFR/N**. It is claimed that **EFR/N** is reflective in **POR/N**. As a first step, define $r(A, P)$ to be the intersection of all euclidean f–rings contained in $\rho(A, P)$ and containing $im(\rho_{(A,P)})$. Let $r_{(A,P)} \colon (A, P) \to r(A, P)$ be the reflection morphism $\rho_{(A,P)}$ with restricted codomain. An intersection of sub–f–rings is a sub–f–ring, hence $r(A, P)$ is an f–ring. In a euclidean f–ring every positive element has a *unique* positive square root. Therefore $r(A, P)$ is also euclidean.

It will be shown that the morphism $r_{(A,P)}$ has the universal property (2.1) of reflections. Observe that $r_{(A,P)}$ is an epimorphism by Proposition 12.14. Pick any reduced poring (A, P) and let $f \colon (A, P) \to (B, Q)$ be a homomorphism into a euclidean f–ring. The real closure reflector yields the diagram

$$
\begin{array}{ccc}
(A, P) & \xrightarrow{\ f\ } & (B, Q) \\
{\scriptstyle \rho_{(A,P)}}\big\downarrow & & \big\downarrow{\scriptstyle \rho_{(B,Q)}} \\
\rho(A, P) & \xrightarrow{\ \rho(f)\ } & \rho(B, Q)
\end{array}
$$

Since $\rho_{(B,Q)}$ is an isomorphism onto a sub–f–ring of $\rho(B, Q)$ belonging to **EFR/N** one checks that $\rho(f)^{-1}(im(\rho_{(B,Q)})) \subseteq \rho(A, P)$ also belongs to **EFR/N**, hence it contributes to the intersection defining $r(A, P)$. This means that $\rho(f)(r(A, P)) \subseteq im(\rho_{(B,Q)})$, hence there is a homomorphism $\overline{f} \colon r(A, P) \to (B, Q)$ such that

$$
\rho_{(B,Q)} f = \rho_{(B,Q)} \overline{f} r_{(A,P)}.
$$

Because $\rho_{(B,Q)}$ is a monomorphism it follows that $f = \overline{f} r_{(A,P)}$. Moreover, $\overline{f}$ is unique with this property since $r_{(A,P)}$ is an epimorphism. Thus, it has been shown that $r_{(A,P)}$ has the universal property (2.1), and r is a reflector.

The reflector r is both H–closed and quotient–closed. For H–closedness, let $f \colon (A, P) \to (B, Q)$ be a surjective homomorphism in

POR/N with (A, P) a euclidean f–ring. Since **FR/N** is H–closed (Example 10.15) one knows that $Q = f(P)$. But then every element of Q is a square, and (B, Q) is a euclidean f–ring. Finally, suppose that (A, P) is a euclidean f–ring and let $S \subseteq A$ be a multiplicative subset. Then (A_S, P_S) is an f–ring (Example 11.13). We must only show that P_S consists of squares. So, let $\frac{a}{s} \in P_S$, i.e., $ast^2 \in P$ for some $t \in S$. Then there are $b, r \in P$ with $|a| = b^2$ and $|s| = r^2$, hence

$$ast^2 = |a||s|t^2 = (brt)^2.$$

Now

$$\frac{a}{s} = \left(\frac{br}{s}\right)^2$$

proves the claim.   □

13 Arities of reflectors and approximations by H–closed reflectors

A general strategy for studying reflectors is to consider subclasses $X \subseteq ob(\mathbf{POR/N})$ and to ask whether a given reflector $r: \mathbf{POR/N} \to \mathbf{D}$ is determined by the reflections $r_{(A,P)}$ with $(A, P) \in X$. If $X = ob(\mathbf{POR/N})$ then the answer is obvious but useless. So we look for, informally speaking, small classes X having this property. For H–closed monoreflectors Theorem 10.6 says that we can choose $X = \{r_{F(n)}; n \in \mathbb{N}\}$ or $X = \{r_{F(\mathbb{N})}\}$. These are rather small sets. But one may be even more ambitious and ask whether a given reflector is already determined by a single reflection $r_{F(n)}$ for some $n \in \mathbb{N}$. As a first step we make the following definition: If $\mathbf{C} \subseteq \mathbf{POR/N}$ is a reflective subcategory and if $r: \mathbf{C} \to \mathbf{D}$ is an H–closed reflector then we define

$$N_{\mathbf{C}}(\mathbf{D}) = N(r) = \min\{n \in \mathbb{N} \cup \{\infty\}; \; ob(\mathbf{D}) = Inj(\{r_{F(m)}; \; m \le n\})\}$$

to be the *arity* of r (or, of $\mathbf{D}$ in $\mathbf{C}$).

Reflectors with small arities are easier to understand and are more readily accessible to the intuition than those with large arities. For example to determine a reflector of arity 1 it suffices to know the reflection of $F(1)$, which is just a ring of semi–algebraic functions on the line. To recognize whether two such reflectors, say, r and s, coincide or whether one of them is stronger than the other one it suffices to compare the rings $r(F(1))$ and $s(F(1))$. It is well–known that, in general, arities are difficult to determine (cf. Hilbert's 13th problem, [80]). So it is no surprise that our results are far from being comprehensive. We show that we know all the reflectors of $\mathbf{POR/N}$ that have arity 0 and that some of the reflectors we have encountered so far have arity 1. Beyond this, the country of arities remains *terra incognita*.

First we note that there exist many H–closed reflectors of finite arity. Let $r: \mathbf{POR/N} \to \mathbf{D}$ be any monoreflector. For each $n \in \mathbb{N} \cup \{\infty\}$, the set $\mathcal{E}_n = \{r_{F(m)}; m \le n\}$ consists of epimorphisms, hence $Inj(\mathcal{E}_n)$ is the object class of some H–closed epireflective subcategory $\mathbf{D}_n$ of $\mathbf{POR/N}$ (Theorem 10.2). The corresponding reflector is denoted by r_n. It is clear that $\mathbf{D}_0 \supseteq \mathbf{D}_1 \supseteq \ldots \supseteq \mathbf{D}_\infty \supseteq \mathbf{D}$ and that $\mathbf{D}_\infty = \mathbf{D}$ if and only if r is H–closed (Theorem 10.6). More precisely, $\mathbf{D} = \mathbf{D}_n \subset \mathbf{D}_{n-1}$ if and only if $\mathbf{D}$ is H–closed and $n = N(r)$. We consider $r_n \le r$ as an approximation of r by an H–closed monoreflector of arity n. It is the strongest H–closed monoreflector of arity n that is weaker than r.

Here are some elementary properties of arities:

Proposition 13.1 Let $\mathbf{E} \subseteq \mathbf{D} \subseteq \mathbf{C} \subseteq \mathbf{POR/N}$ be H–closed monoreflective subcategories.

(a) $N_{\mathbf{C}}(\mathbf{E}) \leq \max\{N_{\mathbf{C}}(\mathbf{D}), N_{\mathbf{D}}(\mathbf{E})\}$.

(b) $N_{\mathbf{D}}(\mathbf{E}) \leq N_{\mathbf{C}}(\mathbf{E})$.

(c) If $N_{\mathbf{C}}(\mathbf{D}) \leq N_{\mathbf{D}}(\mathbf{E})$ then $N_{\mathbf{C}}(\mathbf{E}) = N_{\mathbf{D}}(\mathbf{E})$

Proof (a) Let $r\colon \mathbf{C} \to \mathbf{D}$, $s\colon \mathbf{C} \to \mathbf{E}$, $t\colon \mathbf{D} \to \mathbf{E}$ be the reflectors, let $n = \max\{N_{\mathbf{C}}(\mathbf{D}), N_{\mathbf{D}}(\mathbf{E})\}$. It is trivial that $ob(\mathbf{E}) \subseteq Inj(\{s_{F(m)}; m \leq n\})$. Now suppose that $(A, P) \in Inj(\{s_{F(m)}; m \leq n\})$. Every reflection $s_{F(m)}$ factors uniquely as

$$F(m) \xrightarrow{\ r_{F(m)}\ } r(F(m)) \xrightarrow{\ t_{r(F(m))}\ } t(r(F(m))) = s(F(m)).$$

Pick any morphism $f\colon F(m) \to (A, P)$. Then f extends to $\overline{f}\colon s(F(m)) \to (A, P)$ such that $f = \overline{f}s_{F(m)} = (\overline{f}t_{r(F(m))})r_{F(m)}$. This means that (A, P) belongs to $Inj(\{r_{F(m)}; m \leq n\}) = ob(\mathbf{D})$. Now consider a morphism $g\colon r(F(m)) \to (A, P)$. Then $gr_{F(m)}\colon F(m) \to (A, P)$ extends to $\overline{g}\colon s(F(m)) \to (A, P)$, i.e., $gr_{F(m)} = (\overline{g}t_{r(F(m))})r_{F(m)}$. Since $r_{F(m)}$ is an epimorphism we conclude that $g = \overline{g}t_{r(F(m))}$, hence $(A, P) \in Inj(\{t_{r(F(m))}; m \leq n\}) = ob(\mathbf{E})$.

(b) Let $n = N_{\mathbf{C}}(\mathbf{E})$. Once again it is clear that $ob(\mathbf{E}) \subseteq Inj(\{t_{r(F(m))}; m \leq n\}) \subseteq ob(\mathbf{D})$. Suppose that $(A, P) \in Inj(\{t_{r(F(m))}; m \leq n\})$, pick any morphism $f\colon F(m) \to (A, P)$ and consider the unique extension $\overline{f}\colon r(F(m)) \to (A, P)$ with $f = \overline{f}r_{F(m)}$. Then $\overline{f}$ extends to $\tilde{f}\colon t(r(F(m))) \to (A, P)$, and

$$f = \overline{f}r_{F(m)} = (\tilde{f}t_{r(F(m))})r_{F(m)} = \tilde{f}s_{F(m)}$$

shows that $(A, P) \in Inj(\{s_{F(m)}; m \leq n\}) = ob(\mathbf{E})$.

(c) is an immediate consequence of (a) and (b). $\qquad\qquad\square$

As a first step in our explicit computation of arities we consider H–closed monoreflective subcategories of arity 0 in $\mathbf{POR/N}$. In fact, we have a complete description of these subcategories.

Proposition 13.2 An H–closed monoreflective subcategory $\mathbf{D} \subseteq \mathbf{POR/N}$ with reflector r has arity 0 if and only if $\mathbf{D} = \mathbf{r}(\mathbb{Z})\mathbf{POR/N}$.

Proof Let $\mathbf{D_0}$ be the arity 0 approximation of $\mathbf{D}$. Then $ob(\mathbf{D_0}) = Inj(\{r_\mathbb{Z}\})$ (since $\mathbb{Z}$ is the free object on the empty set in $\mathbf{POR/N}$). The injectivity class of $r_\mathbb{Z}$ is the class of $r(\mathbb{Z})$–algebras, i.e., $\mathbf{D_0} = \mathbf{r}(\mathbb{Z})\mathbf{POR/N}$. This finishes the proof since $\mathbf{D}$ has arity 0 if and only if $\mathbf{D} = \mathbf{D_0}$. $\square$

Beyond arity 0 our results are fragmentary, at best. We are only able to determine arities in a few special cases.

Example 13.3 FR/N *has arity 1 in* **POR/N**
Recall that $\mathbf{FR/N}$ is H–closed by Example 10.15. We claim that $ob(\mathbf{FR/N})$ is the injectivity class of $\{\varphi_{F(1)}\}$. The inclusion $ob(\mathbf{FR/N}) \subseteq Inj(\{\varphi_{F(1)}\})$ is trivial. So, we pick any $(A, P) \in Inj(\{\varphi_{F(1)}\})$. Via $\Delta_{(A,P)}$, (A, P) is mapped into the reduced f–ring $\Pi(A, P)$. It suffices to show that $\Delta_{(A,P)}(a) \vee 0 \in im(\Delta_{(A,P)})$ for each $a \in A$. To prove this we define $f \colon \mathbb{Z}[T] \to A \colon T \to a$ and consider f as a morphism $F(1) \to (A, P)$. Since $(A, P) \in Inj(\{\varphi_{F(1)}\})$ there is an extension $\overline{f} \colon \varphi(F(1)) \to (A, P)$ of f, i.e., $f = \overline{f}\varphi_{F(1)}$. The composition $\Delta_{(A,P)}\overline{f} \colon \varphi(F(1)) \to \Pi(A, P)$ is a $\mathbf{POR/N}$–morphism of reduced f–rings, hence it preserves the lattice operations ([44], Lemma 2.2). But then

$$
\begin{aligned}
\Delta_{(A,P)}(a) \vee 0 &= \Delta_{(A,P)}\overline{f}(\varphi_{F(1)}(T)) \vee 0 \\
&= \Delta_{(A,P)}\overline{f}(\varphi_{F(1)}(T) \vee 0) \in im(\Delta_{(A,P)}). \qquad \square
\end{aligned}
$$

Example 13.4 *The Pierce–Birkhoff Conjecture*
Let r be the reflector $\mathbf{POR/N} \to \mathbf{CPWPFR}$. By an example of Mahé's, r is not H–closed ([89], p. 985). The same example shows that the reflector $s \colon \mathbf{POR/N} \to \mathbf{R_0CPWPFR}$ is not H–closed, either. Thus, we cannot talk about the arity of the reflector s. However, one can consider the sequence $s_1 \leq s_2 \leq \ldots$ of approximations of s. The reflector $\mathbf{POR/N} \to \mathbf{R_0FR/N}$ is denoted by $\overline{\varphi}$. The *Pierce–Birkhoff Conjecture* is concerned with the precise relationship between $\overline{\varphi}$ and the sequence $s_1, s_2, \ldots$. For the classical formulation of the conjecture, see [89] or [85]. Using our setup, the question may be asked in the following form: Is

$\overline{\varphi} = s_1 = s_2 = \ldots$? It is well–known and easy to prove that $\overline{\varphi} = s_1$ (for more general results, see [85], p. 569, Proposition, and [91]). It was proved by Mahé ([89]) that also $\overline{\varphi} = s_2$. (A strenghtening of this result may be found in [43]. It shows that the same result holds when R_0 is replaced by some intermediate field between $\mathbb{Q}$ and and R_0.) The first open case of the Pierce–Birkhoff Conjecture is whether $\overline{\varphi} = s_3$. $\square$

Example 13.5 BIFR/N is H–closed of arity 1 in FR/N

It was shown in Example 10.13 that **BIPOR/N** is not H–closed in **POR/N**. So it is not a priori clear whether the intersection **BIFR/N** = **BIPOR/N** $\cap$ **FR/N** is H–closed. However, factor porings exist in **BIPOR/N** (Example 10.13) and in **FR/N** (Example 10.15, Remark 10.1), hence they exist in the intersection of the categories. But since **BIFR/N** is contained in **FR/N** it follows from Remark 10.1 that **BIFR/N** is H–closed both in **POR/N** and in **FR/N**. We show that the arity of **BIFR/N** in **FR/N** is 1: By Proposition 11.7 and Example 11.13, the reflection of $\varphi(F(1))$ in **BIFR/N** is $s_{\varphi(F(1))}\colon \varphi(F(1)) \to s(\varphi(F(1)))$ (where $s\colon$ **POR/N** $\to$ **BIPOR/N** is the bounded inversion reflector). To prove that $ob(\textbf{BIFR/N}) = Inj(\{s_{\varphi(F(1))}\})$ it suffices to show that every reduced f–ring $(A, P) \in Inj(\{s_{\varphi(F(1))}\})$ belongs to **BIFR/N**. So, suppose that $a \in A$ and $a(\alpha) \geq 0$ for all $\alpha \in Sper(A, P)$. The homomorphism $f\colon \mathbb{Z}[T] \to A\colon T \to a$ can be considered as a morphism $F(1) \to (A, P)$ in **POR/N**. It extends uniquely to $g\colon \varphi(F(1)) \to (A, P)$. By the choice of (A, P) there is a morphism $\overline{g}\colon s(\varphi(F(1))) \to (A, P)$ with $g = \overline{g}s_{\varphi(F(1))}$. Note that $\overline{g}(|T|) = |g(T)| = |a| = a$. Since $1 + |T| \in s(\varphi(F(1)))^{\times}$ we conclude that $1 + a = \overline{g}(1 + |T|) \in A^{\times}$, i.e., $1 + P \subseteq A^{\times}$ as claimed.

Together, Example 13.3 and Proposition 13.1 imply that the arity of **BIFR/N** in **POR/N** is also 1. $\square$

Example 13.6 EFR/N has arity 1 in POR/N

The H–closed monoreflective subcategory of euclidean f–rings was introduced in Example 12.19. Let $r\colon$ **POR/N** $\to$ **EFR/N** be the corresponding reflector. Suppose that (A, P) belongs to the injectivity class of $\{r_{F(1)}\}$. Then (A, P) is a reduced f–ring by Example 13.3. It remains to show that every $a \in P$ is a square: A homomorphism $f\colon F(1) \to (A, P)$ is defined by $T \to a$. Let $\overline{f}\colon r(F(1)) \to (A, P)$ the unique extension of f. Then $\overline{f}(T \vee 0) = \overline{f}(T) \vee 0 = a$. Since $r(F(1))$ is euclidean and $T \vee 0 \in r^{+}(F(1))$ there exists some $t \in r(F(1))$ with $T \vee 0 = t^2$. It follows that $a = \overline{f}(t)^2$, and (A, P) is euclidean. $\square$

Example 13.7 VNRFR *has arity* 1 *in* **POR/N**

According to Example 10.17 the monoreflector r is H–closed. To prove that it has arity 1 in **POR/N** we have to show that each object $(A, P) \in Inj(\{\nu_{F(1)}\})$ belongs to **VNRFR**. First note that (A, P) is a reduced f–ring since $\varphi \leq \nu$ and hence $(A, P) \in Inj(\{\varphi_{F(1)}\})$ (Example 13.3). To show that it is also von Neumann regular, pick $a \in A$ and define $f \colon \mathbb{Z}[T] \to A \colon T \to a$. Again, f is considered as a morphism $F(1) \to (A, P)$. As such it extends uniquely to $\overline{f} \colon \nu(F(1)) \to (A, P)$ (by the choice of (A, P)). In $\nu(F(1))$ there is an element T^* with $T^2 T^* = T$. This implies

$$a = f(T) = f(T^2 T^*) = \overline{f}(T)^2 f(T^*) = a^2 \overline{f}(T^*),$$

i.e., A is von Neumann regular. $\square$

We suspect that the arities of **WBIPOR/N**, **RCR** and **SAFR** in **POR/N** are infinite. However, at this time we do not have a single example of an H–closed reflector for which we were able to prove that the arity is larger than 1!

Remark 13.8 Our notion of the arity of a reflector refers to free objects. It is possible to define arities in a more general category theoretic setup: Let **C** be a complete and wellpowered category with forgetful functor $U \colon \mathbf{C} \to \mathbf{SETS}$. Assume that U preserves limits. Thus, subobjects generated by subsets exist (cf. section 1). An object C is *generated* by a subset $X \subseteq U(C)$ if $\langle X \rangle = C$. In particular, C is *generated by* n *elements* if there is a subset $X \subseteq U(C)$, $|X| \leq n$, with $C = \langle X \rangle$. If $r \colon \mathbf{C} \to \mathbf{D}$ is any epireflector then there is a class $\mathcal{G}$ of reflection morphisms such that $ob(\mathbf{D}) = Inj(\mathcal{G})$ (Theorem 8.7), e.g., the class of all reflection morphisms. For every $n \in \mathbb{N}$, define $\mathcal{G}_n$ to be the class of all reflections $r_C \colon C \to r(C)$ where C is generated by n elements. Then we call

$$n_{\mathbf{C}}(\mathbf{D}) = n(r) = \min\{n \in \mathbb{N} \cup \{\infty\}; \ ob(\mathbf{D}) = Inj(\mathcal{G}_n)\}$$

the *arity* of the reflector r, or of the epireflective subcategory **D**.

If $\mathbf{C} = \mathbf{POR/N}$ and r is H–closed then we have two notions of arity, given by the numbers $n(r)$ and $N(r)$. If $\mathcal{E}_n$ denotes the set of reflections of the free objects on the sets $\{1, \ldots, m\}$, $m \leq n$, then $\mathcal{E}_n \subseteq \mathcal{G}_n$. It follows that $n(r) \leq N(r)$. For, if $n = N(r)$ then $Inj(\mathcal{E}_n) = ob(\mathbf{D})$. Since $Inj(\mathcal{G}) = ob(\mathbf{D})$ and $\mathcal{E}_n \subseteq \mathcal{G}$ it is clear that $Inj(\mathcal{G}') = ob(\mathbf{D})$ for any class $\mathcal{G}'$ with $\mathcal{E}_n \subseteq \mathcal{G}' \subseteq \mathcal{G}$. One such class is $\mathcal{G}_n$, hence $n(r) \leq N(r)$.

We show that $n(r) = N(r)$ by proving the other inequality. Let $n = n(r)$ and let $r_n \colon \mathbf{POR}/\mathbf{N} \to \mathbf{D_n}$ be the approximation of arity n. It is trivial that $ob(\mathbf{D_n}) \supseteq ob(\mathbf{D})$; we claim that these classes are equal. If we know that $(r_n)_{(A,P)} = r_{(A,P)}$ for all $r_{(A,P)} \in \mathcal{G}_n$ then $ob(\mathbf{D_n}) \subseteq Inj(\mathcal{G}_n) = ob(\mathbf{D})$, and the proof is finished. So it suffices to show that $r_n(A, P) = r(A, P)$ whenever (A, P) is generated by at most n elements. We assume that this is false, say $r_n(A, P) \neq r(A, P)$. Since (A, P) is generated by at most n elements there is a surjective morphism $f \colon F(n) \to (A, P)$. By Proposition 10.8(a) the morphism $r(f) \colon r(F(n)) \to r(A, P)$ is surjective as well. Since $r_{F(n)} = (r_n)_{F(n)}$ it follows that $r(f)$ maps $r_n(F(n))$ surjectively onto $r(A, P)$. The diagram

$$
\begin{array}{ccccc}
F(n) & \longrightarrow & r_n(F(n)) & \overset{=}{\longrightarrow} & r(F(n)) \\
{\scriptstyle f}\big\downarrow & & {\scriptstyle r_n(f)}\big\downarrow & \quad , & \big\downarrow {\scriptstyle r(f)} \\
(A, P) & \longrightarrow & r_n(A, P) & \overset{\subseteq}{\longrightarrow} & r(A, P)
\end{array}
$$

shows that the underlying rings of $r_n(A, P)$ and $r(A, P)$ coincide. The difference, if there is any difference, must be in the partial orders. Since r is H–closed and $r_n(A, P) \notin ob(\mathbf{D})$ (by assumption) there exist some $p > n$ and some $\mathbf{POR}/\mathbf{N}$–morphism $g \colon F(p) \to r_n(A, P)$ such that g cannot be extended to a $\mathbf{POR}/\mathbf{N}$–morphism $\bar{g} \colon r(F(p)) \to r_n(A, P)$. (Note that $\bar{g}$ exists as a *ring* homomorphism since the underlying rings of $r_n(A, P)$ and $r(A, P)$ coincide.) Since $r_n(f)$ is surjective there is some $\mathbf{POR}/\mathbf{N}$–morphism $h \colon F(p) \to r_n(F(n))$ such that $r_n(f)h = g$. The universal property (2.1) of the reflection $r_{F(p)}$ yields $\bar{h} \colon r(F(p)) \to r_n(F(n)) = r(F(n))$ with $h = \bar{h}r_{F(p)}$. But then $r_n(f)\bar{h}$ is a $\mathbf{POR}/\mathbf{N}$–morphism and $(r_n(f)\bar{h})r_{F(p)} = g$. This contradicts the choice of g, and the proof is finished. $\square$

14 Epimorphic extensions of reduced porings

Frequently one of the most important and difficult steps needed to identify a construction with porings as epireflective is to recognize whether the presumable reflection maps are epimorphisms. Therefore any result is welcome which says that certain intermediate porings between a reduced poring (A, P) and its reflection $\sigma(A, P)$ are epimorphic over (A, P). For example, according to Proposition 12.14 every f–ring (B, Q) between (A, P) and $\rho(A, P)$ is epimorphic over (A, P). The main result of the present section characterizes the reduced porings (A, P) for which every subporing of $\rho(A, P)$ containing $im(\rho_{(A,P)})$ is epimorphic over (A, P). The criterion uses only the real spectrum (Theorem 14.6).

The most general question we consider in this section is:

(14.1) Given a reduced poring (A, P_A) and a factorization of $\sigma_{(A,P_A)}$ into two monomorphisms $(A, P_A) \xrightarrow{f} (C, P_C) \xrightarrow{g} \sigma(A, P_A)$, when is it true that every subporing (B, P_B) of (C, P_C) containing $im(f)$ is epimorphic over (A, P_A)?

One may ask an even broader question by allowing (B, P_B) to be any subobject of (C, P_C) such that f factors through (B, P_B). However, this generalization would be very costly in terms of the results that one can expect. For example, if (A, P_A) is a totally ordered field then $\sigma(A, P_A)$ is the real closure of (A, P_A) and (C, P_C) is an intermediate field. If every intermediate *partially ordered* field of (A, P_A) and (C, P_C) is epimorphic over (A, P_A) then each such field must be totally ordered (by Theorem 5.2), i.e., P_A can be extended in only one way to a total order of any intermediate field between A and C. The class of field extensions having this property is rather narrow. Correspondingly, the class of poring extensions $(A, P_A) \subseteq (C, P_C)$ for which every intermediate poring is epimorphic over (A, P_A) is also very small. If we add the additional assumption that the intermediate poring is *embedded* in (C, P_C) then the class of extensions becomes large enough to be useful for our purposes.

Compared with the above most general question we have to make one additional assumption, namely that the poring (C, P_C) is not only contained in $\sigma(A, P_A)$, but rather in $\rho(A, P_A)$. This is due to the following result:

Proposition 14.2 Suppose that (A, P_A) is a reduced poring, that $i\colon (C, P_C) \overset{\subseteq}{\longrightarrow} \sigma(A, P_A)$ is a sub–f–ring and that $\sigma_{(A,P_A)}$ factors as $(A, P_A) \overset{f}{\longrightarrow} (C, P_C) \overset{i}{\longrightarrow} \sigma(A, P_A)$. If $Sper(f)$ is not a homeomorphism then there is some subporing $j\colon (B, P_B) \overset{\subseteq}{\longrightarrow} (C, P_C)$ together with a factorization $f = jg$ such that g is not an epimorphism.

Proof If $Sper(f)$ is not injective then we may choose $(B, P_B) = (C, P_C)$ (Theorem 5.2), and we are finished. So, now we suppose that $Sper(f)$ is injective. Then it is even bijective (since $Sper(\sigma_{(A,P_A)})$ is surjective). Therefore the real spectra of (A, P_A) and (C, P_C) can be identified as *sets*. But the topology of $Sper(C, P_C)$ is finer than the one of $Sper(A, P_A)$. Since $Sper(f)$ is a homeomorphism with respect to the *constructible* topologies, the constructible sets in both spectra correspond to each other under $Sper(f)$. There is an open constructible set $U \subseteq Sper(C, P_C)$ such that $Sper(f)(U)$ is not open in $Sper(A, P_A)$. Thus, U is closed under generalization, whereas this is not true for $Sper(f)(U)$. Pick some $\alpha \in Sper(A, P_A)$ which is a closed point of $Sper(A, P_A) \backslash Sper(f)(U)$, but has a specialization in $Sper(f)(U)$; let $\beta = Sper(f)^{-1}(\alpha)$. There is an immediate specialization α' of α in $Sper(f)(U)$; let $\beta' = Sper(f)^{-1}(\alpha')$. We claim that β is a closed point of $Sper(C, P_C)$. Assume that this is not so. Then there is a proper specialization β'' of β. By continuity of $Sper(f)$, $\alpha'' = Sper(f)(\beta'')$ is a proper specialization of α. Since $\beta'' \neq \beta'$ we know that $\alpha'' \neq \alpha'$. The specializations of α form a chain, hence $\alpha \subset \alpha' \subset \alpha''$. By [65], Lemma 2.1, $Sper(f)$ is convex, i.e., there is some $\gamma \in Sper(C, P_C)$ with $\beta \subset \gamma \subset \beta''$ and $Sper(f)(\gamma) = \alpha'$. But then $\gamma = Sper(f)^{-1}(\alpha') = \beta'$, and $\beta \subset \beta' \in U$. Since U is closed under generalization and $\beta \notin U$, this is impossible. We conclude that β is a closed point of $Sper(C, P_C)$, hence $\beta \not\subseteq \beta'$. It is obviously impossible that $\beta' \subseteq \beta$. Thus β and β' are incomparable in $Sper(C, P_C)$, and there exists some $c \in C$ with $c(\beta) > 0$ and $c(\beta') < 0$. Replacing c by $c \wedge 1$ if necessary we may assume that $c(\beta) \leq 1$.

The real closed residue fields $\rho(\alpha)$ and $\rho(\beta)$ are isomorphic; we identify them. Then $\overline{V}(\alpha) \subseteq \rho(\beta)$ is a convex valuation ring. Now we define $B = \rho_\beta^{-1}(\overline{V}(\alpha)) \subseteq C$ and $P_B = B \cap P_C$. Furthermore, $g : (A, P_A) \to (B, P_B)$ is obtained from f by restriction of the codomain, $j : (P, P_B) \to (C, P_C)$ is the inclusion. We claim that g is not an epimorphism. Three points in $Sper(B, P_B)$ will be used for the proof. Let γ be the point defined by $\rho_\beta j$, let γ' be the point defined by $\rho_{\beta'} j$. For the definition of the third point, note that there is a nontrivial prime ideal $p \subseteq \overline{V}(\alpha)$ such that α' is the point of $Sper(A, P_A)$ defined by

$$(A, P_A) \xrightarrow{\rho_\alpha} \overline{V}(\alpha) \longrightarrow \overline{V}(\alpha)_p / p\overline{V}(\alpha)_p.$$

We define $\gamma'' \in Sper(B, P_B)$ to be the point determined by

$$(B, P_B) \xrightarrow{\rho_\beta} \overline{V}(\alpha) \longrightarrow \overline{V}(\alpha)_p / p\overline{V}(\alpha)_p.$$

It is clear that $Sper(g)(\gamma) = \alpha$ and $Sper(g)(\gamma') = \alpha' = Sper(g)(\gamma'')$. It suffices to show that $\gamma' \neq \gamma''$. Since $0 < c(\beta) \leq 1$ we know that $c \in B$ and $c \in \gamma$. Because of $\gamma \subseteq \gamma''$ it follows that $c \in \gamma''$. On the other hand, $c(\gamma') < 0$ shows that $c \notin \gamma'$. We conclude that $Sper(g)$ is not injective, hence g is not an epimorphism. $\square$

Example 14.3 *If the map $Sper(f)$ induced by $f : (A, P_A) \to (C, P_C)$, $(C, P_C) \subseteq \sigma(A, P_A)$ is not a homeomorphism then there is an intermediate poring (B, P_B) between (A, P_A) and (C, P_C) which is not epimorphic over (A, P_A).*
To see an explicit example of the phenomenon observed in the proposition, let R be the real closed formal power series field $\mathbb{R}((\mathbb{Q}))$, let (A, P_A) be the natural valuation ring with its unique total order. This is a real closed ring, the real spectrum consists of two points $\alpha \subset \beta$. More precisely, $\alpha = P_A$, and β is determined by the canonical place onto the residue field. It is then obvious that $\sigma(A, P_A) = \mathbb{R}((\mathbb{Q})) \times \mathbb{R}$ and $\sigma_{(A, P_A)}$ is the homomorphism $a \to (\rho_\alpha(a), \rho_\beta(a))$. Consider the intermediate ring $(A, P_A) \times \mathbb{R}$ of (A, P) and $\sigma(A, P_A)$: Its real spectrum has 3 points, hence it cannot be epimorphic over (A, P_A). $\square$

The proposition suggests that, when looking for positive results, we restrict our attention to monomorphisms $f : (A, P_A) \to (C, P_C)$ for which $Sper(f)$ is a homeomorphism. Since (C, P_C) is an intermediate ring between (A, P_A) and $\sigma(A, P_A)$, Theorem 12.13 says that (C, P_C) is even contained in $\rho(A, P_A)$. Thus, the largest choice we have for the subporing (C, P_C) of $\sigma(A, P_A)$ is $\rho(A, P_A)$. Therefore, the following question is studied:

(14.4) Given a reduced poring (A, P), when is it true that every subporing $(B, Q) \subseteq \rho(A, P)$ containing $im(\rho_{(A,P)})$ is epimorphic over (A, P)?

First we attack the question for the case that (A, P) is a partially ordered field. In this case $\rho(A, P) = \sigma(A, P)$ and $\rho(A, P) = \prod\limits_{\alpha \in Sper(A,P)} \rho(\alpha)$ if the real spectrum is finite. Similarly, $\varphi(A, P) = \nu(A, P)$ and $\varphi(A, P) = \prod\limits_{\alpha \in Sper(A,P)} \kappa(\alpha)$ if $Sper(A, P)$ is finite.

Proposition 14.5 Let (A, P) be a partially ordered field. If every subporing $(B, Q) \subseteq \varphi(A, P)$ containing $im(\varphi_{(A,P)})$ is epimorphic over (A, P) then (A, P) is a SAP–field. Conversely, if (A, P) is a SAP–field then every subporing of $\rho(A, P)$ containing $im(\rho_{(A,P)})$ is epimorphic over (A, P).

Before proving the proposition we need to recall some terminology. The real spectrum of the partially ordered field (A, P) is the same as its *space of orderings* with the *Harrison topology*(cf. [75], Lecture 3; [76], §1; or [92] for a large number of more recent references). If the subbasis of $Sper(A, P)$ consisting of the sets $pos(a), a \in A$, is even a *basis* of the Harrison topology then (A, P) is called a *SAP–field* ([77], §17; [97], §9). There is another characterization of SAP-fields using fans: A subset $X \subseteq Sper(A, P)$ is called a *fan* if and only if for any three total orders $\alpha_1, \alpha_2, \alpha_3 \in X$ the product of the corresponding sign functions $sign_{\alpha_i} : A^\times \to \mathbb{Z}^\times$ is the sign function of some element of X. Any subset $X \subseteq Sper(A, P)$ containing at most two elements is a fan. Such fans are called *trivial*; a fan is *nontrivial* if it contains more than two elements. The notion of fans was first introduced in [14], p. 166. Fans have found a large number of applications in the areas of quadratic forms, real algebra and real geometry. We mention [3], [4] and [92] as general references. The connection with SAP-fields is that (A, P) is a SAP-field if and only if $Sper(A, P)$ does not contain any nontrivial fans ([14], §§4,5).

As in [3], Definition III 3.12, and [92], p. 162, the notion of a fan can be extended from partially ordered fields to porings. So, now let (A, P) be a reduced poring and let $q \subseteq (A, P)$ be a convex prime ideal. We denote the factor poring by $(\overline{A}, \overline{P})$ and its partially ordered quotient field by (K, Q). If $\pi_q : (A, P) \to (K, Q)$ is the canonical homomorphism then $Sper(\pi_q)$ is a homeomorphism onto the subspace $supp_{(A,P)}^{-1}(q) \subseteq Sper(A, P)$. A subset $X \subseteq Sper(A, P)$ is a fan if there is some q such that X is the image of a fan of (K, Q) under $Sper(\pi_q)$. The fan $X \subseteq Sper(A, P)$ is *nontrivial* if and only if the corresponding fan of (K, Q) is nontrivial.

Proof of Proposition 14.5 First suppose that (A, P) is not an SAP–field. Then there is a nontrivial fan $\{\tau_1, \tau_2, \tau_3, \tau_4\} \subseteq Sper(A, P)$ and $Sper(A, P)$ can be partitioned into two constructible sets X, Y with $\tau_1 \in X$, $\tau_2, \tau_3, \tau_4 \in Y$. The characteristic function χ_X of X belongs to $\varphi(A, P)$. We define $B = A[\chi_X]$ and equip it with the restriction Q of the partial order of $\varphi(A, P)$. The idempotent element χ_X splits $\prod\limits_{\alpha \in Sper(A,P)} \kappa(\alpha)$ into a direct product

$$\prod_{\alpha \in Sper(A,P)} \kappa(\alpha) = \prod_{\alpha \in X} \kappa(\alpha) \times \prod_{a \in Y} \kappa(\alpha).$$

Let p_X and p_Y be the projections onto the components. Since χ_X belongs to both B and $\varphi(A, P)$ these porings also split as $(B, Q) = (B_X, Q_X) \times (B_Y, Q_Y)$ and $\varphi_{(A,P)} = \varphi(A, P)_X \times \varphi(A, P)_Y$. As rings, B_X and B_Y are both isomorphic to A. The positive cones are $Q_X = \bigcap\limits_{\alpha \in X} \alpha$ and $Q_Y = \bigcap\limits_{\alpha \in Y} \alpha$. The real spectrum of (B, Q) is the disjoint union of the real spectra of (B_X, Q_X) and (B_Y, Q_Y). In $Sper(B_X, Q_X)$ there is a point β_1 defined by

$$B_X \to \prod_{\alpha \in X} \kappa(\alpha) \xrightarrow{p_{\tau_1}} \kappa(\tau_1).$$

In $Sper(B_Y, Q_Y)$ there are three points $\beta_2, \beta_3, \beta_4$ defined similarly by τ_2, τ_3, τ_4. Since $\{\tau_1, \tau_2, \tau_3, \tau_4\}$ is a fan of (A, P), each one of these total orders contains the intersection of the other three. In the underlying field A of (B_Y, Q_Y), the prime cones β_i coincide with the total orders τ_i, $i = 2, 3, 4$. Therefore the partial order Q_Y of B_Y is contained in the partial order $\tau_2 \cap \tau_3 \cap \tau_4$ of A. But then τ_1 also contains Q_Y, hence τ_1 defines a point $\beta_1' \in Sper(B_Y, Q_Y)$. Since $\beta_1 \neq \beta_1'$ and both restrict to $\tau_1 \in Sper(A, P)$ we see that $(A, P) \to (B, Q)$ is not an epimorphism.

Now we assume that (A, P) is a SAP–field. Let $(A, P) \xrightarrow{f} (B, Q) \xrightarrow{g} \rho(A, P)$ be an intermediate extension with g an embedding. The diagram

$$
\begin{array}{ccccc}
\rho_{(A,P)} : (A, P) & \xrightarrow{\;\;f\;\;} & (B, Q) & \xrightarrow{\;\;g\;\;} & \rho(A, P) \\
{\scriptstyle \rho_{(A,P)}} \downarrow & & {\scriptstyle \rho_{(B,Q)}} \downarrow & & \| \\
id: \rho(A, P) & \xrightarrow{\;\;\rho(f)\;\;} & \rho(B, Q) & \xrightarrow{\;\;\rho(g)\;\;} & \rho(A, P)
\end{array}
$$

shows that $Sper(g)$ is a homeomorphism onto a closed subspace of $Sper(B, Q)$. If it is surjective then f is epimorphic. Assume that this is not the case and pick some $\beta \in Sper(B,Q)\backslash im(Sper(g))$; define $\alpha = Sper(f)(\beta)$, $\gamma = Sper(gf)^{-1}(\alpha)$, $\delta = Sper(g)(\gamma)$. Since $im(Sper(g))$ is a closed subspace of $Sper(B,Q)$, β ist not a specialization of δ, hence there is some $b \in B$ with $b(\beta) < 0$ and $b(\delta) \geq 0$. The sets $U = pos(-g(b))$ and $V = Sper(\rho(A,P))\backslash U$ form a constructible partition of $Sper(\rho(A,P))$. If $U = \emptyset$ then $g(b) \geq 0$ in $\rho(A,P)$, hence $b \geq 0$ in (B,Q). This is impossible, hence $U \neq \emptyset$. Because $\delta \in im(Sper(g))$ and $b(\delta) \geq 0$ it is clear that also $V \neq \emptyset$. The real spectra of (A,P) and $\rho(A,P)$ are homeomorphic, hence there is some $a \in A$ with $U = pos(-(gf)(a))$, $V = pos((gf)(a))$ (here we use the SA–property). Then $g(b \cdot f(a)) \geq 0$ in $\rho(A,P)$, hence $b \cdot f(a) \geq 0$ in (B,Q). This implies $(b \cdot f(a))(\beta) \geq 0$, whereas $b(\beta) < 0$ and $f(a)(\beta) = a(\alpha) = (gf)(a)(\gamma) = f(a)(\delta) > 0$. This contradiction ends the proof. $\square$

Our next step is to extend Proposition 14.5 from partially ordered fields to reduced porings. This is the main result of the section.

Theorem 14.6 For the reduced poring (A,P) the following conditions are equivalent:

(a) If $f : (A,P) \to (B,Q)$ is a monomorphism and $g : (B,Q) \to \rho(A,P)$ is an embedding such that $\rho_{(A,P)} = gf$ then f is an epimorphism.

(b) If $f : (A,P) \to (B,Q)$ is a monomorphism and $g: (B,Q) \to \varphi(A,P)$ is an embedding such that $\varphi_{(A,P)} = gf$ then f is an epimorphism.

(c) In $Sper(A,P)$ every fan is trivial.

Proof The implication (a) $\Rightarrow$ (b) is trivial. For the proof of (b) $\Rightarrow$ (c), suppose that (c) fails, i.e., there is a nontrivial fan $F \subseteq Sper(A,P)$. By the defintion of a fan, the elements of F all have the same support, say p. Let $K = A_p/pA_p$ and $X = Sper(A,P) \cap Sper(K) \subseteq Sper(K)$, let $Q = \bigcap_{\tau \in X} \tau$. Then (K,Q) is not a SAP-field. The prime ideal p is convex in (A,P), and we can form the factor poring $(\overline{A},\overline{P}) = (A/p, P+p/p)$. With $S = A - p$ the underlying ring of the poring $(\overline{A}_S,\overline{P}_S)$ is K, the partial order is Q (because of [77], Theorem 1.6). We apply the reduced f–ring reflector φ to the homomorphisms

$$(A,P) \xrightarrow{\pi} (\overline{A},\overline{P}) \xrightarrow{i} (K,Q)$$

and obtain the diagram

$$(A,P) \xrightarrow{\ \pi\ } (\overline{A},\overline{P}) \xrightarrow{\ i\ } (K,Q)$$

$$\varphi_{(A,P)} \downarrow \qquad\quad \varphi_{(\overline{A},\overline{P})} \downarrow \qquad\qquad\qquad \downarrow \varphi_{(K,Q)}$$

$$\varphi(A,P) \xrightarrow{\varphi(\pi)} \varphi(\overline{A},\overline{P}) \xrightarrow{\varphi(i)} \varphi(K,Q).$$

Note that $\varphi(\pi)$ is surjective since φ is H–closed (Proposition 10.8 (a)) and that $\varphi(i)_S \colon \varphi(\overline{A},\overline{P})_S \to \varphi(K,Q)$ is an isomorphism (Propositon 11.2 and Example 11.13). By the proof of Proposition 14.5 there is an intermediate extension

$$\varphi_{(K,Q)} \colon (K,Q) \xrightarrow{\ f\ } (K[c], P_{K[c]}) \xrightarrow{\ g\ } \varphi(K,Q)$$

such that g is an embedding and f is not epimorphic. Since $\varphi(i)_S$ is an isomorphism there are $b \in \varphi(\overline{A},\overline{P})$ and $s \in S$ with $\varphi(i)_S(\frac{b}{s}) = c$. Let $\overline{s} \in K$ be the canonical image of s. Then $\overline{s} \in K^\times$, hence c can be replaced by $c\overline{s}^2$. Now $\varphi(i)(bs) = c\overline{s}^2$ with $bs \in \varphi(\overline{A},\overline{P})$. Thus, we may assume that $c = \varphi(i)(b)$ for a suitable $b \in \varphi(\overline{A},\overline{P})$. We define $\overline{B} = \overline{A}[b] \subseteq \varphi(\overline{A},\overline{P})$ and let $P_{\overline{B}}$ be the restriction of the partial order of $\varphi(\overline{A},\overline{P})$. Then $\varphi(i)$ restricts to $m \colon (\overline{B}, P_{\overline{B}}) \to (K[c], P_{K[c]})$. Now $B = \varphi(\pi)^{-1}(\overline{B}) \subseteq \varphi(A,P)$ together with the positive cone $P_B = B \cap \varphi^+(A,P)$ is an intermediate extension of (A,P) and $\varphi(A,P)$ which is embedded in $\varphi(A,P)$. Let $l \colon (B,P_B) \to (\overline{B}, P_{\overline{B}})$ be the restriction of $\varphi(\pi)$. Altogether we have the diagram

$$(A,P) \xrightarrow{\ \pi\ } (\overline{A},\overline{P}) \xrightarrow{\ i\ } (K,Q)$$

$$h \downarrow \qquad\qquad \overline{h} \downarrow \qquad\qquad\qquad \downarrow f$$

$$(B,P_B) \xrightarrow{\ l\ } (\overline{B}, P_{\overline{B}}) \xrightarrow{\ m\ } (K[c], P_{K[c]})$$

$$k \downarrow \qquad\qquad \overline{k} \downarrow \qquad\qquad\qquad \downarrow g$$

$$\varphi(A,P) \xrightarrow{\varphi(\pi)} \varphi(\overline{A},\overline{P}) \xrightarrow{\varphi(i)} \varphi(K,Q)$$

in which every horizontal map is an epimorphism on the level of underlying rings, hence also in **POR/N**. Thus, if h is an epimorphism then so is f. But this contradicts the choice of f, and we see that h is not epimorphic.

(c) $\Rightarrow$ (a) We assume that (A,P) satisfies condition (c), but fails (a). Pick an intermediate extension

$$\rho_{(A,P)} \colon (A,P) \xrightarrow{\ f\ } (B,Q) \xrightarrow{\ g\ } \rho(A,P)$$

with g an embedding and f not an epimorphism. Then $im(Sper(g)) \subseteq Sper(B,Q)$ is a proper closed subspace. Pick any

$$\beta \in Sper(B,Q) \setminus im(Sper(g))$$

and let $\alpha = Sper(f)(\beta)$, $\gamma = Sper(gf)^{-1}(\alpha)$. We define $S = A \setminus supp(\alpha)$ and form porings of quotients with denominators in S:

$$(A_S, P_S) \xrightarrow{f_S} (B_S, Q_S) \xrightarrow{g_S} \rho(A,P)_S.$$

It is clear that both f_S and g_S are monomorphisms in **R**, hence also in **POR/N**. Since **RCR** is quotient–closed (Proposition 12.6; or: [111], Theorem I 3.19, Theorem I 4.8; [116], §6) it follows that $\rho(A_S, P_S) \cong \rho(A,P)_S$ (Proposition 11.2). Because g is an embedding it is clear that so is g_S. The real spectra of the three porings can be identified canonically with the subspaces

$$
\begin{aligned}
U &= \{\delta \in Sper(A,P); \ S \cap supp(\delta) = \emptyset\}, \\
V &= \{\varepsilon \in Sper(B,Q); \ f(S) \cap supp(\varepsilon) = \emptyset\}, \\
W &= \{\zeta \in Sper(\rho(A,P)); \ (gf)(S) \cap supp(\zeta) = \emptyset\}.
\end{aligned}
$$

Since $f^{-1}(supp(\beta)) = supp(\alpha)$ one sees that $\beta \in V \setminus Sper(g)(W)$. Thus, f_S is not an epimorphism.

These considerations show that it suffices to deal with the following special case: A is a local $\mathbb{Q}$-algebra with maximal ideal $M = supp(\alpha) = f^{-1}(supp(\beta))$.

In B there are finitely many $b_1, \ldots, b_n$ such that: $b_i(\beta) > 0$ for each i, and for every $\zeta \in Sper(\rho(A,P))$ there is some i with $g(b_i)(\zeta) \leq 0$. The subporing $(C, C \cap Q)$, $C = f(A)[b_1, \ldots, b_n]$, of (B,Q) is not epimorphic over (A,P) either. For, the prime cone $\beta \cap C$ is not the restriction of a prime cone of $\rho(A,P)$. Thus, one may further assume that B is of finite type over A.

The extension $\rho_{(A,P)}$ is always semi–integral ([27], p. 126). For, if $a \in \rho(A,P)$ then for every closed point $\eta \in Sper(A,P)$ there is some $a_\eta \in A$ with $a_\eta(\eta) > |a|(\eta)$ (since $\rho(\eta)$ is the real closure of $\kappa(\eta)$ and A/η is cofinal in $\kappa(\eta)$). The open sets

$$U_\eta = \{\vartheta \in Sper(A,P); \ a_\eta(\vartheta) > |a|(\vartheta)\}$$

cover the compact space of closed points of $Sper(A,P)$, hence there is a finite subcover

$$Sper(A,P) = U_{\eta_1} \cup \ldots \cup U_{\eta_r}.$$

If one defines $b = \sum_i (1 + a_{\eta_i})^2$ then $-\rho_{(A,P)}(b) < a < \rho_{(A,P)}(b)$ in $\rho(A,P)$.

By [27], Proposition 6.4.1, this proves that $\rho_{(A,P)}$ is semi–integral, hence the same is true for the embedding g. The prime ideal $supp(\beta) \subseteq (B,Q)$ is convex. Therefore there exists a convex prime ideal $q \subseteq \rho(A,B)$ with $g^{-1}(q) = supp(\beta)$ ([27], Proposition 6.4.2). Let $\zeta \in Sper\,\rho(A,P)$ be the unique prime cone with support q, let $\delta = Sper(gf)(\zeta)$. From

$$supp(\delta) = (gf)^{-1}(supp(\zeta)) = f^{-1}(supp(\beta)) = M$$

we conclude that δ and ζ are closed points in their spectra. Also, q is a maximal ideal of $\rho(A,P)$, $\rho(A,P)/q$ is the real closure of $\kappa(\delta) = A/\delta$. The ring $B/supp(\beta)$ is an intermediate domain of $\kappa(\delta)$ and $\rho(\delta) = \rho(A,P)/q$. Since $\kappa(\delta) \subseteq \rho(\delta)$ is an algebraic extension, $B/supp(\beta)$ is also a field. Hence $supp(\beta) \subseteq B$ is a maximal ideal, and β is a closed point of $Sper(B,Q)$. As $Sper(g)(\gamma)$ is contained in a closed subspace of $Sper(B,Q)$ which does not include the point β, none of β and $Sper(g)(\gamma)$ can be a specialization of the other one.

We define

$$
\begin{aligned}
X &= supp^{-1}(\{M\}) \subseteq Sper(A,P), \\
Z &= Sper(gf)^{-1}(X) \subseteq Sper(\rho(A,P)), \\
Y &= Sper(g)(Z) \subseteq Sper(B,Q).
\end{aligned}
$$

Each one of these sets is closed in its ambient spectrum. If we reduce A modulo the convex prime ideal M and denote the induced partial order of $\overline{A} = A/M$ by $\overline{P}$ then X may be identified with $Sper(\overline{A}, \overline{P})$. This is a partially ordered field. By hypothesis there is no nontrivial fan in its real spectrum. The canonical homomorphism $k \colon \rho(A,P) \to \rho(\overline{A}, \overline{P})$ is surjective; $Sper(\rho(\overline{A}, \overline{P}))$ is homeomorphic to Z, hence the two spaces will be identified. Finally, we define $\overline{B} = k(B)$. This is an intermediate ring of $\overline{A}$ and $\rho(\overline{A}, \overline{P})$. The kernel of $k|_B$ is convex in (B,Q), so Q induces a partial order $\overline{Q}$ on $\overline{B}$. Also, $\overline{B}$ carries the restriction of the partial order of $\rho(\overline{A}, \overline{P})$, which is denoted by Q'. It is clear that $\overline{Q} \subseteq Q'$, but it is not a priori clear whether the two are equal. To clarify the relationship between them, first note that Proposition 14.5 can be applied to

$$(\overline{A}, \overline{P}) \xrightarrow{\;\overline{f}\;} (\overline{B}, Q') \xrightarrow{\;\overline{g}\;} \rho(\overline{A}, \overline{P}),$$

hence $Sper(\overline{g})$ is surjective. On the other hand, $(gf)^{-1}(q) = M$ implies that $\delta \in X$ and $\zeta \in Z$. From $ker(k) \subseteq supp(\zeta) = q$ it follows that $ker(k|_B) \subseteq supp(\beta)$, hence $\rho_\beta \colon (B,Q) \to \rho(\beta)$ factors through $(\overline{B}, \overline{Q})$.

But then β induces a point $\overline{\beta} \in Sper(\overline{B}, \overline{Q})$ which does not belong to Y, i.e., $Y = Sper(\overline{B}, Q')$ is a proper closed subspace of $Sper(\overline{B}, \overline{Q})$, and $\overline{Q} \subset Q'$.

Every element of $\rho(\overline{A}, \overline{P})$ is integral over $\overline{A}$. For, $\overline{A}$ is a field, $\rho(\overline{A}, \overline{P})$ is a von Neumann regular ring, and for each prime ideal $r \subseteq \rho(\overline{A}, \overline{P})$ the residue field $\rho(\overline{A}, \overline{P})/r$ is an algebraic extension of $\overline{A}$. If $a \in \rho(\overline{A}, \overline{P})$ then $a(r)$ is a root of some monic polynomial with coefficients in $\overline{A}$. By a compactness argument one finds a monic polynomial $F \in \overline{A}[T]$ with $F(a) = 0$, hence a is integral over $\overline{A}$. It follows that the intermediate ring $\overline{B}$ is also integral over $\overline{A}$. Thus, being an A–algebra of finite type, $\overline{B}$ is even a finite $\overline{A}$–algebra. Moreover $\overline{B}$ is reduced, hence it is a finite product of fields, $\overline{B} = \prod_{i=1}^{r} \overline{B}_i$, with each $\overline{B}_i$ a finite algebraic extension of $\overline{A}$. Both $(\overline{B}, \overline{Q})$ and $(\overline{B}, Q')$ are direct products of partially ordered fields: $(\overline{B}, \overline{Q}) = \prod(\overline{B}_i, \overline{Q}_i)$, $(\overline{B}, Q') = \prod(\overline{B}_i, Q'_i)$. The direct decompositions induce partitions of the real spectra:

$$Sper(\overline{B}, \overline{Q}) = \dot{\bigcup} Sper(\overline{B}_i, \overline{Q}_i),$$
$$Sper(\overline{B}, Q') = \dot{\bigcup} Sper(\overline{B}_i, Q'_i).$$

Moreover, $Sper(\overline{B}, Q') \cap Sper(\overline{B}_i, \overline{Q}_i) = Sper(\overline{B}_i, Q'_i)$. As $Q' \not\subseteq \overline{\beta}$ there is some $\overline{b} \in Q' \backslash \overline{\beta}$, i.e., $\overline{b}(\beta) < 0$ and $\overline{b}(\lambda) \geq 0$ for each $\lambda \in Sper(\overline{B}, Q')$. The components of $\overline{b}$ in the various $\overline{B}_i$ are denoted by $\overline{b}_i$. We may assume that $\overline{\beta} \in Sper(\overline{B}_1, \overline{Q}_1)$. Then $\overline{b}_1 \neq 0$, and for each $\lambda \in Sper(\overline{B}_1, Q'_1)$ we have $\overline{b}_1(\lambda) > 0$ (since $\overline{B}_1$ is a field). We set $\overline{c} = (\overline{b}_1, \overline{1}, \ldots, \overline{1}) \in \overline{B}$ and note that $\overline{c}(\overline{\beta}) < 0$ and $\overline{c}(\lambda) > 0$ for each $\lambda \in Sper(\overline{B}, Q')$. Pick some representative $c \in B$ for $\overline{c}$. The set

$$U = \{\lambda \in Sper(\rho(A, P)); \, g(c)(\lambda) < 0\}$$

is open constructible in $Sper(\rho(A, P))$ and the topological closure $\overline{U}$ has empty intersection with Z (by the choice of c). Thus, for every $\lambda \in U$ the maximal specialization $\overline{\lambda}$ of λ in $Sper(\rho(A, P))$ does not belong to Z. With $\overline{\mu} = Sper(gf)(\overline{\lambda})$ we note that $supp(\overline{\mu}) \subset M$ and $A/\overline{\mu}$ is cofinal in $\rho(\overline{\mu}) = \rho(\overline{\lambda})$. The ring $A/\overline{\mu}$ is local with maximal ideal $M/supp(\overline{\mu})$, but the maximal ideal is not convex. Therefore there are $0 < v < u$ in $A/\overline{\mu}$ such that $v \notin M/supp(\overline{\mu})$ and $u \in M/supp(\overline{\mu})$. Since $v \in (A/\overline{\mu})^{\times}$ we may multiply by v^{-1} and get $0 < 1 < uv^{-1} \in M/supp(\overline{\mu})$. We conclude that for every $\lambda \in U$ there is some $0 \leq a_\lambda \in M$ with

$$0 \leq -g(c)(\overline{\lambda}) \; < \; (gf)(a_\lambda)(\overline{\lambda}),$$
$$1 \; < \; (gf)(a_\lambda)(\overline{\lambda}).$$

For $\lambda \in U$ the set

$$U_\lambda = \{\kappa \in U; \; 0 < -g(c)(\kappa) < (gf)(a_\lambda)(\kappa), 1 < (gf)(a_\lambda)(\kappa)\}$$

is open and constructible and contains λ, i.e., these sets form an open constructible cover of U. By compactness there is a finite subcover, say $U = \bigcup_{i=1}^{s} U_{\lambda_i}$. Now $a = a_{\lambda_1} + \ldots + a_{\lambda_s} \in M$, and $(g(c) + (gf)(a))(\kappa) > 0$ for every $\kappa \in U$. The definition of U shows that $(g(c) + (gf)(a))(\kappa) \geq 0$ for all $\kappa \in Sper(\rho(A, P))$, i.e., $g(c) + (gf)(a) \geq 0$ in $\rho(A, P)$. This implies that $c + f(a) \in \overline{Q}$ since g is an embedding. In particular,

$$0 \leq (c + f(a))(\beta) = c(\beta).$$

This contradicts the choice of c, and the proof is finished. $\qquad \square$

First we record a couple of immediate applications of the theorem:

Corollary 14.7 Suppose that the reduced poring (A, P) has an injective support function from the real spectrum to the prime spectrum. Then the equivalent conditions of Theorem 14.6 are satisfied. This is the case, for example, if (A, P) is an f–ring. $\qquad \square$

Corollary 14.8 Let (A, P) be a reduced poring, let $f : (A, P) \to (B, Q)$ be a monomorphism into a von Neumann regular f–ring, let $g : (B, Q) \to \sigma(A, P)$ be an embedding such that $gf = \sigma_{(A,P)}$. Then f is epimorphic.

Proof We use the factorization

$$f : (A, P) \xrightarrow{\nu_{(A,P)}} \nu(A, P) \xrightarrow{\overline{f}} (B, Q)$$

supplied by the universal property of the reflection $\nu_{(A,P)}$. Since $\sigma(A, P)$ is canonically isomorphic to $\rho(\nu(A, P))$ (Example 12.15) we conclude that $\overline{f}$ is an epimorphism. The reflection $\nu_{(A,P)}$ is epimorphic as well, hence so is f. $\qquad \square$

Corollary 14.7 and Proposition 9B.1 together show that there are a lot of cases where the infimum of two monoreflectors is easy to determine:

Corollary 14.9 Suppose that $r : \mathbf{POR}/\mathbf{N} \to \mathbf{D}$ and $s : \mathbf{POR}/\mathbf{N} \to \mathbf{E}$ are monoreflectors stronger than the reduced f-ring reflector φ. If a least one of r and s is weaker than the real closure reflector ρ then $r \wedge s$ is determined by $(r \wedge s)(C) = r(C) \cap s(C)$. $\square$

The following example gives an application of Theorem 14.6 which is not connected with f-rings.

Example 14.10 *Coordinate rings of curves over real closed fields*
Let A be the coordinate ring of some affine curve over a real closed field R, let P be the weak partial order of A. The prime spectrum of A has dimension 1. A maximal ideal $p \subseteq A$ is the support of at most one prime cone since the residue field is either R or its algebraic closure. If $p \subseteq A$ is a minimal prime ideal then A_p/pA_p is a field of transcendence degree 1 over R, hence it has the SAP ([97], Theorem 9.4). We see that condition (b) of Theorem 14.6 is satisfied, hence every intermediate ring of (A, P) and $\rho(A, P)$ is epimorphic over (A, P). $\square$

Example 14.11 *Differentiable semi–algebraic functions*
In Example 10.20 the subporing $(A_n, P_n) \subseteq \sigma(F(n))$ of r-times continuously differentiable semi-algebraic functions on R_0^n was introduced. Obviously, (A_n, P_n) is also a subporing of $\rho(F(n))$. Let $f_n : F(n) \to (A_n, P_n), n \in \mathbb{N}$, be the family of canonical homomorphisms. Theorem 14.6 can be used to show that every f_n is an epimorphism. As a first step we introduce an intermediate poring between $F(n)$ and (A_n, P_n): Let

$$
\begin{aligned}
A'_n &= F(n)[(a \vee 0)^{r+1}; \; a \in F(n)] \subseteq \rho(F(n)); \\
P'_n &= A'_n \cap \rho^+(F(n)).
\end{aligned}
$$

Note that each of the functions $(a \vee 0)^{r+1}$ is r-times continuously differentiable. Therefore the subring of $\rho(F(n))$ generated by these functions is contained in A_n, and (A'_n, P'_n) is a subporing of (A_n, P_n). Or, to put it differently: (A_n, P_n) is an intermediate poring of (A'_n, P'_n) and $\rho(F(n))$. Let $f'_n : F(n) \to (A'_n, P'_n)$ be the canonical homomorphism. It follows from [117], Example 2.4, in connection with Theorem 5.2 that f'_n is an epimorphism. It is also shown in [117], Example 2.4, that the support map $supp_{(A'_n, P'_n)}$ is injective. By Corollary 14.7 this implies that (A_n, P_n) is epimorphic over (A'_n, P'_n), hence f_n is an epimorphism.

With Example 10.20 we conclude that the family (A_n, P_n), $n \in \mathbb{N}$, of porings of r–times continuously differentiable functions defines an H–closed monoreflector of **POR/N**.   □

We close the section with a discussion of rings of Nash functions. In Example 10.19 we introduced the Nash reflector. Another approach to Nash functions was first introduced in [105] and [2]. We discuss a modification here which includes partial orders. First, recall the universal property of the Nash sheaf without partial orders:

A local ring is said to be *strictly real*, if it is Henselian and the residue field is real closed (*real closed local rings* in the terminology of [2], §2; [105], Definition 4.1). With every ring A one associates its *affine real scheme* ([103], Defintion 3.1). It is a locally ringed space whose underlying space is $Sper(A)$ and whose structure sheaf is the sheaf $\mathcal{N}_A$ of Nash functions ([2], §4; [105], §2). The stalks are strictly real. There is a canonical morphism $g\colon (Sper(A), \mathcal{N}_A) \to Spec(A)$ into the affine scheme belonging to A. If $f\colon (X, \mathcal{F}) \to Spec(A)$ is any morphism of locally ringed spaces and if the stalks of $\mathcal{F}$ are strictly real then f factors uniquely as $f = gh$, where $h\colon (X, \mathcal{F}) \to (Sper(A), \mathcal{N}_A)$ is a morphism of locally ringed spaces.

To extend these concepts to reduced porings we introduce the category **POLRSP** of locally ringed spaces with reduced partially ordered stalks. Its objects are locally ringed spaces whose rings of sections and stalks are reduced porings. Restriction maps are order preserving. The morphisms are those maps of locally ringed spaces whose homomorphisms between rings of sections and stalks are order preserving. Given a reduced poring (A, P) we define an object $Spec(A, P)$ of **POLRSP**. As a ringed space it is the affine scheme $Spec(A)$. For any basic open set $D(s) = \{p \in Spec(A); s \notin p\}$ the ring of sections A_s is partially ordered by P_s. One checks easily that, given some object $(X, \mathcal{F})$ of **POLRSP** and some homomorphism $f\colon (A, P) \to \Gamma(\mathcal{F})$ of reduced porings, there is a unique morphism $\varphi\colon (X, \mathcal{F}) \to Spec(A, P)$ such that $\Gamma(\varphi)\colon (A, P) \to \Gamma(\mathcal{F})$ coincides with f. In **POLRSP** we define **SRPOLRSP** to be the full subcategory of locally rings spaces with reduced strictly real partially ordered stalks. The affine scheme belonging to a real closed ring is a member of this category.

Let $i\colon Sper(A, P) \to Sper(A)$ be the inclusion. We define a sheaf $\mathcal{N}_{(A,P)}$ on $Sper(A, P)$: As a first step we consider the presheaf $i^{-1}\mathcal{N}_A$ on $Sper(A, P)$. For every basic open set $U \subseteq Sper(A, P)$ we define $\mathcal{N}'_{(A,P)}$ to be the image of $(i^{-1}\mathcal{N}_A)(U)$ in $\rho(A, P)_u$, where $U = \{\alpha \in Sper(A, P);$

$u(\alpha) > 0\}$, $u \geq 0$. Let $\mathcal{N}_{(A,P)}$ be the sheaf on $Sper(A, P)$ associated with the presheaf $\mathcal{N}'_{(A,P)}$. Thus, $\mathcal{N}_{(A,P)}$ is a subsheaf of the structure sheaf of the affine scheme $Spec(\rho(A, P))$. The stalks are strictly real local rings. There is a canonical morphism $g\colon (Sper(A, P), \mathcal{N}_{(A,P)}) \to Spec(A, P)$. If we equip $\mathcal{N}_{(A,P)}$ with the weakest partial order such that g is a morphism in the category **POLRSP** then $(Sper(A, P), \mathcal{N}_{(A,P)})$ is a member of **SRPOLRSP**. One can check that g has the following universal property: If $(X, \mathcal{F}) \in ob(\mathbf{SRPOLRSP})$ and $f\colon (X, \mathcal{F}) \to Spec(A, P)$ is a morphism in **POLRSP** then there is a unique morphism $h\colon (X, \mathcal{F}) \to (Sper(A, P), \mathcal{N}_{(A,P)})$ such that $f = gh$.

In particular, consider a real ring A and let P be its weak partial order, hence $Sper(A, P) = Sper(A)$. Assume furthermore that every strict real localization A_α ([2], §3; [105], §3) is real. Then, as locally ringed spaces, $(Sper(A), \mathcal{N}_A)$ and $(Sper(A, P), \mathcal{N}_{(A,P)})$ coincide (cf. [37], Chapitre V, §2).

It is a major open question whether the construction of the Nash sheaf is idempotent ([102]), i.e., whether $(Sper(N_A), \mathcal{N}_{N_A})$ is isomorphic to $(Sper(A), \mathcal{N}_A)$. The same question can be asked about the partially ordered Nash sheaf: Setting $N_{(A,P)} = \Gamma(\mathcal{N}_{(A,P)})$, is the canonical morphism $(Sper(N_{(A,P)}), \mathcal{N}_{N(A,P)}) \to (Sper(A, P), \mathcal{N}_{(A,P)})$ an isomorphism? Note that the canonical morphism arises in the following way: The canonical morphism $g\colon (Sper(A, P), \mathcal{N}_{(A,P)}) \to Spec(A, P)$ factors as $g = g_2 g_1$ where $g_1\colon (Sper(A, P), \mathcal{N}_{(A,P)}) \to Spec(N_{(A,P)})$ is induced by the identity homomorphism $N_{(A,P)} \to \Gamma(\mathcal{N}_{(A,P)})$ and $g_2\colon Spec(N_{(A,P)}) \to Spec(A, P)$ is induced by the homomorphism $(A, P) \to N_{(A,P)}$ between porings of global sections. Let $h\colon (Sper(N_{(A,P)}), \mathcal{N}_{N(A,P)}) \to Spec(N_{(A,P)})$ be the canonical morphism belonging to $N_{(A,P)}$. Then $g_2 h = gf$ with a unique morphism $f\colon (Sper(N_{(A,P)}), \mathcal{N}_{N(A,P)}) \to (Sper(A, P), \mathcal{N}_{(A,P)})$. This is the canonical morphism referred to in the above question. The universal property of the morphism h implies that $g_1 = hk$ with a unique morphism $k\colon (Sper(A, P), \mathcal{N}_{(A,P)}) \to (Sper(N_{(A,P)}), \mathcal{N}_{N(A,P)})$. Because of

$$gfk = g_2 hk = g_2 g_1 = g$$

the universal property of g implies that $fk = id$. So the question about the idempotency of the construction is reduced to showing that $kf = id$. It is true that

(14.12) $$g_2 hkf = g_2 g_1 f = g_2 h.$$

If this implies that $hkf = h$ then the universal property of the morphism

h yields the desired identity $kf = id$. However, it is not clear at all whether g_2 can always be cancelled from (14.12). A sufficient condition for cancellation is that g_2 is a monomorphism in **POLRSP**, or, equivalently, that the corresponding homomorphism $\Gamma(g_2)\colon (A, P) \to N_{(A,P)}$ between rings of global sections is an epimorphism in **POR/N**. Since (by construction) $N_{(A,P)}$ is an intermediate ring between (A, P) and $\rho(A, P)$ we can use Theorem 14.6 to obtain idempotency at least in some cases:

Proposition 14.13 The canonical homomorphism

$$f\colon (Sper(N_{(A,P)}), \mathcal{N}_{N_{(A,P)}}) \to (Sper(A, P), \mathcal{N}_{(A,P)})$$

is an isomorphism if $Sper(A, P)$ does not contain any nontrivial fan. This is the case if (A, P) is a reduced f–ring, or if (A, P) is the partially ordered coordinate ring of some affine curve over a real closed field.   □

We conjecture that the construction of the sheaf of partially ordered rings of Nash functions is always idempotent. Obviously, Proposition 14.13 is only a small step in this direction. Theorem 14.6 shows that obstructions to the idempotency property are connected with the fans of $Sper(A, P)$.

15 Essential monoreflectors

A monoreflector is said to be *essential* if every reflection morphism is an essential monomorphism. Our main result about essential monoreflectors is Theorem 15.5. It says that the real closure reflector is the strongest essential monoreflector of $\mathbf{FR/N}$, another remarkable extremal property of real closures.

Following [121], Definition 8.1, we call a monomorphism $f : C \to C'$ *essential* if a composition $f'f : C \to C' \to C''$ is monomorphic only if f' is a monomorphism. The definition applies to arbitrary categories. Since we consider many different subcategories $\mathbf{C} \subseteq \mathbf{POR/N}$ it is necessary to exercise some caution when using the term essential. In general, $\mathbf{C}$–morphisms that are essential in $\mathbf{POR/N}$ can only be expected to be essential in $\mathbf{C}$ if $\mathbf{C}$–monomorphisms are injective. Note that this is the case whenever $\mathbf{C}$ is reflective in $\mathbf{POR/N}$ (Proposition 2.5). On the other hand, a $\mathbf{C}$–morphism may be essential in $\mathbf{C}$ without being so in $\mathbf{POR/N}$. But the next result shows that this is not a problem in the situation which is most interesting for us.

Lemma 15.1 Let $r\colon \mathbf{POR/N} \to \mathbf{D}$ be a monoreflector. If $f\colon (A, P_A) \to (B, P_B)$ is an essential morphism in $\mathbf{D}$ then it is also essential as a $\mathbf{POR/N}$–morphism.

Proof Let $g\colon (B, P_B) \to (C, P_C)$ be a $\mathbf{POR/N}$–morphism such that gf is monomorphic in $\mathbf{POR/N}$. Then $r_{(C,P_C)}gf$ is also monomorphic in $\mathbf{POR/N}$. This is even a $\mathbf{D}$–morphism, hence it is a $\mathbf{D}$–monomorphism. Since f is essential in $\mathbf{D}$ one concludes that $r_{(C,P_C)}g$ is monomorphic in $\mathbf{D}$. By Proposition 2.5 every $\mathbf{D}$–monomorphism is a $\mathbf{POR/N}$–monomorphism. Being an initial factor of the monomorphism $r_{(C,P_C)}g$, g is monomorphic. $\square$

Suppose that $\mathbf{D} \subseteq \mathbf{C} \subseteq \mathbf{POR/N}$ are monoreflective subcategories; let r be the reflector of $\mathbf{POR/N}$ belonging to $\mathbf{D}$. The lemma says that the reflection maps of $r|_{\mathbf{C}}$ are essential in $\mathbf{C}$ if and only if they are essential in $\mathbf{POR/N}$. So, the subtle dependence of the notion of essentiality on the subcategory is of no importance for us.

In [117], §2, the question is discussed under what conditions about a poring (A, P) the real closure reflection $\rho_{(A,P)}$ is an essential monomorphism: If the support map of (A, P) is assumed to be injective then $\rho_{(A,P)}$ is essential if and only if the set of those prime cones whose supports are minimal prime ideals is dense in $Sper(A, P)$ ([117], Proposition 2.13). This is the case, for example, if the support map is a homeomorphism onto its image ([117], Corollary 2.14). All reduced f–rings satisfy this condition, so the restriction of ρ to $\mathbf{FR/N}$ is an essential monoreflector. The same is true for the restriction of ρ to any intermediate category $\mathbf{FR/N} \supseteq \mathbf{C} \supseteq \mathbf{RCR}$.

Before starting the systematic investigation of essential monoreflectors we mention that essential monomorphisms may be helpful in identifying epimorphic extensions.

Proposition 15.2 Suppose that $r: \mathbf{POR/N} \to \mathbf{D}$ is a monoreflector. Let $f : (A, P) \to (B, Q)$, $g : (B, Q) \to r(A, P)$ be monomorphisms such that $r_{(A,P)} = gf$. If the reflection $r_{(B,Q)}$ is an essential extension then f is an epimorphism.

Proof Consider the diagram

$$
\begin{array}{ccccc}
r_{(A,P)}: (A, P) & \xrightarrow{\;f\;} & (B, Q) & \xrightarrow{\;g\;} & r(A, P) \\
{\scriptstyle r_{(A,P)}}\big\downarrow & & {\scriptstyle r_{(B,Q)}}\big\downarrow & & \big\downarrow{\scriptstyle r_{r(A,P)}} \\
r(r_{(A,P)}): r(A, P) & \xrightarrow{\;r(f)\;} & r(B, Q) & \xrightarrow{\;r(g)\;} & r(r(A, P)).
\end{array}
$$

By Proposition 2.4 it suffices to show that $r(f)$ is an epimorphism. First note that $r(r_{(A,P)})$ is an isomorphism. Since $r_{(B,Q)}$ is essential and $r(g)r_{(B,Q)} = r_{r(A,P)}g$ is a monomorphism, $r(g)$ is a monomorphism as well. But then $r(g)$ is even an isomorphism, being a final monomorphic factor of the isomorphism $r(r_{(A,P)})$. Thus, $r(f)$ is also an isomorphism, in particular, $r(f)$ is epimorphic as claimed. $\qquad\square$

Suppose that $\mathbf{C}$ is an intermediate category of $\mathbf{RCR}$ and $\mathbf{POR/N}$ and that $\mathbf{C}$ is H–closed and quotient–closed in $\mathbf{POR/N}$. We look for conditions about $\mathbf{C}$ implying that the reflector $\rho|_{\mathbf{C}}$ is essential. Note that the representable reflection of (A, P) is the image of $\rho_{(A,P)}$ together with the restriction of the partial order of $\rho(A, P)$. Thus, being H–closed, $\mathbf{C}$ contains the representable reflection of each of its objects.

Proposition 15.3 For the category $\mathbf{C}$ the following conditions are equivalent:

(a) Every partially ordered field in $\mathbf{C}$ is totally ordered.
(b) For every $\mathbf{C}$–object (A, P) the support map $supp_{(A,P)} : Sper(A, P) \to Spec(A, P)$ is injective.

Proof (a) $\Rightarrow$ (b) Suppose that $\alpha, \beta \in Sper(A, P)$ have the same support p. Since $\mathbf{C}$ is H–closed the domain $(\bar{A}, \bar{P}) = (A/p, P + p/p)$ belongs to $\mathbf{C}$; since $\mathbf{C}$ is quotient–closed the partially ordered quotient field (K, T) of $(\bar{A}, \bar{P})$ belongs to $\mathbf{C}$, hence it is totally ordered. In $Sper(K, T)$ there are prime cones γ and δ which are induced by α and β. Since T is a total order, γ and δ coincide, hence so do α and β. – (b) $\Rightarrow$ (a) If (A, P) is a partially ordered field then (b) implies $|Sper(A, P)| \leq |Spec(A, P)| = 1.\square$

Proposition 15.4 For the category $\mathbf{C}$ the following conditions are equivalent:

(a) $\rho|_{\mathbf{C}}$ is essential.
(b) Every representable integral domain in $\mathbf{C}$ is totally ordered.
(c) For every $\mathbf{C}$–object the support map is a homeomorphism onto its image.

Proof The implication (c) $\Rightarrow$ (a) follows from [117], Corollary 2.14. To prove (a) $\Rightarrow$ (b), let (A, P) be a representable integral domain in $\mathbf{C}$. If (A, P) is not totally ordered then there are two generic points $\alpha, \beta \in Sper(A, P)$. Since some generic point has support (0) we may assume that this is the case for α. There is some $a \in \rho(A, P)$ with $a(\alpha) = 0$ and $a(\beta) \neq 0$. (Note, that we may identify the real spectra of (A, P) and $\rho(A, P)$ by Proposition 12.5.) Thus, $a \neq 0$ and it belongs to the kernel of the evaluation map $e_\alpha : \rho(A, P) \to \rho(\alpha) : f \to f(\alpha)$. So, e_α is not a monomorphism, but $e_\alpha \rho_{(A,P)}$ is monomorphic. This contradicts the hypothesis that $\rho_{(A,P)}$ is essential. – (b) $\Rightarrow$ (c) Partially ordered fields are always representable ([77], Theorem 1.6), so Proposition 15.3 shows that support maps are always injective in $\mathbf{C}$. Since they are morphisms of spectral spaces it only remains to show that $\alpha \subseteq \beta$ whenever $supp(\alpha) \subseteq supp(\beta)$. To prove this, let $p = supp(\alpha)$ and form the factor poring $(\bar{A}, \bar{P}) = (A/p, P + p/p)$. There are unique points

$\overline{\alpha}, \overline{\beta} \in Sper(\overline{A}, \overline{P})$ corresponding to α and β in $Sper(A, P)$. Moreover, $supp(\overline{\alpha}) = (0) \subseteq supp(\overline{\beta})$. So, we may assume that (A, P) is a domain and that $supp(\alpha) = (0)$. The representable poring reflection does not change the real spectrum. Therefore it suffices to discuss the case that (A, P) is representable. Then P is a total order according to (b). Also, α is a total order and $P \subseteq \alpha$. This implies $\alpha = P \subseteq \beta$, and the proof is finished. □

There is a largest subcategory $\mathbf{C} \subseteq \mathbf{POR/N}$ satisfying the hypotheses and the equivalent conditions of Proposition 15.4, namely the full subcategory of the reduced porings whose support map is a homeomorphism onto the image. We know that $\mathbf{FR/N}$ is contained in $\mathbf{C}$; that the containment is proper follows from [117], Example 2.4. Another example of a poring belonging to $\mathbf{C}$, but not to $\mathbf{FR/N}$, is the subring $\mathbb{Z}(1,1)[(2,0)] \subseteq \mathbb{Z} \times \mathbb{Z}$ with the restriction of the componentwise lattice–order of $\mathbb{Z} \times \mathbb{Z}$. We do not know whether $\mathbf{C}$ is a reflective subcategory of $\mathbf{POR/N}$, but we would rather conjecture that it is not. We do not even know whether $\mathbf{C}$–monomorphisms are injective.

So far we have only discussed the question for which categories the restriction of the real closure reflector is essential. Of course, there are other interesting essential reflectors, for example the representable poring reflector $Rep: \mathbf{POR/N} \to \mathbf{REPPOR}$. Individual porings may also have essential extensions that are much larger than the real closure, e.g. [117], Theorem 4.4, Proposition 4.5, shows that every reduced f–ring has an epimorphic hull. The epimorphic hull is always a ring of semi–algebraic functions, hence in general it is a proper extension of the real closure. However, the construction of epimorphic hulls is not a functorial operation, so it does not contribute to the investigation of reflectors.

In [55], 6.4, it is shown that in $\mathbf{FR/N}$ there is a smallest essential monoreflective subcategory. Our main result about essential reflectors identifies this subcategory as $\mathbf{RCR}$

Theorem 15.5 Let $r : \mathbf{FR/N} \to \mathbf{D}$ be an essential monoreflector. Then r is weaker than the real closure reflector ρ.

Proof First we form the supremum of r and ρ. As in Section 9A we construct a functorial extension t together with a natural transformation

$Id_{\mathbf{FR/N}} \to t$: We define $t(A, P)$ to be the direct limit of the sequence

$$(A, P) \longrightarrow r(A, P) \longrightarrow \rho(r(A, P)) \longrightarrow r(\rho(r(A, P))) \to \dots .$$

The definition of t on the morphisms uses the universal property of the reflectors and the direct limit. The canonical map $t_{(A,P)}: (A, P) \to t(A, P)$ is the component of the natural transformation at (A, P). Every map in the sequence is essential, the underlying ring of $t(A, P)$ in the direct limit of the underlying rings in the sequence. Thus, $t_{(A,P)}$ is essential, i.e., the functorial extension operator is essential. The reflector $r \vee \rho$ is the direct limit of the transfinite sequence of iterations of the functorial extension opertor (cf. Section 9A). Therefore each reflection morphism $(r \vee \rho)_{(A,P)}$ is essential. So, by forming the supremum every essential monoreflector of $\mathbf{FR/N}$ yields another one which is even stronger than ρ. To finish, we shall show that there is no essential monoreflector that is properly stronger than ρ. So, assume that $\mathbf{D} \subset \mathbf{RCR}$. There is some real closed ring (A, P) outside $\mathbf{D}$. Then $r(A, P)$ is a real closed intermediate ring of (A, P) and $\sigma(A, P)$, properly containing (A, P). By Theorem 12.13 it suffices to show that $Sper(r_{(A,P)})$ is a homeomorphism.

First note that $r_{(A,P)}$ is an epimorphism, hence $Sper(r_{(A,P)})$ is bijective and is a homeomorphism with respect to the constructible topology. To show that it is also a homeomorphism for the spectral topology we have to prove that, given $\gamma, \delta \in Sper(r(A, P))$, $\alpha = r(A, P)^{-1}(\gamma) \subseteq r_{(A,P)}^{-1}(\delta) = \beta$ implies $\gamma \subseteq \delta$. Using the commutative diagram

$$
\begin{array}{ccc}
(A, P) & \xrightarrow{\ \pi_\alpha\ } & A/\alpha \\
{\scriptstyle r_{(A,P)}}\Big\downarrow & & \Big\downarrow{\scriptstyle r_{A/\alpha}} \\
r(A, P) & \xrightarrow{\ r(\pi_\alpha)\ } & r(A/\alpha).
\end{array}
$$

one reduces the problem to dealing with the real closed domain A/α. Thus, we may assume that (A, P) is a real closed domain and that $\alpha \in Sper(A, P)$ is the generic point. Assume by way of contradiction that $\gamma \not\subseteq \delta$. Since $Sper(r_{(A,P)})$ is bijective and $\alpha \subseteq \beta$ it follows that γ and δ are incomparable. Any two generic points $\gamma_0 \subseteq \gamma$ and $\delta_0 \subseteq \delta$ must be incomparable as well. In fact, $r_{(A,P)}^{-1}(\gamma_0) \subseteq \alpha$ implies that $r_{(A,P)}^{-1}(\gamma_0) = \alpha$ (since α is generic), hence $\gamma_0 = \gamma$. From $\delta_0 \neq \gamma$ it follows that $\alpha \subset r_{(A,P)}^{-1}(\delta_0)$. Replacing δ by δ_0 we may assume that δ is another generic point of $Sper(r(A, P))$. Then there is some $a \in r(A, P)$ with $a(\gamma) = 0$, $a(\delta) \neq 0$. The evaluation $\rho_\gamma: r(A, P) \to \rho(\gamma)$ is not a monomorphism since $a \in ker(\rho_\gamma)$ and $a \neq 0$. On the other hand, the composition $\rho_\gamma r_{(A,P)} = \rho_\alpha$ (recall that $\rho(\alpha) \cong \rho(\gamma)$) is a monomorphsim as (A, P) is a domain and $\alpha \in Sper(A, P)$ is generic. But then $r_{(A,P)}$ is not essential, i.e., the reflector r is not essential. This contradiction finishes the proof. $\square$

16 Reflections of totally ordered fields

In the next section we shall try to classify the reflective subcategories of **VNRFR**. Since von Neumann regular rings are very close to fields, our approach will be to find out to what an extent a reflector of **VNRFR** is determined by its effect on totally ordered fields. As a first step, the present section is devoted to reflectors of the category **TOF** of totally ordered fields. The main results are characterizations of the epireflective (Theorem 16.6) and monoreflective (Theorem 16.7) subcategories of **TOF**. The first question is whether a reflector of **POR** or **POR/N** can always be restricted to a reflector of **TOF**. In complete generality we do not know the answer. But, in the case of epireflectors we have a positive result which is based on the following characterization of epimorphisms whose domain is a totally ordered field:

Proposition 16.1 Let **C** be either **POR** or **POR/N**, let (A, P) be a totally ordered field. If $f : (A, P) \to (B, Q)$ is an epimorphism in **C** then (B, Q) is a totally ordered field. It is the zero ring or an intermediate field of (A, P) and its real closure. Conversely, if (B, Q) is the zero ring or an intermediate field of (A, P) and its real closure then $f : (A, P) \to (B, Q)$ is an epimorphism in **C**.

Proof First let $\mathbf{C} = \mathbf{POR/N}$. Then $|Sper(B, Q)| \leq |Sper(A, P)| = 1$ (Theorem 5.2). If $Sper(B, Q) = \emptyset$ then (B, Q) is the zero field. (Otherwise (B, Q) has a convex prime ideal, which must belong to the image of the support map, cf. Proposition 4.3.) Now suppose that $Sper(B, Q) = \{\beta\}$. The intersection of the supports of all prime cones of (B, Q) is the nilradical (section 4), hence $supp(\beta) = (0)$. This means that (B, β) is a totally orderd domain. By Theorem 5.2 it is an intermediate domain of the field (A, P) and its real closure. The real closure is an algebraic extension, hence every intermediate ring is a field. Altogether this shows that (B, β) is a totally ordered field. The partial order Q of B is an intersection of total orders ([73], p. 2, Theorem 1). Because B has only one total order containing Q, namely β, it follows that $Q = \beta$ is a total order.

Now suppose that $\mathbf{C} = \mathbf{POR}$. If (B, Q) is the zero ring then there is nothing to prove. So assume that $1 \neq 0$ in B. Let $(\overline{B}, \overline{Q})$ be the reduction of (B, Q) modulo the nilradical. The canonical map $\pi \colon (B, Q) \to (\overline{B}, \overline{Q})$ is surjective, hence it is epimorphic in **POR**. Then πf is epimorphic in **POR** and also in **POR/N**. The first part of the proof shows that $(\overline{B}, \overline{Q})$ is an intermediate field of (A, P) and its real closure. We see that (B, Q) is a local ring of dimension 0. By Zorn's Lemma there is a subfield $K \subseteq B$ which contains $f(A)$ and is maximal with this property. We claim that $K \subseteq B \xrightarrow{\pi} \overline{B}$ is an isomorphism. If not, then there is some $\overline{b} \in \overline{B} \backslash \pi(K)$. Pick any representative $b \in B$ of $\overline{b}$. Since $\pi(K) \subseteq \overline{B}$ is an algebraic extension, b satisfies a relation

$$b^n + a_{n-1} b^{n-1} + \ldots + a_0 \in Nil(B),$$

where $F = T^n + a_{n-1}T^{n-1} + \ldots + a_0 \in K[T]$ is an irreducible polynomial. Because of $F(b)^r = 0$ for some $1 \leq r \in \mathbb{N}$ the ring $K[b]$ is a local Artinian ring; its residue field is isomorphic to $\pi(K)[\overline{b}]$. By [23], Chapitre IX, §3, No. 2, Proposition 1, there is a field of representatives $L \subseteq K[b]$ which contains K. But then $K \subset L \subseteq B$ violates the maximality of K. This contradiction shows that $\pi(K) = \overline{B}$.

If the field K is endowed with the partial order $K \cap Q$ then $\pi|_K \colon (K, K \cap Q) \to (\overline{B}, \overline{Q})$ is a homomorphism of partially ordered fields. We claim that $\pi|_K$ is an isomorphism. The map is an isomorphism of the underlying fields. It remains to show that $K \cap Q$ is a total order. So, pick $a \in K$ with $\pi(a) > 0$. Then there are some $b \in B^\times \cap Q$ and some $x \in Nil(B)$ sucht hat $a = b + x$. Note that $c^{-1} \in Q$ for every $c \in B^\times \cap Q$. For, $(c^{-1})^2 \in Q$ by definition of a partial order, hence $c^{-1} = c(c^{-1})^2 \in Q$. Because of $a = (ab^{-1})b = (1 + xb^{-1})b$ it suffices to show that $1 + y \in Q$ for all $y \in Nil(B)$. Pick $r \in \mathbb{N}$ such that $y^{2^r} = 0$. Then

$$1 = 1 - y^{2^r} = (1 - y^2)(1 + y^2)(1 + y^4) \cdot \ldots \cdot (1 + y^{2^{(r-1)}}) \in Q$$

and $1 + y^{2^s} \in B^\times \cap Q$ for all $s = 1, \ldots, r-1$. This implies that $(1 + y^{2^s})^{-1} \in Q$ for all s, hence $1 - y^2 \in Q$. Finally

$$1 + y = \frac{1}{2}(2 + 2y) = \frac{1}{2}((1 + y)^2 + (1 - y^2)) \in Q$$

proves the claim. Let $i \colon (K, K \cap Q) \to (B, Q)$ be the inclusion. Because f is an epimorphism, the identity

$$id_{(B,Q)}f = f = i(\pi|_K)^{-1}\pi f$$

implies that $id_{(B,Q)} = i(\pi|_K)^{-1}\pi$. But then π is an isomorphism, and (B, Q) is a totally ordered field, as claimed.

For the converse, suppose that (B, Q) is the zero field or that (B, Q) is an intermediate field of (A, P) and its real closure $\rho(A, P)$. If $\mathbf{C} = \mathbf{POR/N}$ then Theorem 5.2 shows that f is an epimorphism. Now suppose that $\mathbf{C} = \mathbf{POR}$. If (B, Q) is the zero field there is nothing to prove. We are left with the case that $\rho_{(A,P)} = if$ where $i\colon (B, Q) \to \rho(A, P)$ is the inclusion. Let $g, h\colon (B, Q) \to (C, P_C)$ be $\mathbf{POR}$–morphisms with $gf = hf$. If $\pi\colon (C, P_C) \to (\overline{C}, \overline{P_C})$ is the reduction modulo the nilradical then $\pi g f = \pi h f$. Since πg and πh are $\mathbf{POR/N}$–morphisms and f is epimorphic in $\mathbf{POR/N}$ it follows that $\pi g = \pi h$. Assume by way of contradiction that $g \neq h$. Then there is some $b \in B$ with $g(b) \neq h(b)$, and we may assume that $B = A[b]$. We consider (C, P_C) as an (A, P)–algebra via $gf = hf$. Because it suffices to deal with the subalgebra generated by $g(b)$ and $h(b)$ over (A, P) we assume that (C, P_C) coincides with this subalgebra. Thus, (C, P_C) is a local (because of $\pi g = \pi h$) Artinian (A, P)–algebra, and both g and h map (B, Q) onto maximal subfields of (C, P_C). The element $x = g(b) - h(b) \in C$ is nilpotent. Let $F \in A[X]$ be the minimal polynomial of b over A, say of degree d. Then

$$
\begin{aligned}
0 &= F(g(b)) = F(h(b) + x) \\
&= F(h(b)) + F_1(h(b))x + \ldots + F_d(h(b))x^d \\
&= F_1(h(b))x + \ldots + F_d(h(b))x^d
\end{aligned}
$$

with polynomials $F_i \in A[X]$ of degree $d - i$. Since F is the minimal polynomial of b it follows that $F_i(b) \neq 0$ for $i = 1, \ldots, d$. This means that $F_i(b) = h(F_i(b)) \in C^\times$, hence

$$
x = -F_1(h(b))^{-1}(F_2(h(b)) + \ldots + F_d(h(b))x^{d-2})x^2.
$$

Because x is nilpotent we conclude that $x = 0$ and $g(b) = h(b)$, contrary to the assumption. $\square$

Corollary 16.2 Let $\mathbf{C} = \mathbf{POR}$ or $= \mathbf{POR/N}$, let $r\colon \mathbf{C} \to \mathbf{D}$ be an epireflector. Then r restricts to an epireflector $r'\colon \mathbf{TOF} \to \mathbf{D} \cap \mathbf{TOF}$. If $\mathbf{C} = \mathbf{POR}$ and r is a monoreflector then so is r'. If $\mathbf{C} = \mathbf{POR/N}$ then r is a monoreflector if and only if so is r'.

Proof The first statement is obvious from the proposition; the second one is trivial. For the third claim note that $\mathbf{RCF} \subseteq \mathbf{D} \cap \mathbf{TOF}$ if r' is a monoreflector. $\square$

Here are a few examples illustrating the corollary:

Example 16.3 *Archimedean f–rings*
A poring (A, P) is called archimedean if $na < b$ for all $n \in \mathbb{Z}$ implies
$a = 0$ ([47], p. 24). For f–rings this condition is equivalent to: $na \leq b$
for all $n \in \mathbb{N}$ implies $a \leq 0$ (loc. cit.). The category of archimedean
reduced f–rings is denoted by **ARCHFR/N**. It is claimed that this
is an epireflective subcategory of **FR/N**. To define the reflection of an
FR/N–object (A, P) recall that **FR/N** is co–wellpowered. Therefore
the epimorphisms with domain (A, P) and codomain in **ARCHFR/N**
are represented by a set; let $\{f_i \colon (A, P) \to (A_i, P_i); i \in I\}$ be a representing set. The product $\prod(A_i, P_i)$ belongs to **ARCHFR/N**. Let
$f \colon (A, P) \to \prod(A_i, P_i)$ be the homomorphism defined by the f_i's. The
image $(f(A), f(P)) \subseteq \prod(A_i, P_i)$ is an archimedean reduced f–ring, the
homomorphism $f \colon (A, P) \to (f(A), f(P))$ is an epimorphism. One checks
easily that f has the universal property (2.1), hence f is the reflection of
(A, P) for a reflector $r \colon$ **FR/N** $\to$ **ARCHFR/N**. Since f is surjective
it is an epireflector. The composition of r with the reduced f–ring reflector φ is an epireflector $r\varphi \colon$ **POR/N** $\to$ **ARCHFR/N**. The restriction
to **TOF** reflects archimedean totally ordered fields onto themselves, all
others onto the zero field. $\qquad\qquad\qquad\qquad\qquad\qquad\qquad\qquad\qquad\square$

The construction of section 9D can be used to determine a large class
of epireflective subcategories of **POR/N** that are not monoreflective:
Let X be a class of totally ordered fields containing every totally ordered subfield of each of its elements. Examples of such classes are the
archimedean fields; the totally ordered fields having at most a given cardinality κ; the totally ordered fields that are isomorphic to some subfield
of a fixed totally ordered field. Let **Y** $\subseteq$ **TOF** be the full subcategory
whose objects do not belong to X. Note that **Y** satisfies condition (9D.1).
Let $r \colon$ **TOF** $\to$ **TOF** be the identity reflector; let **E** $\subseteq$ **TOF** be the reflective subcategory consisting only of the zero field, let $s \colon$ **TOF** $\to$ **E** be
the reflector. The construction of section 9D can be applied using the
reflectors r and s and the subcategory **Y** to determine a new reflector t of
TOF. If (K, T) is a totally ordered field then $t(K, T) = (K, T)$ whenever
$(K, T) \in X$, $t(K, T) = 0$ for all $(K, T) \notin X$. This reflector is the restriction of an epireflector of **POR/N**. Namely, let **D** $\subseteq$ **POR/N** be the
full subcategory whose objects are the sub–porings $(A, P) \subseteq \prod_{i \in I}(K_i, T_i)$,

where $(K_i, T_i)_{i \in I}$ is any family in X. It follows immediately from Theorem 8.3 that this subcategory is epireflective. Moreover, it is clear that $ob(\mathbf{D} \cap \mathbf{TOF}) = X \cup \{0\}$.

Example 16.4 *Strongly archimedean porings*
In the construction just explained, let X be the class of all archimedean totally ordered fields, i.e., the totally ordered subfields of $\mathbb{R}$. The objects of the category $\mathbf{D}$ are called *strongly archimedean porings*. They are the sub–porings of rings $\mathbb{R}^M$, M varying in the category of sets. Both $\mathbf{D}$ and $\mathbf{FR/N}$ are epireflective subcategories of $\mathbf{POR/N}$, hence so is $\mathbf{D} \cap \mathbf{FR/N}$. It is obvious that every object of this category is archimedean. Therefore $\mathbf{D} \cap \mathbf{FR/N} \subseteq \mathbf{ARCHFR/N}$ is a subcategory. It is well–known that it is a proper subcategory. To see a specific example of a reduced archimedean f–ring that is not strongly archimedean, let $\mathcal{E}$ be the set of finite subsets of $[0,1]$. For every $E \in \mathcal{E}$ let $(A_E, P_E) = C([0,1]\backslash E, \mathbb{R})$. If $E \subseteq E'$ then the restriction provides a homomorphism $\rho_{E'E} : (A_E, P_E) \to (A_{E'}, P_{E'})$. Let (A, P) be the direct limit of the rings (A_E, P_E) with these transition maps. It is easy to check that (A, P) belongs to $\mathbf{ARCHFR/N}$, but has no homomorphism into $\mathbb{R}$ whatsoever. $\square$

Example 16.5 *The real closure*
The real closure reflector $\rho : \mathbf{POR/N} \to \mathbf{RCR}$ is a monoreflector, hence ρ restricts to a monoreflector on $\mathbf{TOF}$. The restriction maps any totally ordered field to its real closure. This is the strongest monoreflector of $\mathbf{TOF}$. Therefore, if $r : \mathbf{POR/N} \to \mathbf{D}$ is any monoreflector stronger than ρ then r restricts to the same monoreflector of $\mathbf{TOF}$. This applies to $\sigma : \mathbf{POR/N} \to \mathbf{SAFR}$, for example. $\square$

The next two results provide characterizations of the epireflective subcategories and the monoreflective subcategories of $\mathbf{TOF}$. We shall consider the following condition about a subcategory $\mathbf{X} \subseteq \mathbf{TOF}$: The subcategory is *closed under intersections* if, given any totally ordered field (K, T) and any nonempty family of totally ordered subfields $(K_i, T_i) \in ob(\mathbf{X})$, the intersection $\bigcap_{i \in I} (K_i, T_i)$ is also a member of $\mathbf{X}$.

Theorem 16.6 For a subcategory $\mathbf{X} \subseteq \mathbf{TOF}$ the following conditions (a) and (b) are equivalent:

(a) **X** is epireflective.
(b)   (i) **X** contains the zero field; and
      (ii) given any extension $f\colon (K, P_K) \to (L, P_L)$ in **TOF** with
      $(L, P_L) \in ob(\mathbf{X})$ there is a commutative diagram

$$
\begin{array}{ccc}
(K, P_K) & \xrightarrow{\;f\;} & (L, P_L) \\
\Big\downarrow{\scriptstyle g} & & \Big\downarrow{\scriptstyle g'} \\
(M, P_M) & \xrightarrow{\;f'\;} & (N, P_N)
\end{array}
$$

      of extensions in **TOF** where $(M, P_M) \in ob(\mathbf{X})$ and g is an
      epimorphism.
      (iii) **X** is closed under intersections.

Proof (a) $\Rightarrow$ (b) Let r be the reflector belonging to **X**. The reflection of the zero field is the zero field, hence (i) holds. For (ii), let
$\overline{f}\colon r(K, P_K) \to (L, P_L)$ be the unique extension of f. Then one obtains the desired diagram by setting $(M, P_M) = r(K)$, $(N, P_N) = (L, P_L)$
and $g = r_{(K,P_K)}$, $g' = id_{(L,P_L)}$, $f' = \overline{f}$. To prove (iii), let (L, S) be a
totally ordered field, let $(K_i, T_i)_{i \in I}$ be a family of totally ordered subfields belonging to **X** and set $(K, T) = \bigcap_{i \in I}(K_i, T_i)$. It is claimed that
$r_{(K,T)}\colon (K, T) \to r(K, T)$ is an isomorphism. For each i the inclusion
$j_i\colon (K, T) \to (K_i, T_i)$ extends uniquely to $\overline{j_i}\colon r(K, T) \to (K_i, T_i)$. Let
$k_i\colon (K_i, T_i) \to (L, S)$ be the inclusion. Since the compositons $k_i \overline{j_i} r_{(K,T)}$
all coincide with the inclusion $l\colon (K, T) \to (L, S)$ and since $r_{(K,T)}$ is an
epimorphism, the homomorphisms $k_i \overline{j_i}$ also coincide; we set $k = k_i \overline{j_i}$.
From $k(r(K, T)) \subseteq K_i$ it follows that $k(r(K, T)) = K$, i.e., $r_{(K,T)}$ is surjective. This means that $r_{(K,T)}$ is even an isomorphism, and the proof of
this implication is finished.
(b) $\Rightarrow$ (a) We define the reflection of every totally ordered field and
show that it has the universal property (2.1). If there is no homomorphism $f\colon (K, T_K) \to (L, T_L)$ into a nonzero field belonging to **X** then
$r(K, T_K)$ is defined to be the zero field. The universal property holds
trivially. Now suppose that $f\colon (K, T_K) \to (L, T_L)$ is a homomorphism
into a nonzero field of **X**. Because of (ii) we may assume that f is
an epimorphism. Using the unique embedding $i\colon (L, T_L) \to \rho(K, T_K)$
over (K, T_K) we consider (L, T_L) as a subfield of the real closure of
(K, T_K). Inside the real closure, the intersection of all totally ordered
subfields belonging to **X** is also a member of **X** (condition (iii)). We define $r(K, T_K)$ to be this intersection, $r_{(K,T_K)}\colon (K, T_K) \to r(K, T_K)$ and

$m\colon r(K, T_K) \to \rho(K, T_K)$ are the inclusion homomorphisms. For the universal property, let $g\colon (K, T) \to (M, T_M)$ be any homomorphism into a member of **X**. We consider a diagram as in (ii), extended by real closures:

$$
\begin{array}{ccc}
(K, T_K) & \xrightarrow{\ g\ } & (M, T_M) \\
h \downarrow & & \downarrow h' \\
(N, T_N) & \xrightarrow{\ g'\ } & (P, T_P) \\
j \downarrow & & \downarrow k \\
\rho(K, T_K) & \xrightarrow{\ l\ } & \rho(P, T_P).
\end{array}
$$

Inside $\rho(P, T_P)$, $kh'(M, T_M) \cap lj(N, T_N) \in ob(\mathbf{X})$ (condition (iii)). Since l is an isomorphism onto its image, $(R, T_R) = l^{-1}(kh'(M, T_M) \cap lj(N, T_N)) \cong kh'(M, T_M) \cap lj(N, T_N)$ is a totally ordered subfield of $\rho(K, T_K)$ belonging to **X**; hence (R, T_R) contributes to the intersection defining $r(K, T_K)$. If $n\colon kh'(M, T_M) \to (M, T_M)$ is the inverse homomorphism of kh' then the composition

$$
p = nlm\colon r(K, T_K) \to (M, T_M)
$$

is well-defined and $pr_{(K, T_K)} = g$. Finally, p is unique because $r_{(K, T_K)}$ is an epimorphism. $\qquad\square$

Theorem 16.7 For a subcategory $\mathbf{X} \subseteq \mathbf{TOF}$ the following conditions are equivalent:

(a) **X** is monoreflective in **TOF**.
(b)   (i) **X** contains the zero field; and
    (ii) **X** contains every real closed field; and
    (iii) **X** is closed under intersections.

Proof (a) $\Rightarrow$ **(b)** A monoreflective subcategory is also epireflective, hence (i) and (iii) follow from Theorem 16.6. For (ii), let R be a real closed field and consider the reflection $r_R\colon R \to r(R)$. This is a monomorphism and an epimorphism. But the real closed field R does not have any proper epimorphic extensions, hence r_R is an isomorphism, and $R \in ob(\mathbf{X})$. – **(b)** $\Rightarrow$ **(a)** It is clear from Theorem 16.6 that **X** is epireflective in **TOF**. To show that it is monoreflective, pick any totally ordered field (K, T) and consider the real closure $\rho_{(K,T)}\colon (K, T) \to \rho(K, T)$. Because $\rho(K, T) \in$

$ob(\mathbf{X})$, $\rho_{(K,T)}$ extends uniquely to $\overline{\rho_{(K,T)}}\colon r(K,T) \to \rho(K,T)$. But then $r_{(K,T)}$ is an initial factor of the monomorphism $\rho_{(K,T)} = \overline{\rho_{(K,T)}}r_{(K,T)}$, hence $r_{(K,T)}$ is a monomorphism. $\qquad\qquad\qquad\qquad\qquad\qquad\qquad\square$

As a consequence of Theorem 16.7 we note that for any class Y of totally ordered fields there is a smallest monoreflective subcategory of **TOF** that contains Y. For, let X_0 be the class of totally fields that are either real closed or are isomorphic to some element of Y or are the zero field. The class X_0 is enlarged to a class X satisfying the equivalent conditions of Theorem 16.7 by adding all totally ordered fields that are isomorphic to an intersection $\bigcap\limits_{(K,T)\in\mathcal{K}} (K,T)$ where $\mathcal{K} \subseteq X_0$ is a set of subfields of some real closed field.

We give a few examples of reflective subcategories of **TOF**:

Example 16.8 *Epireflective subcategories of* **TOF**

(a) The subcategory of all archimedean fields.
(b) Given some cardinal κ, the subcategory of all totally ordered fields of cardinality at most κ.
(c) Given some totally ordered field (K,T), the subcategory of all totally ordered fields isomorphic to a subfield of (K,T), together with the zero field. $\qquad\qquad\qquad\qquad\qquad\qquad\square$

Example 16.9 *Monoreflective subcategories of* **TOF**

(a) The category of real closed fields is monoreflective in **TOF**.
(b) A field is *euclidean* if the squares are the positive cone of a total order ([13]). The euclidean fields form a monoreflective subcategory of **TOF**.
(c) A field is called *pythagorean* if every sum of squares is a square. In **TOF**, the subcategory of totally ordered pythagorean fields is monoreflective.
(d) The same is true for the class of totally ordered fields whose natural valuation ring is Henselian. This is a consequence of Hensel's Lemma (cf. [104], p. 185, Theorème 4) or of the universal property of Henselizations (cf. [46], p. 131).
(e) Finally, let X be the class of totally ordered fields whose natural valuation ring is Henselian with real closed residue field. These

fields are studied under the name *HRC–fields* in [15] and under the name *almost real closed field* in [42]. The members of X can also be characterized as those totally ordered fields which have some convex Henselian subring with real closed residue field. The full subcategory of **TOF** with objects in X is monoreflective. The reflection of a totally ordered field (K, T) with natural valuation v is the maximal unramified extension of K in its real closure (cf. [46], §19). $\qquad\qquad\square$

We finish the section with model theoretic considerations arising in connection with H–closed monoreflectors of **POR/N**.

Theorem 16.10 Suppose that $r: \mathbf{POR/N} \to \mathbf{D}$ is an H–closed monoreflector; let $r': \mathbf{TOF} \to \mathbf{D'}$ be its restriction. Then the objects of $\mathbf{D'}$ are an elementary class of fields. It can be axiomatized by the usual axioms of totally ordered fields together with a set of formulas of the following type:

$$\forall\, X_1, \dots, X_n\, [\, \exists\, Y : \phi(X_1, \dots, X_n, Y)\ \&$$
$$\forall\, Z, T\, (\phi(X_1, \dots, X_n, Z)\ \&\ \phi(X_1, \dots, X_n, T) \to Z = T)],$$

where $\phi(X_1, \dots, X_n, Y)$ is a functional formula.

Proof Theorem 10.6 shows that $ob(\mathbf{D})$ is the injectivity class of the reflection $r_{F(\mathbb{N})} : F(\mathbb{N}) \to r(F(\mathbb{N}))$. For each $f \in r(F(\mathbb{N}))$ we choose some quantifier free functional formula defining f, say $\Theta_f = \Theta_f(X_1, \dots, X_{n(f)}, Y)$. Inside the class of totally ordered fields we consider the elementary subclass X consisting of all fields satisying the sentences

$$\phi_f \ \equiv\ \forall\, X_1, \dots, X_{n(f)}\, [\, \exists\, Y : \Theta_f(X_1, \dots, X_{n(f)}, Y)\ \&$$
$$\forall\, Z, T\, (\Theta_f(X_1, \dots, X_{n(f)}, Z)\ \&\ \Theta_f(X_1, \dots, X_{n(f)}, T) \to Z = T)]$$

Now suppose that $(K, T) \in ob(\mathbf{D'})$, pick one of the formulas Θ_f and let $a_1, \dots, a_{n(f)} \in K$. The free object $F(\mathbb{N})$ is the polynomial ring $\mathbb{Z}[T_1, T_2, \dots]$ with its weak partial order. Let $h : F(\mathbb{N}) \to (K, T)$ be any homomorphism with $h(T_i) = a_i$ for $i = 1, \dots, n(f)$. The universal property of reflections supplies a unique extension $\bar{h} : r(F(\mathbb{N})) \to (K, T)$ of h. Define $b = \bar{h}(f)$. Then $\Theta_f(a_1, \dots, a_{n(f)}, b)$ holds in (K, T). Since Θ_f is functional, b is the only element $c \in (K, T)$ with $\Theta_f(a_1, \dots, a_{n(f)}, c)$.

Thus, (K,T) is a model of $\{\phi_f; f \in r(F(\mathbb{N}))\}$. Conversely, let (K,T) be a totally ordered field which is a model of this set of sentences. We must show that every homomorphism $h\colon F(\mathbb{N}) \to (K,T)$ extends uniquely to $\overline{h}\colon r(F(\mathbb{N})) \to (K,T)$. If R is the real closure of (K,T) then there is a unique extension $\overline{h}\colon r(F(\mathbb{N})) \to R$ (since R belongs to $\mathbf{D}$). We claim that $im(\overline{h}) \subseteq K$. To prove this, pick any $f \in r(F(\mathbb{N}))$. Then $\overline{h}(f) \in R$ is the unique element y such that $\Theta_f(h(T_1),\dots,h(T_{n(f)}),y)$ holds. The formula ϕ_f is satisfied by (K,T), hence $y \in K$, i.e., $\overline{h}(f) \in K$. $\qquad\square$

In the next section we shall show that the converse of Theorem 16.10 also holds: Any elementary class of totally ordered fields that can be axiomatized as in the theorem is the class of totally ordered fields of some H–closed monoreflective subcategory of $\mathbf{POR/N}$. Note that such an elementary class of totally ordered fields satisfies the equivalent conditions of Theorem 16.7: Elementary classes are always isomorphism closed ([98], Satz 2.3). If Θ is a functional formula then

$$\phi \equiv \forall\, X_1,\dots,X_n\, [\, \exists\, Y : \Theta(X_1,\dots,X_n,Y)\ \&$$
$$\forall\, Z,T\, (\Theta(X_1,\dots,X_n,Y)\ \&\ \Theta(X_1,\dots,X_n,T) \to Z = T)]$$

is satisfied by every real closed field (by the definition of functionality). Thus, all real closed fields belong to the class. Now let (F,T_F) be a totally ordered field, let (K_i,T_i), $i \in I$, be a family of subfields, all the fields belonging to the elementary class. Then $(K,T) = \bigcap(K_i,T_i)$ is a member of this class as well. For, let Θ and ϕ be as above, let $a_1,\dots,a_n \in K$. For each $i \in I$ there is a unique $a_i \in K_i$ with $\Theta(a_1,\dots,a_n,a_i)$. There is at most one element $a \in F$ satisfying $\Theta(a_1,\dots,a_n,Y)$. Since each a_i also belongs to F we conclude that $a = a_i$ for every $i \in I$, which means that $a \in K$. Thus, $\Theta(a_1,\dots,a_n,a)$ holds in (K,T). As uniqueness of a is clear it follows that (K,T) is a model of ϕ.

For future reference we record the essence of this discussion in the next proposition. These considerations are continued in Theorem 17.6 and Theorem 17.7.

Proposition 16.11 Let $\mathbf{D} \subseteq \mathbf{POR/N}$ be an H–closed monoreflective subcategory, let X be the class of totally ordered fields belonging to $\mathbf{D}$. Then the class X is closed under formation of ultraproducts and intersections.

Proof It suffices to note that X, being an elementary class of totally ordered fields, is closed under formation of ultraproducts, cf. [30], Theorem 4.1.12.   □

17 von Neumann regular f-rings

It is a general experience in commutative algebra that von Neumann regular rings have properties that are particularly close to those of fields. Similarly, von Neumann regular f-rings are close to totally ordered fields. This has been recognized a long time ago, in particular by model theorists, and led to the results in [79]; [81]; [82], §6; [123], §4; [122]. In this section we undertake a systematic investigation of the monoreflective subcategories of **VNRFR**. Our results will reaffirm the above mentioned general experience. Our strongest results are about H-closed monoreflectors of **VNRFR**. These are completely determined by their restrictions to **TOF** (Theorem 17.6). The question which subcategories of **TOF** belong to some H-closed monoreflector of **VNRFR** is answered completely in Theorem 17.7. We start with a few useful alternative characterizations of von Neumann regular f-rings.

Proposition 17.1 Let (A, P) be a reduced poring. The following properties of (A, P) are equivalent:

  (a) (A, P) is a von Neumann regular f-ring.
  (b) $\sigma_{(A,P)} : (A, P) \to \sigma(A, P)$ is an isomorphism onto the subring

$$\{a \in \sigma(A, P); \ \forall \, \alpha \in Sper(A, P) \colon a(\alpha) \in \kappa_{(A,P)}(\alpha)\}.$$

  (c) A is von Neumann regular and the support map $supp :$ $Sper(A, P) \to Spec(A, P)$ is injective.

Proof (a) $\Rightarrow$ (c) A is von Neumann regular by hypothesis. The support map is injective for all f-rings. – (c) $\Rightarrow$ (b) By Proposition 4.4 the minimal prime ideals (for a von Neumann regular ring that is: all prime ideals) belong to the image of the support. So, the support is bijective. Both spectra are 0-dimensional. Therefore the support is a homeomorphism. Also, $Sper(\sigma_{(A,P)}) : Sper(\sigma(A, P)) \to Sper(A, P)$ is a homeomorphism (Proposition 7.9). We shall identify all these spectra. It is clear that $\sigma_{(A,P)}$ maps into the ring described in (b). We only need to show that this ring is the exact image. So, pick $a \in \sigma(A, P)$ with $a(\alpha) \in \kappa_{(A,P)}(\alpha)$ for every $\alpha \in Sper(A, P)$. Since $\kappa_{(A,P)}(\alpha) = A/\alpha$ (by von Neumann regularity) we find some $a_\alpha \in A$ with $a_\alpha(\alpha) = a(\alpha)$. The sets

$$C_\alpha = \{\beta \in Sper(A,P); \ a_\alpha(\beta) = a(\beta)\}$$

are constructible (Lemma 7.6) and cover $Sper(A,P)$. By compactness there is a finite subcover, say $Sper(A,P) = C_{\alpha_1} \cup \ldots \cup C_{\alpha_r}$. Without loss of generality one may assume that the covering sets are pairwise disjoint. Since A is von Neumann regular there is some $c \in A$ such that $c(\alpha) = a_{\alpha_i}(\alpha)$ for each i and each $\alpha \in C_{\alpha_i}$. But then $\sigma_{(A,P)}(c) = a$.   –
(b) $\Rightarrow$ (a) One checks easily that the ring of (b) is von Neumann regular and is a sub–f–ring of $\sigma(A,P)$. $\square$

Proposition 17.2 Suppose that (A,P) is a von Neumann regular f–ring. Let $f : A \to B$, $g : B \to \sigma(A,P)$ be monomorphisms of *rings* with $gf = \sigma_{(A,P)}$. If B is equipped with the restriction, say Q, of the partial order of $\sigma(A,P)$ then (B,Q) is a von Neumann regular f–ring and f is a **POR/N**–epimorphism.

Proof First we show that (B,Q) is an f–ring. It suffices to prove that for each $b \in B$ there is some $c \in B$ with $g(c) = g(b) \vee 0$: The set $C = \{\alpha \in Sper(A,P); \ g(b)(\alpha) \geq 0\}$ is constructible. Therefore the characteristic function χ_C belongs to A. The element $c = f(\chi_C)b \in B$ has the desired property, i.e., B is an f–ring.

By Corollary 14.7, f is an epimorphism in **POR/N**, hence $Sper(f)$ is injective (Theorem 5.2). Since $Sper(gf)$ is a homeomorphism it follows that $Sper(f)$ is also a homeomorphism. Because (B,Q) is an f–ring, the support map $supp: Sper(B,Q) \to Spec(B)$ is a homeomorphism onto the image. If it is surjective then $Spec(B)$ is homeomorphic to $Sper(A,P)$, hence B is reduced and has Krull dimension 0, which means that B is von Neumann regular. So it remains to show that $supp$ is surjective.

First pick some minimal prime ideal $p \subseteq B$. Since g is an extension of rings there is a minimal prime ideal $q \subseteq \sigma(A,P)$ with $p = g^{-1}(q)$. There is a unique $\gamma \in Sper(\sigma(A,P))$ with $q = supp(\gamma)$. But then $p = supp(g^{-1}(\gamma))$, as claimed. Now let $p \subseteq B$ be any prime ideal and pick a minimal prime ideal $p_0 \subseteq p$, say $p_0 = supp(\beta)$, $\beta \in Sper(B,Q)$. Let $\alpha = f^{-1}(\beta)$, let $\gamma \in Sper(\sigma(A,P))$ be such that $g^{-1}(\gamma) = \beta$. Then the extension $\kappa_{(A,P)}(\alpha) \subseteq \kappa_{\sigma(A,P)}(\gamma) = \rho(\alpha)$ is an algebraic field extension and B/p_0 is an intermediate ring, hence is a field as well. The prime ideal $p/p_0 \subseteq B/p_0$ must be trivial, i.e., $p = p_0$ is a minimal prime ideal, hence

$p \in im(supp)$ as claimed.   □

The discussion (in section 9A) of the supremum of two monoreflectors of **POR/N** shows that the construction can be very simple (Proposition 9A.1), but is complicated in general. It is a consequence of Proposition 17.2 that the situation is always simple if one of the reflectors is the von Neumann regularization ν.

Corollary 17.3 Suppose that $r: \mathbf{POR/N} \to \mathbf{D}$ is a monoreflector. Then $(r \vee \nu)(A, P) = r(\nu(A, P))$ for every reduced poring (A, P).

Proof Considering all the rings involved as subrings of $\sigma(A, P)$ we have

$$(A, P) \subseteq \nu(A, P) \subseteq r(\nu(A, P)) \subseteq (r \vee \nu)(A, P) \subseteq \sigma(A, P).$$

From Proposition 17.2 we know that $r(\nu(A, P))$ is von Neumann regular. The support map of $\nu(A, P)$ is injective. The same is true for $r(\nu(A, P))$ since the real spectra of the two rings are bijective to each other. Proposition 17.1 implies that $r(\nu(A, P))$ is a von Neumann regular f–ring. But then $(r \vee \nu)(A, P) \subseteq r(\nu(A, P))$, and the two rings coincide.   □

If $\mathbf{X} \subseteq \mathbf{TOF}$ is a monoreflective subcategory then there is a smallest monoreflective subcategory $\mathbf{M(X)} \subseteq \mathbf{POR/N}$ containing $\mathbf{X}$ (section 8). Let $f_\mathbf{X}$ be the monoreflector of $\mathbf{TOF}$ belonging to $\mathbf{X}$, let $r_\mathbf{X}$ be the monoreflector of $\mathbf{POR/N}$ belonging to $\mathbf{M(X)}$.

Proposition 17.4 The reflection of (A, P) in $\mathbf{M(X)}$ is

$$(r_\mathbf{X})(A, P) = \sigma(A, P) \cap \prod_{\alpha \in Sper(A,P)} f_\mathbf{X}(\kappa_{(A,P)}(\alpha)).$$

The reflector $r_\mathbf{X}$ is stronger than ν; $\mathbf{X} = \mathbf{M(X)} \cap \mathbf{TOF}$.

Proof For every reduced poring (A, P) we define

$$F(A, P) = \prod_{\alpha \in Sper(A,P)} f_\mathbf{X}(\kappa_{(A,P)}(\alpha))$$

and let $F_{(A,P)}\colon (A,P) \to F(A,P)$ be the canonical homomorphism. Obviously, F is a subfunctor of the complete rings of functions functor Π. Now let $s(A,P) = \sigma(A,P) \cap F(A,P)$, $s_{(A,P)}\colon (A,P) \to s(A,P)$ the canonical homomorphism. We wish to show that s is a reflector into some subcategory of **POR/N**. It is clear that $s(A,P)$ is a von Neumann regular f–ring. According to Proposition 9C.1 we need to check that every $s_{(A,P)}$ is an epimorphism (which follows from Corollary 14.8) and that $F(s_{(A,P)})$ is an isomorphism. The real spectra of $s(A,P)$ and $\sigma(A,P)$ are homeomorphic to each other (Proposition 17.2). Let $\alpha \in Sper(s(A,P))$, $\beta = s_{(A,P)}^{-1}(\alpha)$. Then

$$\kappa_{(A,P)}(\beta) \to \kappa_{s(A,P)}(\alpha) \to f_{\mathbf{X}}(\kappa_{(A,P)}(\beta))$$

are **TOF**–morphisms. The universal property of reflections implies that the canonical homomorphism

$$f_{\mathbf{X}}(\kappa_{(A,P)}(\beta)) \to f_{\mathbf{X}}(\kappa_{s(A,P)}(\alpha))$$

is an isomorphism. Thus, $F(s_{(A,P)})$ is an isomorphism.

So far we have shown that s is a monoreflector which is stronger than ν. If (A,P) is a totally ordered field then the definition of $s(A,P)$ shows that the reflection is $f_{\mathbf{X}}(A,P) \in ob(\mathbf{X})$. Therefore $ob(\mathbf{X})$ is the class of fields belonging to the reflective subcategory corresponding to s. The reflector s is weaker than $r_{\mathbf{X}}$ because $r_{\mathbf{X}}$ is the strongest monoreflector preserving the fields in $\mathbf{X}$. On the other hand, the image of the canonical homomorphism $r_{\mathbf{X}}(A,P) \to \sigma(A,P)$ is always contained in $s(A,P)$. Together this implies that $r_{\mathbf{X}} = s$, and the proof is finished. $\square$

In Corollary 16.2 it was shown that $\mathbf{D} \cap \mathbf{TOF}$ is monoreflective in **TOF** for all monoreflective subcategories $\mathbf{D} \subseteq \mathbf{POR/N}$. As a consequence of Proposition 17.4 we record the following improvement of this result:

Corollary 17.5 The class of monoreflective subcategories of **TOF** coincides with the class of subcategories $\mathbf{D} \cap \mathbf{TOF}$, $\mathbf{D}$ varying among the monoreflective subcategories of **POR/N** (or **VNRFR**). $\square$

In **POR/N** there are many monoreflective subcategories restricting to the same monoreflective subcategory $\mathbf{X} \subseteq \mathbf{TOF}$. For the two extreme cases that $\mathbf{X} = \mathbf{TOF}$ or $\mathbf{X} = \mathbf{RCF}$ this can be seen as follows: The monoreflective subcategories $\mathbf{POR/N} \subseteq \mathbf{POR/N}$ and $\mathbf{VNRFR} \subseteq \mathbf{POR/N}$

both restrict to **TOF**, **RCR** and **SAFR** restrict to **RCF**. Let α be some ordinal, let R_α be a real closed field of cardinality $\aleph_\alpha$, let $\mathbf{D}_\alpha \subseteq \mathbf{POR}$ be the subcategory of R_α-algebras. As in the proof of Theorem 9D.3, one finds a proper class of monoreflective subcategories between **VNRFR** and **POR/N**, and between **SAFR** and **RCR**. All of these restrict to the subcategories **TOF** and **RCF** of **TOF**, resp.

The related question whether the restriction from the monoreflective subcategories of **VNRFR** to the monoreflective subcategories of **TOF** is injective is much more delicate. In fact, it is so delicate that the answer has eluded us so far. An equivalent way to phrase the question is whether every monoreflective subcategory of **VNRFR** is of the form $\mathbf{M}(\mathbf{X})$ for some monoreflective subcategory $\mathbf{X} \subseteq \mathbf{TOF}$. There is one special case in which the answer to these questions is accessible, namely if the subcategory is H–closed.

Theorem 17.6 Let $\mathbf{D} \subseteq \mathbf{VNRFR}$ be a monoreflective subcategory with reflector r; let X be the class of totally ordered fields in $\mathbf{D}$. Then $\mathbf{D}$ is H–closed if and only if for all $(A, P) \in ob(\mathbf{D})$ and all $\alpha \in Sper(A, P)$ the totally ordered field $\kappa_{(A,P)}(\alpha)$ belongs to X. If the equivalent conditions hold then $\mathbf{D}$ is the monoreflective subcategory generated by X.

Proof First suppose that $\mathbf{D}$ is H–closed. Pick some **D**-object (A, P) and some $\alpha \in Sper(A, P)$. The image of the canonical homomorphism $\rho_\alpha : (A, P) \to \rho(\alpha)$ is $\kappa_{(A,P)}(\alpha)$, hence, by H–closedness, $\kappa_{(A,P)}(\alpha)$ is a member of **D**. – Conversely, assume the condition about the residue fields $\kappa_{(A,P)}(\alpha)$. Let $s \colon \mathbf{POR/N} \to \mathbf{E}$ be the monoreflector generated by the class X. The reflection of (A, P) under s was described explicitly in Proposition 17.4. Since **D** and **E** both contain the same fields it is clear that $\mathbf{D} \supseteq \mathbf{E}$. For each **D**-object (A, P), the reflection $s_{(A,P)} : (A, P) \to s(A, P)$ is an epimorphic extension. The map $Sper(s_{(A,P)})$ is a homeomorphism. Suppose that $\beta \in Sper(s(A, P))$ and let $\alpha = s_{(A,P)}^{-1}(\beta)$. From $\kappa_{(A,P)}(\alpha) \in X$ and Proposition 17.4 it follows that the canonical homomorphism $\kappa_{(A,P)}(\alpha) \to \kappa_{s(A,P)}(\beta)$ is an isomorphism. But then (7.13) implies that $s_{(A,P)}$ is an isomorphism, i.e., (A, P) belongs to **E**. Thus, **D** is the monoreflective subcategory generated by X. The proof of H–closedness can be done via same the scheme theoretic argument: Suppose that $(A, P) \in ob(\mathbf{D})$ and that $I \subseteq A$ is a (radical) ideal. Then I is an l–ideal, and $(\overline{A}, \overline{P}) = (A/I, P+I/I)$ is in **VNRFR**. We consider $Sper(\overline{A}, \overline{P})$ as a closed subspace of $Sper(A, P)$. If $\alpha \in Sper(\overline{A}, \overline{P})$

then $\kappa_{(\overline{A},\overline{P})}(\alpha) = \kappa_{(A,P)}(\alpha)$ belongs to **D**. Therefore $Sper(r_{(\overline{A},\overline{P})})$ is a homeomorphism and the maps $\kappa_{(\overline{A},\overline{P})}(\alpha) \to \kappa_{r(\overline{A},\overline{P})}(\beta)$ are isomorphisms (with $\beta \in Sper(r(\overline{A},\overline{P}))$ and $\alpha = r_{(\overline{A},\overline{P})}^{-1}(\beta)$). Once again, (7.13) can be applied, and $r_{(\overline{A},\overline{P})}$ is an isomorphism.   □

Corollary 17.5 in connection with Theorem 16.7 gives a complete descrip-
tion of those monoreflective subcategories $\mathbf{X} \subseteq \mathbf{TOF}$ that are of the form
$\mathbf{D} \cap \mathbf{TOF}$, **D** monoreflective in $\mathbf{POR/N}$. We consider the same question
for the H–closed monoreflective subcategories $\mathbf{D} \subseteq \mathbf{POR/N}$.

Theorem 17.7 Let $\mathbf{X} \subseteq \mathbf{TOF}$ be a monoreflective subcategory with
reflector $f_{\mathbf{X}}$. The reflector of the monoreflective subcategory $\mathbf{M(X)} \subseteq$
$\mathbf{POR/N}$ is denoted by $r_{\mathbf{X}}$. The following conditions are equivalent:

(a) $\mathbf{M(X)}$ is H–closed;
(b) $\mathbf{X}$ is closed under the formation of ultraproducts and under formation of intersections.

Proof The implication (a) $\Rightarrow$ (b) is Proposition 16.11. For (b) $\Rightarrow$ (a),
assume that $\mathbf{M(X)}$ is not H–closed. Then there are some (A,P) in
$\mathbf{M(X)}$ and some $\alpha \in Sper(A,P)$ such that $(f_{\mathbf{X}})_{\kappa_{(A,P)}(\alpha)} : \kappa_{(A,P)}(\alpha) \to$
$f_{\mathbf{X}}(\kappa_{(A,P)}(\alpha))$ is not surjective (Theorem 17.6). So, pick an element

$$x \in f_{\mathbf{X}}(\kappa_{(A,P)}(\alpha)) \backslash im((f_{\mathbf{X}})_{\kappa_{(A,P)}(\alpha)}).$$

There is some $a \in \sigma(A,P)$ with $a(\alpha) = x$. Since

$$(A,P) = r_{\mathbf{X}}(A,P) = \sigma(A,P) \cap \prod_{\beta \in Sper(A,P)} f_{\mathbf{X}}(\kappa_{(A,P)}(\beta))$$

(Proposition 17.4) this element does not belong to $\prod_{\beta} f_{\mathbf{X}}(\kappa_{(A,P)}(\beta))$. Let
U be any open constructible neighborhood of α. Then also $a|_U \notin$
$\prod_{\beta \in U} f_{\mathbf{X}}(\kappa_{(A,P)}(\beta))$, hence

$$F(U) = \{\beta \in U; \ a(\beta) \notin f_{\mathbf{X}}(\kappa_{(A,P)}(\beta))\}$$

is nonempty. By the choice of x and a, $\alpha \notin F(U)$. The set $\{F(U); U\}$
is a filter basis on $Sper(A,P)$ converging to α. Pick some ultrafilter $\mathcal{F}$

containing $\{F(U); U\}$. Then $\mathcal{F}$ converges to α as well. This can also be phrased as follows: Let $\Delta = \Delta_{(A,P)} : (A, P) \to \Pi(A, P)$ be the canonical homomorpism into the complete ring of functions. The ultrafilter $\mathcal{F}$ may be considered as point of $Sper(\Pi(A, P))$. Then $\Delta^{-1}(\mathcal{F}) = \alpha$ and $\rho(\alpha)$ may be identified with a subfield of $\rho(\mathcal{F})$. Under the identification x corresponds to $a(\mathcal{F})$. We want to show that $a(\mathcal{F}) \notin (\prod_\beta f_{\mathbf{X}}(\kappa_{(A,P)}(\beta)))/\mathcal{F}$;

assume this to be false, by way of contradiction. Then there is some $F \in \mathcal{F}$ such that $a|_F \in \prod_{\beta \in F} f_{\mathbf{X}}(\kappa_{(A,P)}(\beta))$. The choice of the ultrafilter implies that $F(U) \cap F \neq \emptyset$ for every open neighborhood U of α. But the definition of $F(U)$ shows that this is impossible. We conclude that $x = a(\mathcal{F}) \in \rho(\mathcal{F}) \backslash (\prod_\beta f_{\mathbf{X}}(\kappa_{(A,P)}(\beta)))/\mathcal{F}$, proving the claim. Because $\mathbf{X}$ is closed under ultraproducts, $(\prod_\beta f_{\mathbf{X}}(\kappa_{(A,P)}(\beta)))/\mathcal{F}$ belongs to $\mathbf{X}$. The intersection property implies that the intersection of $\rho(\alpha)$ and $(\prod_\beta f_{\mathbf{X}}(\kappa_{(A,P)}(\beta)))/\mathcal{F}$ inside $\rho(\mathcal{F})$ is a member of $\mathbf{X}$ as well. This intersection does not contain x. On the other hand, $f_{\mathbf{X}}(\kappa_{(A,P)}(\alpha))$ is a subfield of the intersection, hence $x \notin f_{\mathbf{X}}(\kappa_{(A,P)}(\alpha))$. This contradicts the choice of x, and the proof is finished. $\qquad\square$

To close the section, here are some examples satisfying the equivalent conditions of the theorem.

Example 17.8 *Elementary classes of totally ordered fields that are closed under intersections*
  (a) Real closed fields.
  (b) Euclidean fields.
  (c) Totally ordered pythagorean fields.
  (d) Let $N \subseteq \mathbb{N}$ be some set of odd primes. For each $n \in N$, let ϕ_n be the sentence $\forall\, a\, \exists\, b : a = b^n$. Then the elementary class of totally ordered fields satisfying all the sentences ϕ_n has the intersection property.
  (e) The subcategory of almost real closed totally ordered fields is monoreflective in **TOF** (Example 16.9(e)). By [15], Corollary 3.2, or [42], Proposition 2.8, this class of fields is elementary. $\qquad\square$

There are other interesting classes of fields for which we do not know whether they meet the conditions of Theorem 17.7. Notable examples are the hereditarily euclidean fields and the totally ordered hereditarily pythagorean fields. Both are elementary classes (cf. [100], Satz 4.1; [67], Corollary 1). But we do not know whether the intersection property holds.

On the other hand, the subcategory of totally ordered fields with Henselian natural valuation ring is monoreflective in **TOF** (Example 16.9), but it is not an elementary class. This was pointed out to us by R. Berr. We give a modification of his argument. Let Th be the theory of this class of totally ordered fields in the language $\mathcal{L}_{po}$. Obviously, $\mathbb{Q}$ is a model of Th. For every natural number $0 < n$ the following formula can be satisfied in $\mathbb{Q}$:

$$\phi_n \equiv 0 < nX < 1 \;\&\; \forall\, Y(1 + X \neq Y^2).$$

(Namely, choose some square free natural number $l > n$, and subsitute $\frac{1}{l}$ for X.) Every finite subset of $\{\phi_n; n \in \mathbb{N}\}$ can be realized in $\mathbb{Q}$, hence $\{\phi_n; n \in \mathbb{N}\}$ is a *type* ([98], p. 122; [106], Section 15). There are an elementary extension (K, T) of $\mathbb{Q}$ and an element $c \in K$ realizing the type. Being an elementary extension of $\mathbb{Q}$, (K, T) is elementarily equivalent to $\mathbb{Q}$, hence (K, T) is also a model of Th. On the other hand, c belongs to the maximal ideal of the natural valuation ring of (K, T), hence $1 + c$ is mapped to a square in the residue field without being a square itself. Thus, the natural valuation ring of (K, T) is not Henselian.

18 Totally ordered domains

Similar to sections 16 and 17, this section and the next one form a unit. Section 19 is devoted to reflective subcategories of **FR/N**. It will turn out that totally ordered domains are an important ingredient for their classification. As a first step we study the category **TOD** of totally ordered domains. The monoreflective subcategories of **TOD** are characterized by the conditions that they contain every real closed domain and are closed both under intersections and under inverse images (Theorem 18.3). They are exactly the restrictions of monoreflective subcategories of **POR/N**.

Proposition 18.1 The category **RCD** of real closed domains is the smallest monoreflective subcategory of **TOD**. The corresponding reflector is the restriction of ρ.

Proof If (A, P) is a totally ordered domain then the real spectrum is a chain with respect to specialization. The functorial map $Sper(\rho_{(A,P)})$ is a homeomorphism (Proposition 12.5 or [111], Proposition 3.11, Corollary 3.20), hence $Sper(\rho(A, P))$ is a chain as well. This is equivalent to $\rho(A, P)$ being a totally ordered domain. Thus, the functor ρ restricts to a functor $\rho' : \mathbf{TOD} \to \mathbf{RCD}$. Since ρ is a reflector the same is true for ρ'. Now let $r : \mathbf{TOD} \to \mathbf{D}$ be any monoreflector. If (A, P) is a totally ordered field then $r_{(A,P)}$ must be an epimorphic extension, hence $r(A, P)$ is contained in the real closure of (A, P) (cf. Remark 5.3). In particular, **D** contains every real closed field. Now pick any $(A, P) \in ob(\mathbf{TOD})$ and let $\alpha \in Sper(A, P)$. The canonical homomorphism $\rho_\alpha : (A, P) \to \rho(\alpha)$ factors uniquely as

$$(A, P) \xrightarrow{r_{(A,P)}} r(A, P) \xrightarrow{\overline{\rho_\alpha}} \rho(\alpha).$$

Therefore the functorial map $Sper(r_{(A,P)})$ is surjective. By Remark 5.3 it is also injective. Because the real spectra of (A, P) and $r(A, P)$ are both totally ordered by specialization, $Sper(r_{(A,P)})$ is even a homeomorphism. Let $\beta \subseteq r(A, P)$ be the prime cone defined by $\overline{\rho_\alpha}$. Then the homomorphism $\rho(\alpha) \to \rho(\beta)$ induced by $r_{(A,P)}$ is an isomorphism. Identifying the real spectra of both rings we now consider (A, P) and $r(A, P)$ as subrings of $\Pi(A, P) = \Pi(r(A, P))$. Given a function $a \in \Pi(A, P)$, constructiblity of a with respect to (A, P) is the same as constructibility with respect to

$r(A, P)$. The same is true for compatibility (Lemma 12.8). Thus, the real closures of (A, P) and $r(A, P)$ agree. In particular, $r(A, P) \subseteq \rho(A, P)$. $\square$

Not every monoreflector of **POR/N** restricts to a monoreflector of **TOD**. This is due to the fact that the bijection $Sper(r_{(A,P)})$ is not a homeomorphism in general. For example, if (A, P) is a totally ordered domain then $Sper(A, P)$ is totally ordered by specialization, whereas $Sper(\sigma(A, P))$ is a Boolean space. Proposition 18.1 suggests an answer to the question which monoreflectors of **POR/N** can be restricted to **TOD**:

Proposition 18.2 Let $r : \textbf{POR/N} \to \textbf{D}$ be a monoreflector. Then r restricts to a monoreflector of **TOD** if and only if $\textbf{RCR} \subseteq \textbf{D}$.

Proof First suppose that $\textbf{RCR} \subseteq \textbf{D}$. If $(A, P) \in ob(\textbf{TOD})$ then $\rho_{(A,P)}$ factors through $r(A, P)$:

$$\rho_{(A,P)} : (A, P) \xrightarrow{r_{(A,P)}} r(A, P) \xrightarrow{\overline{\rho_{(A,P)}}} \rho(A, P).$$

The map $Sper(r_{(A,P)})$ is a homeomorphism, i.e., $Sper(r(A, P))$ is a chain. This means that $r(A, P)$ is a totally ordered domain. Thus, r restricts to a functor $\textbf{TOD} \to \textbf{D} \cap \textbf{TOD}$. It is clear that the restriction is a monoreflector. Conversely, suppose that $\textbf{RCR} \not\subseteq \textbf{D}$. Then there is a real closed ring (A, P) for which $r_{(A,P)}$ is not an isomorphism. The complete rings of functions $\Pi(A, P)$ and $\Pi(r(A, P))$ may be identified. If $Sper(r_{(A,P)})$ is a homeomorphism then for $a \in \Pi(A, P)$ to be constructible and compatible over (A, P) is the same as to be constructible and compatible over $r(A, P)$. But this means that the real closure of $r(A, P)$ agrees with the real closure of (A, P), a contradiction. Thus, $Sper(r_{(A,P)})$ is not a homeomorphism. Since it is a homeomorphism *with respect to the constructible topology* there must exist prime cones $\beta, \beta' \in Sper(r(A, P))$ which are not comparable, but such that $r_{(A,P)}^{-1}(\beta) \subseteq r_{(A,P)}^{-1}(\beta')$. Now define $\alpha = r_{(A,P)}^{-1}(\beta)$ and consider the totally ordered domain A/α. It follows from the commutative diagram

$$
\begin{array}{ccc}
(A, P) & \xrightarrow{r_{(A,P)}} & r(A, P) \\
\downarrow & & \downarrow \\
A/\alpha & \xrightarrow{r_{A/\alpha}} & r(A/\alpha)
\end{array}
$$

that $Sper(r_{A/\alpha})$ cannot be a homeomorphism, either. So, replacing (A, P) by A/α we may assume that (A, P) is a real closed domain. If $Sper(r(A, P))$ is totally ordered then $Sper(r_{(A,P)})$ is a homeomorphism, and we get a contradiction. Thus, $Sper(r(A, P))$ has a least two different generic points, hence $r(A, P)$ is not a totally ordered domain. We conclude that r cannot be restricted to **TOD**. $\qquad\qquad\square$

Theorem 18.3 A subcategory $\mathbf{X} \subseteq \mathbf{TOD}$ is monoreflective if and only if it contains all real closed domains and the following conditions hold:

(18.4)   If $(A, P) \in ob(\mathbf{X})$ and if $(A_i, P_i)_{i \in I}$ is a family of $\mathbf{X}$–subobjects of (A, P) then $\bigcap_{i \in I} (A_i, P_i)$ belongs to $\mathbf{X}$.

(18.5)   If $f: (A, P) \to (B, Q)$ is an $\mathbf{X}$–morphism and if $(B', Q') \subseteq (B, Q)$ is an $\mathbf{X}$–subobject then $(f^{-1}(B'), f^{-1}(B') \cap P)$ belongs to $\mathbf{X}$.

Every monoreflective subcategory of **TOD** is of the form $\mathbf{D} \cap \mathbf{TOD}$ for some monoreflective subcategory $\mathbf{D} \subseteq \mathbf{POR/N}$.

Proof Suppose that $\mathbf{X} \subseteq \mathbf{TOD}$ is monoreflective with reflector r. By Proposition 18.1, $\mathbf{RCD} \subseteq \mathbf{X}$. To check the intersection property (18.4), set $(B, Q) = \bigcap_{i \in I}(A_i, P_i)$ and consider the reflection $r_{(B,Q)}: (B, Q) \to r(B, Q)$. For each i, the inclusion $f_i: (B, Q) \to (A_i, P_i)$ is a morphism into an $\mathbf{X}$–object, hence it factors uniquely as

$$ f_i: (B, Q) \xrightarrow{\ r_{(B,Q)}\ } r(B, Q) \xrightarrow{\ \overline{f_i}\ } (A_i, P_i). $$

If $g_i: (A_i, P_i) \to (A, P)$ denotes the inclusion then $g_i \overline{f_i}$ is independent of i. Thus, $\overline{f_i}(r(B, Q))$ is independent of i as well. We conclude that $\overline{f_i}(r(B, Q)) \subseteq B$, i.e., there is a unique morphism $\overline{f} : r(B, Q) \to (B, Q)$ such that $\overline{f} r_{(B,Q)} = id_{(B,Q)}$. Being an initial epimorphic factor of an isomorphism, $r_{(B,Q)}$ is an isomorphism as well, and $(B, Q) \in ob(\mathbf{X})$ as claimed. Next we check condition (18.5). Let $h: (f^{-1}(B'), f^{-1}(B') \cap P) \to (A, P)$ be the inclusion and let $g: (f^{-1}(B'), f^{-1}(B') \cap P) \to (B', Q')$ be the restriction of f. The universal property of reflections yields both $\overline{g}$ and $\overline{h}$ in the diagram

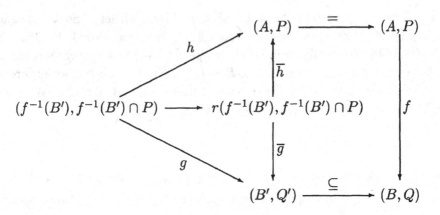

The diagram is commutative since $r_{(f^{-1}(B'),f^{-1}(B')\cap P)}$ is an epimorphism. In particular, this implies that $im(\overline{h}) \subseteq f^{-1}(B')$, and we obtain a homomorphism

$$\overline{h}' : r(f^{-1}(B'),\ f^{-1}(B') \cap P) \to (f^{-1}(B'),\ f^{-1}(B') \cap P)$$

by restricting the codomain of $\overline{h}$. It is clear that $\overline{h}' r_{(f^{-1}(B'),f^{-1}(B')\cap P)} = id_{(f^{-1}(B'),f^{-1}(B')\cap P)}$. Now $r_{(f^{-1}(B'),f^{-1}(B')\cap P)}$ is an initial epimorphic factor of an isomorphism, hence, $r_{(f^{-1}(B'),f^{-1}(B')\cap P)}$ is an isomorphism, and $(f^{-1}(B'),f^{-1}(B') \cap P)$ is a member of **X**.

Conversely, assuming that **X** satisfies the conditions (18.4) and (18.5) we want to show that **X** is monoreflective. For every (A,P) in **TOD**, let (A_i, P_i), $i \in I$, be the set of all totally ordered subrings of $\rho(A, P)$ that contain $\rho_{(A,P)}(A, P)$ and belong to **X**. The intersection $\bigcap(A_i, P_i)$ is denoted by $r(A, P)$; according to (18.4), it is a menber of **X**. By restriction of the codomain, $\rho_{(A,P)}$ yields a homomorphism $r_{(A,P)} : (A, P) \to r(A, P)$. It is claimed that $r_{(A,P)}$ has the universal property of reflections. So, let $f: (A, P) \to (B, Q)$ be a homomorphism into an **X**–object. Then $\rho(f): \rho(A, P) \to \rho(B, Q)$ is an **X**–morphism, and (B, Q) can be considered as a subobject of $\rho(B, Q)$. By (18.5), $\rho(f)^{-1}(B, Q) \subseteq \rho(A, P)$ belongs to **X**. Containing $\rho_{(A,P)}(A, P)$, it contributes to the intersection that defines $r(A, P)$. Therefore, $r(A, P) \subseteq \rho(f)^{-1}(B, Q)$, and f factors as

$$(A,P) \xrightarrow{\ r_{(A,P)}\ } r(A,P) \xrightarrow{\ \overline{f}\ } (B,Q).$$

According to Proposition 12.14, $r_{(A,P)}$ is an epimorphism in **POR/N**, hence it is also epimorphic in **TOD**. Therefore the factorization is unique. This finishes the proof that r is a reflector.

Now let **X** $\subseteq$ **TOD** be any monoreflective subcategory; let d be the corresponding reflector. For every $(A, P) \in ob(\mathbf{POR/N})$ we form the subring

$$D(A,P) = \prod_{\alpha \in Sper(A,P)} d(A/\alpha) \subseteq \Pi(A,P).$$

With the obvious definition of D on the morphisms of **POR/N**, D is a subfunctor of Π. Also, the homomorphism $\Delta_{(A,P)}\colon (A,P) \to \Pi(A,P)$ restricts to $D_{(A,P)}\colon (A,P) \to D(A,P)$. We define $r(A,P) = D(A,P) \cap \rho(A,P)$ and let $r_{(A,P)}\colon (A,P) \to r(A,P)$ be the restriction of $\Delta_{(A,P)}$. Proposition 9C.1 will be used to show that r is a reflector: The map $r_{(A,P)}$ is epimorphic by Proposition 12.14 (note that $D(A,P)$ and $\rho(A,P)$ are both sub-f-rings of $\Pi(A,P)$, hence so is $r(A,P)$). This implies that $Sper(r_{(A,P)})$ is a homeomorphism. Let $\beta \in Sper(r(A,P))$, $\alpha = r_{(A,P)}^{-1}(\beta)$. It remains to show that the canonical homomorphism $d(A/\alpha) \to d(r(A,P)/\beta)$ is an isomorphism. Composing the inclusion $r(A,P) \to D(A,P)$ with the projection onto $d(A/\alpha)$ and applying the universal property of reflections one obtains the diagram

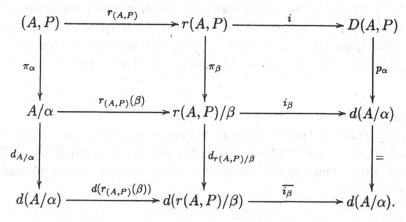

From $i_\beta r_{(A,P)}(\beta) = d_{A/\alpha}$ it follows that $\overline{i_\beta} d(r_{(A,P)}(\beta)) = id_{d(A/\alpha)}$. The maps $r_{(A,P)}$ and π_β are epimorphic in **POR/N**, hence so is $r_{(A,P)}(\beta)$. Then the latter map is also epimorphic in **TOD**. As $d_{r(A,P)/\beta}$ is epimorphic in **TOD** it follows that $d(r_{(A,P)}(\beta))$ is epimorphic in **TOD**. But then it is an initial epimorphic factor in an isomorphism, hence is an isomorphism itself.

Altogether, if $\mathbf{D} \subseteq \mathbf{POR/N}$ is the subcategory of those reduced porings (A,P) for which $r_{(A,P)}$ is an isomorphism then $\mathbf{D}$ is monoreflective and $\mathbf{D} \cap \mathbf{TOD} = \mathbf{X}$. □

Example 18.6 *RCD is monoreflective in* **TOD**
One extreme possibility in Theorem 18.3 is to choose $\mathbf{X} = \mathbf{RCD}$. It

is already known from Proposition 18.1 that **RCD** $\subseteq$ **TOD** is monoreflective. But it can also be deduced from Theorem 18.3. For, the class
of all real closed domains satisfies the conditions of the theorem: (18.4)
follows from [111], Theorem I 4.12, and (18.5) is [96], p. 18, Korollar.
The monoreflective subcategory **D** with **D** $\cap$ **TOD** = **RCD** constructed
in the proof of the theorem is **RCR**. $\square$

Example 18.7 *TOD is monoreflective in itself*
Besides **X** = **RCD** (Example 18.6) the other extreme possibility for **X** is
TOD. The conditions of the theorem are trivially satisfied. The monoreflective subcategory **D** $\subseteq$ **POR/N** with **D** $\cap$ **TOD** = **TOD** constructed
in the proof is the category **CPWPFR** of Example 12.17. $\square$

Example 18.8 *Totally ordered integrally closed domains*
Let **X** $\subseteq$ **TOD** be the subcategory of totally ordered integrally closed
domains. Every real closed domain belongs to **X** ([109], Proposition 9).
The conditions (18.4) and (18.5) can be checked directly (or, see [22],
Chapitre 5, p. 12, Proposition 12, and p. 16, Corollaire). $\square$

Example 18.9 *Totally ordered local domains*
Let **X** $\subseteq$ **TOD** be the subcategory of totally ordered local domains with
convex maximal ideal. The real closed domains belong to **X** ([111], Theorem I 3.10). Once again, conditions (18.4) and (18.5) can be checked
directly. $\square$

As with totally ordered fields, one may ask whether the categories **D** $\cap$
TOD, with **D** $\subseteq$ **POR/N** H–closed monoreflective, have any special
properties beyond those of Theorem 18.3. Here is a necessary condition:

Proposition 18.10 Let **X** $\subseteq$ **TOD** be a monoreflective subcategory.
If **X** = **D** $\cap$ **TOD** for some H–closed monoreflective subcategory **D** $\subseteq$
POR/N then **X** is closed under formation of ultraproducts.

Proof Setting **X** = **D** $\cap$ **TOD** let $(A_i, P_i)_{i \in I}$ be any famliy in **X** and
let $\mathcal{F}$ be an ultrafilter on the set I. The category **D** is complete (Proposition 2.7), hence $\prod(A_i, P_i)$ exists in **D**. The ultraproduct $\prod(A_i, P_i)/\mathcal{F}$

is a surjective image of $\prod(A_i, P_i)$, hence belongs to **D** by H–closedness. General properties of ultraproducts imply that $\prod(A_i, P_i)/\mathcal{F}$ is a totally ordered domain ([30], Theorem 4.1.9). $\qquad\qquad\qquad\qquad\qquad\qquad\qquad\qquad$ $\square$

We do not know whether the converse of the proposition is also true. As a first attempt one may try to show that the category **D** constructed in the proof of Theorem 18.3 (with the property that **D**∩**TOD** = **X**) is H–closed whenever **X** is closed under ultraproducts. However, we know already that this is not true in general. For, if **X** = **TOD** then **D** = **CPWPFR** (Example 18.7). But **CPWPFR** is not H–closed according to Example 12.17. The natural thing to do now is to replace **D** by the smallest H–closed monoreflective subcategory **E** ⊆ **POR/N** containing **D**. However, we are unable to decide whether **E** ∩ **TOD** = **X**.

19 Reduced f–rings

In section 17 we discussed the question of how much information is required to determine a monoreflective subcategory $\mathbf{D} \subseteq \mathbf{VNRFR}$. The
class of totally ordered fields contained in $\mathbf{D}$ always carries a great deal of
information about the reflector. If the monoreflector is H–closed then $\mathbf{D}$
is completely determined by the fields. In the present section the monoreflectors stronger than the reduced f–rings reflector φ will be studied in
the same spirit. It is clear that in general these monoreflectors are not
completely determined by their effect on totally ordered fields, even if
they are H–closed. For example, both $\mathbf{FR/N}$ and $\mathbf{VNRFR}$ contain
every totally ordered field. So, we will have to look for more powerful invariants. Instead of studying the restriction of the monoreflector to $\mathbf{TOF}$
we shall look at the restriction to $\mathbf{TOD}$. From Proposition 18.2 we know
that such a restriction exists only if the monoreflector is weaker than the
real closure reflector. Therefore we shall concentrate our efforts on the
class $[\varphi, \rho] = \{r; \varphi \leq r \leq \rho\}$ of reflectors. The main result says that
two such H–closed monoreflectors coincide if and only if the reflections of
certain fibre products of totally ordered domains agree (Theorem 19.8).

First we consider the class of monoreflectors having a given restriction
to $\mathbf{TOD}$. So, let $d : \mathbf{TOD} \to \mathbf{X}$ be a monoreflector, and let

$$[\varphi, \rho]_d = \{r \in [\varphi, \rho];\ r|\mathbf{TOD} = d\}.$$

Proposition 19.1 The class $[\varphi, \rho]_d$ is a complete lattice.

Proof There is a smallest monoreflective subcategory $\mathbf{D}_1$ with $\mathbf{RCR} \subseteq$
$\mathbf{D}_1 \subseteq \mathbf{POR/N}$ that contains $\mathbf{X}$. It is a consequence of Theorem 18.3 that
$\mathbf{X} = \mathbf{D}_1 \cap \mathbf{TOD}$. (Note that $\mathbf{D}_1$ is the category constructed in the proof
of Theorem 18.3.) For each $(A, P) \in ob(\mathbf{POR/N})$,

$$\{r(A, P);\ r \in [\varphi, \rho]_d\}$$

is a set of intermediate rings of $\varphi(A, P)$ and $\rho(A, P)$. We define $m(A, P) = \bigcap r(A, P)$ and let $m_{(A,P)} : (A, P) \to m(A, P)$ be the homomorphism
$\rho_{(A,P)}$ with restricted codomain. We claim that m is the smallest element
of $[\varphi, \rho]_d$. First note that $m(A, P)$ is a reduced f–ring, hence $m_{(A,P)}$ is

an epimorphism (Proposition 12.14) and $r(m_{(A,P)})$ is an epimorphism as well (Proposition 2.4). For every $r \in [\varphi, \rho]_d$ the diagram

$$
\begin{array}{ccc}
(A,P) \longrightarrow & m(A,P) & \longrightarrow r(A,P) \\
\downarrow & \downarrow & \downarrow = \\
r(A,P) \longrightarrow & r(m(A,P)) & \longrightarrow r(A,P)
\end{array}
$$

(obtained from the universal property of reflections) shows that $r(A,P) \to r(m(A,P))$ is an isomorphism, hence $r(m(A,P))$ can be identified with the subring $r(A,P)$ of $\rho(A,P)$. But then $m(A,P) = m(m(A,P))$, and $m(A,P)$ belongs to the subcategory $\mathbf{D_0} \subseteq \mathbf{POR/N}$ whose objects are the reduced porings (A,P) for which $m_{(A,P)}$ is an isomorphism. Since the composition of m with the inclusion functor $\mathbf{D_0} \to \mathbf{POR/N}$ is an idempotent functorial extension operator (section 2) we conclude that m is a reflector. To finish the proof of the proposition we remark that $m(A,P) = d(A,P)$ if $(A,P) \in ob(\mathbf{TOD})$ (because $r(A,P) = d(A,P)$ for every $r \in [\varphi, \rho]_d$). Thus, m is the weakest reflector in $[\varphi, \rho]_d$. Being an interval in a complete lattice, $[\varphi, \rho]_d$ is a complete lattice. $\square$

In general, the class $[\varphi, \rho]_d$ has more than one element and must be expected to be quite large. For example, both $\mathbf{FR/N}$ and $\mathbf{CPWPFR}$ have the same intersection with $\mathbf{TOD}$, namely the entire category $\mathbf{TOD}$. We do not know whether there exists some reflector d such that $[\varphi, \rho]_d$ has just one element. Also, it is an open question whether an H–closed monoreflector in $[\varphi, \rho]$ is determined by its restriction to $\mathbf{TOD}$. But, concerning this latter question, we shall show that there is a slightly larger class of rings than $\mathbf{TOD}$ which can be used to compare H–closed monoreflectors. To work in this direction we need a tool to recognize whether a given morphism of f–rings is an isomorphism. Therefore we discuss a sheaf representation of reduced f–rings.

    Sheaf representations of f–rings were first studied by K. Keimel ([71]; see also [17], Chapitre 10). With every reduced f–ring (A,P) one associates a spectral space which we shall call the *Keimel spectrum*, following the usage in [70], p. 212. The notation will be $SpeK(A,P)$. The points of $SpeK(A,P)$ are the irreducible l–ideals of (A,P) ([17], §8.4). These are the l–ideals $I \subseteq A$ for which $(A/I, P + I/I)$ is totally ordered. For $a \in A$ one defines

$$
\begin{array}{rcl}
S(a) & = & \{I \in SpeK(A,P);\ a \notin I\}, \\
H(a) & = & \{I \in SpeK(A,P);\ a \in I\}.
\end{array}
$$

Note that $S(a) = S(|a|)$ and $H(a) = H(|a|)$. Therefore, when working with these sets one may assume most of the time that $a \geq 0$. If $0 \leq a, b$ then

$$\begin{aligned}
S(a) \cup S(b) &= S(a + b) &= S(a \vee b), \\
S(a) \cap S(b) &= S(ab) &= S(a \wedge b), \\
H(a) \cup H(b) &= H(ab) &= H(a \wedge b), \\
H(a) \cap H(b) &= H(a + b) &= H(a \vee b).
\end{aligned}$$

The sets $S(a), a \in A$, are the basis of a topology of $SpeK(A, P)$ such that $SpeK(A, P)$ is a spectral space in the sense of Hochster ([17], §10.1; [71], §1). By [64], Proposition 8, the sets $H(a)$ are the basis of another topology with which $SpeK(A, P)$ is also a spectral space. In [73], p. 118, Definition 7, and in [114] this topology was called the *inverse topology* of the spectral space. We shall write $SpeK^*(A, P)$ if we use the inverse topology. The specialization relation with respect to the inverse topology is the opposite of the specialization relation with respect to the usual topology. In particular, the generalizations of any point of $SpeK^*(A, P)$ form a chain. Every convex prime ideal of (A, P) is a member of $SpeK^*(A, P)$, i.e., the Brumfiel spectrum is a subset of the Keimel spectrum. The support map is a homeomorphism from $Sper(A, P)$ onto the Brumfiel spectrum, with respect to the usual topology or with respect to the inverse topology. So, $Sper^*(A, P)$, i.e., $Sper(A, P)$ with the inverse topology, may be considered as a subspace of $SpeK^*(A, P)$ via the support map. Because of the different topologies, our sheaf representation of (A, P) is not the same as the one in [17] and [71]. But it is essentially the same as the sheaf representation of abelian l–groups discussed in [118]. It is briefly discussed in [70], V 4.8, in a form which, however, is not very explicit. Therefore we include a complete and detailed discussion of the sheaf.

The category $\mathbf{FR}$ is complete; in particular, arbitrary projective limits exist in $\mathbf{FR}$. Therefore, presheaves and sheaves of f–rings on $SpeK^*(A, P)$ can be defined on a basis of the topology. For $U = H(a)$ we set $I_U = \bigcap_{I \in U} I$ and $\mathcal{F}_{(A,P)}(U) = (A/I_U, P + I_U/I_U)$. If $V \subseteq U$ then $I_U \subseteq I_V$. With the canonical homomorphisms $\mathcal{F}_{(A,P)}(U) \to \mathcal{F}_{(A,P)}(V)$ as restriction maps we obtain a presheaf of f–rings.

Proposition 19.2 The presheaf $\mathcal{F}_{(A,P)}$ is a sheaf.

Proof The stalk of $\mathcal{F}_{(A,P)}$ at $I \in SpeK^*(A, P)$ is $(A/I, P + I/I)$. If $a + I_U \in \mathcal{F}_{(A,P)}(U)$ is mapped to 0 in every stalk $I \in U$ then $a \in \bigcap_{I \in U} I =$

I_U, i.e., $a + I_U = 0$. Thus, the presheaf is separated. It remains to prove the following: If $U = \bigcup_{i \in I} U_i$ (with $U = H(a)$, $U_i = H(a_i)$) and if $x_i \in \mathcal{F}_{(A,P)}(U_i)$ with $x_i|_{U_i \cap U_j} = x_j|_{U_i \cap U_j}$ for all i, j then there is some $x \in \mathcal{F}_{(A,P)}(U)$ with $x|_{U_i} = x_i$ for each i. Since U is quasi–compact we may assume that I is finite, say $I = \{1, \dots, n\}$. By separatedness of the presheaf we may even assume that $n = 2$. Then we pick $c_1, c_2 \in A$ such that $c_i|_{U_i} = x_i$. Replacing c_1 and c_2 by $c_1 - c_2$ and $c_2 - c_2$ it suffices to deal with the case that $x_2 = 0$. Taking into account all these reductions we have to prove the following

Claim: If $U = U_1 \cup U_2$ and if $c \in A$ has the property that $c|_{U_1 \cap U_2} = 0$ then there is some $d \in A$ such that $d|_{U_1} = c|_{U_1}$, $d|_{U_2} = 0$.
Proof Pick any $I \in U_1 \setminus U_2$ and let $Gen(I)$ be the chain of generalizations of I. If $J \in Gen(I)$ then $a_2 \in J$ implies $c \in J$. We conclude that there must be some $a_I \in A$ with $|a_I a_2| + I > |c| + I$. We replace a_I by $|a_I| \vee 1$ and may then assume that $1 \leq a_I$. The sets

$$V_I = \{J \in SpeK^*(A, P);\ a_I a_2 + J > |c| + J\}$$

form a constructible cover of the constructible set $U_1 \setminus U_2$. By compactness there is a finite subcover $U_1 \setminus U_2 \subseteq V_{I_1} \cup \dots \cup V_{I_k}$. With $b = a_{I_1} \vee \dots \vee a_{I_k}$ we have $ba_2 + J > |c| + J$ for every $J \in U_1 \setminus U_2$. Now we set

$$d_1' = ba_2 \vee c, \qquad d_2' = ba_2 \vee (-c),$$
$$d' = d_1' - d_2', \qquad d = c - d'.$$

If $I \in U_1$ then $d_1' + I = ba_2 + I$ and $d_2' + I = ba_2 + I$; if $I \in U_2$ then $d_1' + I = c^+ + I$ and $d_2' + I = c^- + I$. Together this implies

$$d' + I = \begin{cases} 0 & \text{if } I \in U_1 \\ c + I & \text{if } I \in U_2. \end{cases}$$

It is clear now that $d + I_U \in \mathcal{F}_{(A,P)}(U)$ is the desired glueing of $c|_{U_1}$ with $0|_{U_2}$.   □

The Keimel spectrum is a contravariant functor from **FR/N** to the category of spectral spaces. We even get a contravariant functor into the category of ringed spaces with totally ordered stalks if we associate $(SpeK^*(A, P), \mathcal{F}_{(A,P)})$ with $(A, P) \in ob(\mathbf{FR/N})$.

Corollary 19.3 Let $f : (A, P) \to (B, Q)$ be an **FR/N**–morphism. Suppose that $Sper(f)$ is a homeomorphism and that the induced homomorphism $A/\alpha \to B/\beta$ (with $\beta \in Sper(B, Q)$, $\alpha = f^{-1}(\beta)$) is an isomorphism for all β. Then f is an isomorphism if and only if the Keimel spectra are homeomorphic with respect to the inverse topology.

Proof One direction is trivial. So, assume that $SpeK^*(f)$ is a homeomorphism. We identify the Keimel spectra with each other, as well as the real spectra. The Keimel spectra are denoted by X, the real spectra by Y. It suffices to show that the morphism $(X, \mathcal{F}_{(B,Q)}) \to (X, \mathcal{F}_{(A,P)})$ of ringed spaces is an isomorphism. For, then the rings of global sections are isomorphic. We have to prove that every homomorphism $f_x : \mathcal{F}_{(A,P),x} \to \mathcal{F}_{(B,Q),x}$ between stalks is an isomorphism. By [17], Théorème 9.3.2, every closed point of X belongs to Y. Pick any $y \in Y$ with $y \in \overline{\{x\}}$. By hypothesis, $f_y : \mathcal{F}_{(A,P),y} \to \mathcal{F}_{(B,Q),y}$ is an isomorphism. But then so is f_x since this is a homomorphism of factor rings of $\mathcal{F}_{(A,P),y}$ and $\mathcal{F}_{(B,Q),y}$ modulo corresponding convex ideals. $\square$

To apply the corollary, one needs to check whether the map between the Keimel spectra is a homeomorphism. A simplification is offered by the next result.

Lemma 19.4 With the hypotheses of Corollary 19.3, the following statements about $SpeK^*(f) : SpeK^*(B, Q) \to SpeK^*(A, P)$ are equivalent:

(a) $SpeK^*(f)$ is a homeomorphism;
(b) $SpeK^*(f)$ is bijective;
(c) $SpeK^*(f)$ is injective.

Proof The implications (a) $\Rightarrow$ (b) and (b) $\Rightarrow$ (c) are trivial. For (c) $\Rightarrow$ (a) first note that every closed point of the Keimel spectrum belongs to the real spectrum ([17], Théorème 9.3.2). The same is true for every generic point: Suppose that $I \in SpeK^*(A, P)$. The nilradical of the totally ordered ring $(A/I, P+I/I)$ is convex, hence $\sqrt{I}$ is convex in (A, P). The ring $A/\sqrt{I}$ is reduced and every minimal prime ideal is convex. Since $(A/\sqrt{I}, P+\sqrt{I}/\sqrt{I})$ is a totally ordered ring its convex ideals form a chain, so there is only one minimal prime ideal, i.e., $\sqrt{I} \subseteq A$ is a prime ideal and is identified with a point of the real spectrum. In particular, if I is a maximal convex ideal then $I = \sqrt{I}$ belongs to $Sper(A, P)$. Thus, every $I \in SpeK^*(A, P)$ lies between the supports of two prime cones. To show bijectivity of $SpeK^*(f)$, pick $I \in SpeK^*(A, P)$ and let $\alpha, \alpha' \in Sper(A, P)$ be such that $supp(\alpha) \subseteq I \subseteq supp(\alpha')$. There are unique points $\beta, \beta' \in Sper(B, Q)$ with $f^{-1}(\beta) = \alpha$ and $f^{-1}(\beta') = \alpha'$. Exactly as with the real spectrum, the map $SpeK^*(f)$ is convex (cf. [65], Lemma 2.1; [111], Proposition I 3.1.9), hence there exists some $J \in SpeK^*(B, Q)$ with $supp(\beta) \subseteq J \subseteq supp(\beta')$ and $f^{-1}(J) = I$. We conclude that $SpeK^*(f)$ is bijective. The map is always continuous, by compactness even a homeomorphism, with respect to the constructible topology and also continuous with respect to the inverse topology. Finally, we show that $SpeK^*(f)$ is a homeomorphism also with respect to the inverse topology: Supposing that $I, J \in SpeK^*(B, Q)$ and $f^{-1}(I) \subseteq f^{-1}(J)$ we claim that $I \subseteq J$. Again, pick $\alpha, \alpha' \in Sper(A, P)$ with $supp(\alpha) \subseteq f^{-1}(I) \subseteq f^{-1}(J) \subseteq supp(\alpha')$. Defining $\beta, \beta' \in Sper(B, Q)$ as above we know that there are $I', J' \in SpeK^*(B, Q)$ such that $supp(\beta) \subseteq I' \subseteq J' \subseteq supp(\beta')$ and $f^{-1}(I') = f^{-1}(I), f^{-1}(J') = f^{-1}(J)$. Since $SpeK^*(f)$ is injective it follows that $I' = I$ and $J' = J$, i.e., $I \subseteq J$.   □

Corollary 19.3 will be applied now to compare monoreflectors in the interval $[\varphi, \rho]$:

Theorem 19.5 Let $r: \mathbf{POR/N} \to \mathbf{D}, s: \mathbf{POR/N} \to \mathbf{E}$ be monoreflectors, $\varphi \leq r \leq s \leq \rho$. Then $r = s$ if and only if for every $(A, P) \in ob(\mathbf{POR/N})$ the homomorphism $\overline{s_{(A,P)}}$ with $s_{(A,P)} = \overline{s_{(A,P)}} r_{(A,P)}$ has the following properties:

(a) $SpeK^*(\overline{s_{(A,P)}})$ is a homeomorphism;

(b) identifying $Sper(A, P) = Sper(r(A, P)) = Sper(s(A, P))$, the canonical homomorphism $r(A, P)/\alpha \to s(A, P)/\alpha$ is an isomorphism for each $\alpha \in Sper(A, P)$.

Proof Trivially, if $r = s$ then conditions (a) and (b) are satisfied. Conversely, suppose that (a) and (b) hold. The conditions of Corollary 19.3 are satisfied by $\overline{s_{(A,P)}}$, hence $\overline{s_{(A,P)}}$ is an isomorphism. Then $s_{r(A,P)} : r(A, P) \to s(r(A, P))$ is an isomorphism as well. This means that $r(A, P)$ belongs to $\mathbf{E}$.   □

Corollary 19.6 With the hypotheses of Theorem 19.5, suppose in addition that r and s are H–closed. Then $r = s$ if and only if

(a) $\mathbf{D} \cap \mathbf{TOD} = \mathbf{E} \cap \mathbf{TOD}$;
(b) for each $(A, P) \in ob(\mathbf{POR/N})$ the functorial map $SpeK^*(\overline{s_{(A,P)}})$ is a homeomorphism.

Proof If $r = s$ then there is nothing to show. Conversely, if (a) and (b) hold then we must check condition (b) of Theorem 19.5. So, let $\alpha \in Sper(A, P)$. With $d = r|_{\mathbf{TOD}} = s|_{\mathbf{TOD}}$ we have the canonical monomorphisms

$$A/\alpha \to r(A, P)/\alpha \to s(A, P)/\alpha \to d(A/\alpha).$$

H–closedness of r and s implies that $r(A, P)/\alpha$ and $s(A, P)/\alpha$ belong to $\mathbf{D} \cap \mathbf{TOD}$. But then the last two maps are isomorphisms. $\qquad\square$

The Pierce–Birkhoff Conjecture is concerned with the comparison between the categories $\mathbf{FR/N}$ and $\mathbf{CPWPFR}$ (cf. Example 13.4). Because of $\mathbf{RCR} \subseteq \mathbf{CPWPFR} \subseteq \mathbf{FR/N}$, both are in the scope of Theorem 19.5. In the following example we use the techniques of this section to deepen our understanding of the difference between the reflectors.

Example 19.7 *A comparison of the categories* $\mathbf{FR/N}$ *and* $\mathbf{CPWPFR}$
Let r be the monoreflector belonging to $\mathbf{CPWPFR}$. Then $\varphi \leq r \leq \rho$, and Theorem 19.5 can be used to compare φ and r. Or, on a more individual level, Corollary 19.3 helps to compare $\varphi(A, P)$ and $r(A, P)$ for any reduced poring (A, P). The real spectra of $\varphi(A, P)$ and $r(A, P)$ are known to be homeomorphic (see also [115], Corollary 9), so we identify them with each other, and also with $Sper(A, P)$. Moreover, the canonical homomorphisms $A/\alpha \to \varphi(A, P)/\alpha \to r(A, P)/\alpha$ are isomorphisms ([115], Proposition 2). Now Corollary 19.3 says that $\varphi(A, P) \to r(A, P)$ is an isomorphism if and only if the functorial map of the Keimel spectra is a homeomorphism (cf. [115], Proposition 14). As pointed out earlier (Example 12.13), φ and r do not coincide. Having pinpointed the difference between φ and r via the Keimel spectrum it is easy to give explicit examples of reduced porings that are not *Pierce–Birkhoff rings* (in the terminology of [85]): Let K be a nonarchimedean totally ordered field with natural valuation ring V. The maximal ideal of V is denoted by

M. Let $I \subset M$ be some M–primary convex ideal. The fibre product $V \times_{V/I} V$ is a subring of $V \times V$. It is an f–ring with the restriction of the componentwise lattice–order. One checks that $r(V \times_{V/I} V) = V \times_{V/M} V$. Thus, $V \times_{V/I} V$ is not a Pierce–Birkhoff ring.

There is another related way to describe the difference between the reflectors φ and r. Let (A, P) be a reduced f–ring, let $i \colon Sper^*(A, P) \to SpeK^*(A, P)$ be the inclusion. We have shown that $\mathcal{F}_{(A,P)}$ is a sheaf on $SpeK^*(A, P)$ (Proposition 19.2); the global ring of sections is (A, P), by construction. Now let $i^* \mathcal{F}_{(A,P)}$ be the inverse image sheaf under i. It is claimed that $r(A, P)$ is the ring of global sections of $i^* \mathcal{F}_{(A,P)}$. First, suppose that $a \in r(A, P)$. Then there is an open constructible cover $Sper^*(A, P) = C_1 \cup \ldots \cup C_r$ and there are $a_1, \ldots, a_r \in A$ such that $a(\alpha) = a_j(\alpha)$ for all $\alpha \in C_j$. It is clear that $a_j|_{C_j} \in i^* \mathcal{F}_{(A,P)}(C_j)$. These sections satisfy the glueing condition, hence they provide a global section, which is exactly a. Conversely, if $a \in i^* \mathcal{F}_{(A,P)}(Sper^*(A, P))$ then a is obtained by glueing together local sections $a_j \in i^* \mathcal{F}_{(A,P)}(C_j)$, $j = 1, \ldots, r$, where C_j is open constructible and a_j is the image of some $b_j \in \mathcal{F}_{(A,P)}(D_j)$, $D_j \subseteq SpeK^*(A, P)$ open constructible, $C_j \subseteq D_j$. Thus $b_j = c_j|_{D_j}$ with $c_j \in A$. Now $a(\alpha) = c_j(\alpha)$ for every $\alpha \in C_j$ shows that $a \in r(A, P)$.

It is also possible to use *separating ideals* ([85]; also see [115], §3) to study the relationship between a reduced f–ring (A, P) and its reflection $r(A, P)$. We identify the real spectra of both rings. If $\alpha, \beta \in Sper(A, P)$ then the separating ideals $\langle \alpha, \beta \rangle_{(A,P)}$ in (A, P) and $\langle \alpha, \beta \rangle_{r(A,P)}$ in $r(A, P)$ are elements of the respective Keimel spectra, whenever they are defined. If $Sper(A, P)$ has only finitely many generic points then it is clear from the sheaf theoretic discussion that $\langle \alpha, \beta \rangle_{r(A,P)}$ has to be a prime ideal, which is not necessarily true for $\langle \alpha, \beta \rangle_{(A,P)}$.

Now let (A, P) be any reduced poring again. We identify the real spectra of (A, P), $\varphi(A, P)$ and $r(A, P)$. Let $\alpha, \beta \in Sper(A, P)$ be points having a separating ideal, say $I = \langle \alpha, \beta \rangle_{(A,P)}$. The fibre product $A/\alpha \times_{A/I} A/\beta$ is a subring of $A/\alpha \times A/\beta$. With the componentwise lattice–order, $A/\alpha \times A/\beta$ is a reduced f–ring; the restriction to the fibre product gives $A/\alpha \times_{A/I} A/\beta$ the structure of a reduced f–ring. The real spectrum of $A/\alpha \times_{A/I} A/\beta$ has two generic points, which are denoted by $\overline{\alpha}$ and $\overline{\beta}$. Their separating ideal is $\langle \overline{\alpha}, \overline{\beta} \rangle = I/supp(\alpha) \times I/supp(\beta)$. The inverse image of this separating ideal under the canonical homomorphism $f \colon (A, P) \to A/\alpha \times_{A/I} A/\beta$ is I. By the universal property of reflections, f factors as

$$f \colon (A, P) \xrightarrow{\varphi(A,P)} \varphi(A, P) \xrightarrow{\overline{f}} A/\alpha \times_{A/I} A/\beta.$$

It follows that $\varphi_{(A,P)}^{-1}(\langle \alpha, \beta \rangle_{\varphi(A,P)}) = I$. We express this fact by saying that *the reflection $\varphi_{(A,P)}$ preserves separating ideals.* The reflection of $A/\alpha \times_{A/I} A/\beta$ with respect to r is the ring $A/\alpha \times_{A/\sqrt{I}} A/\beta$. The same arguments as above show that $r_{(A,P)}^{-1}(\langle \alpha, \beta \rangle_{r(A,P)}) = \sqrt{I}$, i.e., the reflection $r_{(A,P)}$ does not preserve separating ideals.

For a specific example, let (A,P) be the weakly ordered coordinate ring of an affine algebraic curve over a real closed field R. For such rings the Pierce–Birkhoff conjecture was studied in [91]. From our discussion we recover [91], Corollary 3.2: Nontrivial separating ideals arise only at singular points of the curve. Let x be singular and let C, D be the germs of two half branches meeting at x. Then C and D determine two generic points $\alpha, \beta \in Sper(A, P)$. These points have a separating ideal which is contained in the maximal ideal m_x determined by x. The separating ideal is m_x if the half branches meet transversally at x or if they approach x along the same tangent, but from different directions. (A general discussion of tangent directions associated with points of the real spectrum may be found in [87], §1.) In all other cases the separating ideal is properly contained in m_x. Then $\varphi_{(A,P)}^{-1}(\langle \alpha, \beta \rangle_{\varphi(A,P)}) = \langle \alpha, \beta \rangle_{(A,P)} \subset r_{(A,P)}^{-1}(\langle \alpha, \beta \rangle_{r(A,P)})$, i.e., (A, P) is not a Pierce–Birkhoff ring. $\square$

In the example we saw that the comparison of the reflectors φ and r hinged largely on studying the effect which r has on a reduced f–ring of the form $(A, P_A) \times_{(C,P_C)} (B, P_B)$ where (A, P_A) and (B, P_B) are totally ordered domains, (C, P_C) is a totally ordered ring, and the fibre product is formed with respect to two **POR**–morphisms $\varphi : (A, P_A) \to (C, P_C)$, $\psi : (B, P_B) \to (C, P_C)$. In fact, for H–closed monoreflectors, these rings can always be used as a *test class* to compare reflectors:

Theorem 19.8   Let $r, s \in [\varphi, \rho]$ be H–closed monoreflectors of **POR/N** and assume that the reflections $r((A, P_A) \times_{(C,P_C)} (B, P_B))$ and $s((A, P_A) \times_{(C,P_C)} (B, P_B))$ are isomorphic over $(A, P_A) \times_{(C,P_C)} (B, P_B)$ where (A, P_A) and (B, P_B) are two totally ordered domains and (C, P_C) is a totally ordered ring. Then r and s coincide.

Proof   Let $\mathbf{D}, \mathbf{E} \subset \mathbf{POR/N}$ be the reflective subcategories belonging to r, s. Then $\mathbf{D} \cap \mathbf{E}$ is the reflective subcategory corresponding to $r \vee s$. It follows from the hypothesis that the reflections of $(A, P_A) \times_{(C,P_C)} (B, P_B)$ with respect to r and s belong to $\mathbf{E} \cap \mathbf{D}$. Therefore they are also isomorphic to $(r \vee s)((A, P_A) \times_{(C,P_C)} (B, P_B))$ over $(A, P_A) \times_{(C,P_C)} (B, P_B)$. Thus it suffices to prove the assertion of the theorem if $r \leq s$. For, in the general situation this implies that $r = r \vee s = s$. So, from now on we assume that $r \leq s$. The proof will be done by using the criterion of Corollary 19.6.

If (A, P) is a reduced poring then each morphism in the sequence

$$(A, P) \longrightarrow r(A, P) \longrightarrow s(A, P) \longrightarrow \rho(A, P)$$

is an epimorphism. Thus, the real spectra of $r(A, P)$ and $s(A, P)$ are both canonically homeomorphic to $Sper(A, P)$. We identify them with $Sper(A, P)$. As before, the homomorphism $r(A, P) \to s(A, P)$ is denoted by $\overline{s_{(A,P)}}$. H–closedness of r and s implies that, given any $\alpha \in Sper(A, P)$, the canonical homomorphisms

$$\begin{aligned} r(A, P)/\alpha &\to r(A/\alpha), \\ s(A, P)/\alpha &\to s(A/\alpha) \end{aligned}$$

are isomorphisms. The totally ordered domain A/α is a fibre product in a trivial way, namely: $A/\alpha = A/\alpha \times_{A/\alpha} A/\alpha$. Using the hypothesis of the theorem we conclude that $r(A, P)/\alpha \to s(A, P)/\alpha$ is an isomorphism. This means that $\mathbf{D} \cap \mathbf{TOD} = \mathbf{E} \cap \mathbf{TOD}$. It remains to check that the functorial map $SpeK^*(\overline{s_{(A,P)}})$ is a homeomorphism. In fact, it suffices to show that the map is injective (Lemma 19.4). Assume this to be false. Then there are two different $I, J \in SpeK^*(s(A, P))$ with the same restriction $K \in SpeK^*(r(A, P))$. First suppose that I and J are comparable, say $I \subset J$. There is some $\alpha \in Sper(A, P)$ with $\alpha \subseteq I \subset J$. Because $r(A, P)/\alpha \to s(A, P)/\alpha$ is an isomorphism the restrictions of I/α and J/α to $r(A, P)/\alpha$ cannot coincide. But then the restrictions of I and J to $r(A, P)$ cannot agree either. This contradiction shows that I and J must be incomparable. Let $\alpha \subseteq I$ and $\beta \subseteq J$ be generic points of $Sper(A, P)$. The separating ideal $\langle \alpha, \beta \rangle_{r(A,P)}$ is defined and is contained in K. Therefore $\overline{\{\alpha, \beta\}}$ has only one closed point, say γ, and the separating ideal of α and β is defined also in (A, P) and in $s(A, P)$. In both porings it is a generalization of γ. The restriction of $\langle \alpha, \beta \rangle_{s(A,P)}$ to $r(A, P)$ contains $\langle \alpha, \beta \rangle_{r(A,P)}$ properly since $I, J \subset \langle \alpha, \beta \rangle_{s(A,P)}$. On the other hand, Proposition 19.2 implies that the canonical homomorphism

$$f : r(A, P) \to (B, Q) = r(A, P)/\alpha \times_{r(A,P)/\langle \alpha, \beta \rangle_{r(A,P)}} r(A, P)/\beta$$

is surjective. From H–closedness of r it follows that $(B, Q) \in ob(\mathbf{D})$. Being a fibre product of totally ordered domains over a totally ordered ring, (B, Q) is also a member of **E**. The homomorphism $r(A, P) \to (B, Q)$ factors through $s(A, P)$:

$$f : r(A, P) \xrightarrow{\overline{s(A,P)}} s(A, P) \xrightarrow{\overline{f}} (B, Q).$$

The separating ideal of the generic points of $Sper(B, Q)$ is the kernel of $(B, Q) \to r(A, P)/\langle \alpha, \beta \rangle_{r(A,P)}$; its inverse image under f is $\langle \alpha, \beta \rangle_{r(A,P)}$, its inverse image under $\overline{f}$ contains $\langle \alpha, \beta \rangle_{s(A,P)}$. But then

$$\overline{s(A,P)}^{-1}(\langle \alpha, \beta \rangle_{s(A,P)}) \subseteq \langle \alpha, \beta \rangle_{r(A,P)},$$

a contradiction. $\qquad\qquad\qquad\qquad\qquad\qquad\qquad\qquad\qquad\qquad\qquad\square$

20 Rings of continuous piecewise polynomial functions

In the preceding section it was shown that one test class to compare H–closed monoreflectors in the interval $[\varphi, \rho]$ consists of the fibre products $(A, P_A) \times_{(C,P_C)} (B, P_B)$, where (A, P_A) and (B, P_B) are totally ordered domains and (C, P_C) is a totally ordered ring. If $\mathbf{RCR} \subseteq \mathbf{C} \subseteq \mathbf{FR/N}$ then, of course, these same rings can be used as a test class to compare any two H–closed monoreflective subcategories between $\mathbf{RCR}$ and $\mathbf{C}$. In general, the test class can be made even smaller if the category $\mathbf{C}$ is small: The smaller the set of H–closed monoreflectors, the smaller the test class that is needed to distinguish between different monoreflective subcategories in the interval $[\mathbf{RCR}, \mathbf{C}]$. In this section we prove the following result:

Theorem 20.1 Let $\mathbf{D}$, $\mathbf{E}$ be H–closed monoreflective subcategories of $\mathbf{POR/N}$ such that $\mathbf{RCR} \subseteq \mathbf{D}, \mathbf{E} \subseteq \mathbf{CPWPFR}$. Then $\mathbf{D} = \mathbf{E}$ if and only if $\mathbf{D} \cap \mathbf{TOD} = \mathbf{E} \cap \mathbf{TOD}$.

Proof Let r and s be the reflectors corresponding to $\mathbf{D}$ and $\mathbf{E}$, resp. One direction of the equivalence is trivial. For the other one we use Theorem 19.8: It suffices to show that the canonical images of $r((A, P_A) \times_{(C,P_C)} (B, P_B))$ and $s((A, P_A) \times_{(C,P_C)} (B, P_B))$ in $\rho((A, P_A) \times_{(C,P_C)} (B, P_B))$ coincide where $(A, P_A) \times_{(C,P_C)} (B, P_B)$ is a fibre product of two totally ordered domains $(A, P_A), (B, P_B)$ over a totally ordered ring (C, P_C), formed in the category $\mathbf{POR}$. We may assume that both $f \colon (A, P_A) \to (C, P_C)$ and $g \colon (B, P_B) \to (C, P_C)$ are surjective. If t denotes the reflector belonging to $\mathbf{CPWPFR}$ then $r((A, P_A) \times_{(C,P_C)} (B, P_B)) \cong r(t((A, P_A) \times_{(C,P_C)} (B, P_B)))$ and $s((A, P_A) \times_{(C,P_C)} (B, P_B)) \cong s(t((A, P_A) \times_{(C,P_C)} (B, P_B)))$. The nilradical of (C, P_C) is convex (section 1 or [17], Théorème 9.2.6). Since (C, P_C) is totally ordered its nilradical is a prime ideal. Let $\overline{C} = C/Nil(C)$ and let $P_{\overline{C}}$ be the induced total order. We claim that the canonical homomorphism $(A, P_A) \times_{(C,P_C)} (B, P_B) \to (A, P_A) \times_{(\overline{C}, P_{\overline{C}})} (B, P_B)$ is the reflection in $\mathbf{CPWPFR}$.

First note that $(A, P_A) \times_{(\overline{C}, P_{\overline{C}})} (B, P_B) \in ob(\mathbf{CPWPFR})$ since $\mathbf{TOD} \subseteq \mathbf{CPWPFR}$ and since $\mathbf{CPWPFR}$ is complete. Next we compare the real spectra. Since $Sper(p_{(A,P_A)})$ and $Sper(p_{(B,P_B)})$ (where

$p_{(A,P_A)}\colon (A,P_A) \times_{(C,P_C)} (B,P_B) \to (A,P_A)$ and $p_{(B,P_B)}\colon (A,P_A) \times_{(C,P_C)}$ $(B,P_B) \to (B,P_B)$ are the canonical projections) are homeomorphisms onto closed subspaces of $Sper((A,P_A) \times_{(C,P_C)} (B,P_B))$ we identify $Sper(A,P_A)$ and $Sper(B,P_B)$ with these subspaces. One checks that $Sper((A,P_A) \times_{(C,P_C)} (B,P_B)) = Sper(A,P_A) \cup Sper(B,P_B)$ and $Sper(A,P_A) \cap Sper(B,P_B) = Sper(C,P_C)$. The same holds if (C,P_C) is replaced by $(\overline{C},P_{\overline{C}})$. From $Sper(C,P_C) = Sper(\overline{C},P_{\overline{C}})$ it follows that the canonical map

$$Sper((A,P_A) \times_{(\overline{C},P_{\overline{C}})} (B,P_B)) \to Sper((A,P_A) \times_{(C,P_C)} (B,P_B))$$

is a homeomorphism. We identify these spaces. For $\alpha \in$ $Sper((A,P_A) \times_{(C,P_C)} (B,P_B))$ it is clear that $(A,P_A) \times_{(C,P_C)} (B,P_B)/\alpha$ $= (A,P_A) \times_{(\overline{C},P_{\overline{C}})} (B,P_B)/\alpha$. But then $\rho((A,P_A) \times_{(C,P_C)} (B,P_B)) =$ $\rho((A,P_A) \times_{(\overline{C},P_{\overline{C}})}(B,P_B))$ and

$$\prod_\alpha (A,P_A) \times_{(C,P_C)} (B,P_B)/\alpha = \prod_\alpha (A,P_A) \times_{(\overline{C},P_{\overline{C}})} (B,P_B)/\alpha.$$

This means that $t((A,P_A)\times_{(C,P_C)}(B,P_B)) = t((A,P_A)\times_{(\overline{C},P_{\overline{C}})}(B,P_B)) =$ $(A,P_A)\times_{(\overline{C},P_{\overline{C}})} (B,P_B)$ (cf. the definition of rings of continuous piecewise polynomial functions in Example 12.17).

From now on it suffices to deal with the case that (C,P_C) is a totally ordered domain. By Proposition 10.8 the diagrams

$$
\begin{array}{ccc}
r((A,P_A) \times_{(C,P_C)} (B,P_B)) & \xrightarrow{\;r(p_{(A,P_A)})\;} & r(A,P_A) \\
{\scriptstyle r(p_{(B,P_B)})}\Big\downarrow & & \Big\downarrow{\scriptstyle r(f)} \\
r(B,P_B) & \xrightarrow{\;r(g)\;} & r(C,P_C)
\end{array}
$$

$$
\begin{array}{ccc}
s((A,P_A) \times_{(C,P_C)} (B,P_B)) & \xrightarrow{\;s(p_{(A,P_A)})\;} & s(A,P_A) \\
{\scriptstyle s(p_{(B,P_B)})}\Big\downarrow & & \Big\downarrow{\scriptstyle s(f)} \\
s(B,P_B) & \xrightarrow{\;s(g)\;} & s(C,P_C)
\end{array}
$$

consist entirely of surjective homomorphisms. The reflectors r and s do not change real spectra. In particular, we identify $Sper(r((A,P_A)\times_{(C,P_C)}(B,P_B))) = Sper((A,P_A)\times_{(C,P_C)}(B,P_B))$. The first part of the proof shows that this space can also be identified with $Sper(r(A,P_A)\times_{r(C,P_C)}r(B,P_B))$. The real closed residue fields of $(A,P_A)\times_{(C,P_C)}(B,P_B)$, $r((A,P_A)\times_{(C,P_C)}(B,P_B))$ and $r(A,P_A)\times_{r(C,P_C)}r(B,P_B)$ at corresponding prime cones agree. Therefore Lemma 12.8 implies that $\rho(r((A,P_A)\times_{(C,P_C)}(B,P_B))) = \rho(r(A,P_A)\times_{r(C,P_C)}r(B,P_B))$. In connection with

$$\prod_{\alpha\in Sper((A,P_A)\times_{(C,P_C)}(B,P_B))} r((A,P_A)\times_{(C,P_C)}(B,P_B))/\alpha$$

$$= \prod_{\alpha\in Sper((A,P_A)\times_{(C,P_C)}(B,P_B))} r(A,P_A)\times_{r(C,P_C)}r(B,P_B)/\alpha$$

one concludes that

$$r((A,P_A)\times_{(C,P_C)}(B,P_B)) = t(r((A,P_A)\times_{(C,P_C)}(B,P_B)))$$
$$= t(r(A,P_A)\times_{r(C,P_C)}r(B,P_B)) = r(A,P_A)\times_{r(C,P_C)}r(B,P_B).$$

The same arguments apply to the reflector s. By hypothesis we know that $r(A,P_A) = s(A,P_A)$, $r(B,P_B) = s(B,P_B)$, and $r(C,P_C) = s(C,P_C)$. Altogether this yields

$$r((A,P_A)\times_{(C,P_C)}(B,P_B)) = r(A,P_A)\times_{r(C,P_C)}r(B,P_B)$$
$$= s(A,P_A)\times_{s(C,P_C)}s(B,P_B) = s((A,P_A)\times_{(C,P_C)}(B,P_B)).$$

$$\square$$

21 Rings of continuous piecewise rational functions

The category **CPWRFR** of rings of continuous piecewise rational functions was introduced in Example 12.16 as the monoreflective subcategory of **POR/N** belonging to the infimum of the reflectors ν (von Neumann regular f–rings) and ρ (real closure). In this section we study the monoreflective subcategories between **RCR** and **CPWRFR**. Since **CPWRFR** $\subseteq$ **CPWPFR** it follows from Theorem 20.1 that two H–closed monoreflective subcategories in this range can be distinguished by comparing their restrictions to **TOD**. The most important tool is to relate the reflectors in the interval $[\rho \wedge \nu, \rho]$ to the reflectors in the interval $[\nu, \sigma]$ by using the lattice operations. These provide two maps

$$\nu^{\vee} \;:\; [\rho \wedge \nu, \rho] \to [\nu, \sigma] : r \to \nu \vee r,$$
$$\rho^{\wedge} \;:\; [\nu, \sigma] \to [\rho \wedge \nu, \rho] : r \to \rho \wedge r.$$

By the results of section 17, the monoreflectors in the interval $[\nu, \sigma]$ are comparably well understood in terms of their restrictions to the category **TOF**. Therefore, in relating the intervals $[\rho \wedge \nu, \rho]$ and $[\nu, \sigma]$ to each other our main question is under what conditions a reflector between $\rho \wedge \nu$ and ρ is determined by its restriction to **TOF**. We show, as part of our main result (Theorem 21.4), that this is the case for reflectors that are both H–closed and quotient–closed. These results lead to another characterization of the real closure reflector by an extremal property (Theorem 21.10).

We start with an alternative description of the reflector $\rho \wedge \nu$. If (A, P) is a reduced poring then

$$(\rho \wedge \nu)(A, P) = \{a \in \rho(A, P); \; \forall \, \alpha \in Sper(A, P) : a(\alpha) \in \kappa_{(A,P)}(\alpha)\}$$

(Example 9C.3, Example 12.16). Given elements $a \in (\rho \wedge \nu)(A, P)$ and $\alpha \in Sper(A, P)$ we choose $b_\alpha, c_\alpha \in A$ with $c_\alpha(\alpha) \neq 0$ and $a(\alpha) = \frac{b_\alpha(\alpha)}{c_\alpha(\alpha)} \in \kappa_{(A,P)}(\alpha)$. The sets

$$K_\alpha = \{\beta \in Sper(A, P); \; c_\alpha(\beta)a(\beta) = b_\alpha(\beta) \; \& \; c_\alpha(\beta) \neq 0\}$$

are constructible and cover $Sper(A, P)$. There is a finite subcover, $Sper(A, P) = K_{\alpha_1} \cup \ldots \cup K_{\alpha_r}$. Thus, a is a continuous semi–algebraic function (since it belongs to $\rho(A, P)$, section 12) which is a quotient of functions in A on constructible pieces of the real spectrum. Conversely, every function having these properties obviously belongs to $(\rho \wedge \nu)(A, P)$. This description explains the name "continuous piecewise rational functions."

It is clear that **CPWRFR** is contained in **CPWPFR** and in **BIPOR/N**. The following example shows that **CPWRFR** is properly contained in the intersection of these two categories.

Example 21.1 CPWRFR $\subset$ CPWPFR $\cap$ BIPOR/N
Every totally ordered domain belongs to **CPWPFR**. If the domain is also local with convex maximal ideal then it is a member of **BIPOR/N**. It is easy to exhibit scores of such rings. The following ring of this type does not belong to **CPWRFR**: Let $\mathbb{R}((\mathbb{Q}))$ be the formal power series field with coefficients in $\mathbb{R}$ and exponents in $\mathbb{Q}$ – a real closed field; let $v \colon \mathbb{R}((\mathbb{Q})) \to \mathbb{Q} \cup \{\infty\}$ be the natural valuation, with valuation ring V. Then $I = \{x \in V; v(x) \geq 2\} \subseteq V$ is a convex ideal. The subring $A = \mathbb{R} + I$ of V carries a unique total order, say P, because $qf(A) = \mathbb{R}((\mathbb{Q}))$. It is a local poring with convex maximal ideal I. Thus, $(A, P) \in ob(\textbf{CPWPFR} \cap \textbf{BIPOR/N})$. The real spectrum of (A, P) has two points $\alpha \subset \beta$ with residue fields $\kappa_{(A,P)}(\alpha) = \mathbb{R}((\mathbb{Q}))$ and $\kappa_{(A,P)}(\beta) = \mathbb{R}$. Therefore $\nu(A, P) = \sigma(A, P)$, and the reflection of (A, P) in **CPWRFR** is $\rho(A, P) = \rho(A, P) \cap \nu(A, P)$. It is easy to check that $\rho(A, P) = V$. We conclude that $(A, P) \notin ob(\textbf{CPWRFR})$. $\square$

The maps $\nu^{\vee}$ and $\rho^{\wedge}$ are our main tools for studying the interval $[\rho \wedge \nu, \rho]$ of reflectors. To work comfortably with $\rho^{\wedge}$ we note that the reflectors $\rho^{\wedge}(r)$, $r \in [\nu, \sigma]$, have an explicit description via $\rho^{\wedge}(r)(A, P) = \rho(A, P) \cap r(A, P)$. For, according to Proposition 9B.1 it suffices to know that the canonical homomorphism $(A, P) \to \rho(A, P) \cap r(A, P)$ is always an epimorphism. Since $\rho(A, P) \cap r(A, P)$ is an intermediate f–ring of (A, P) and $\rho(A, P)$ this is guaranteed by Proposition 12.14.

Proposition 21.2 $\nu^{\vee}\rho^{\wedge}$ is the identity of $[\nu, \sigma]$.

Proof If $r \in [\nu, \sigma]$ then r is stronger than both ν and $\rho^{\wedge}(r)$, hence $r \geq \nu^{\vee}\rho^{\wedge}(r)$. Assume by way of contradiction that r is properly stronger than $\nu^{\vee}\rho^{\wedge}(r)$. Then there is a reduced poring (A, P) such that $\nu^{\vee}\rho^{\wedge}(r)_{(A,P)}$ is an isomorphism, but $r_{(A,P)}$ is not an isomorphism. Since (A, P) is a von Neumann regular f–ring it follows that $\rho(A, P) = \sigma(A, P)$. Considering all the rings involved as subrings of $\sigma(A, P)$ we have

$$
\begin{aligned}
(A, P) &= \nu^\vee \rho^\wedge(r)(A, P) \\
&= \rho(A, P) \cap r(A, P) \\
&= \sigma(A, P) \cap r(A, P) \\
&= r(A, P).
\end{aligned}
$$

This contradicts the choice of (A, P), and the proof is complete. $\square$

We are most interested in the behavior of H–closed and quotient–closed monoreflectors under $\nu^\vee$ and $\rho^\wedge$. Recall that H–closedness and quotient–closedness are equivalent for reflectors $r \in [\nu, \sigma]$ (Proposition 11.4). Also, if $r \in [\rho \wedge \nu, \rho]$ is H–closed or quotient–closed then the same is true for $\nu^\vee(r)$ (Theorem 10.2, Example 10.17 and Section 11). The behavior of these properties under $\rho^\wedge$ is less obvious.

Proposition 21.3 Suppose that $r \in [\nu, \sigma]$ is an H–closed (= quotient–closed) monoreflector. Then $\rho^\wedge(r)$ is quotient–closed.

Proof Let $\mathbf{D} \subseteq \mathbf{POR/N}$ be the subcategory belonging to $\rho^\wedge(r)$. Pick (A, P) in $\mathbf{D}$ and let $S \subseteq A$ be a multiplicative subset. We consider $\rho(A_S, P_S)$ and $r(A_S, P_S)$ as sub–f–rings of $\Pi(A_S, P_S)$. It is claimed that (A_S, P_S) belongs to $\mathbf{D}$, i.e., that the canonical homomorphism $(A_S, P_S) \to \rho(A_S, P_S) \cap r(A_S, P_S)$ is an isomorphism. Injectivity being clear, only surjectivity needs to be checked. Therefore, pick any element $x \in \rho(A_S, P_S) \cap r(A_S, P_S)$. According to Proposition 11.2 there are $a \in \rho(A, P)$, $b \in r(A, P)$ and $s, t \in S$ such that the canonical isomorphisms

$$
\begin{aligned}
\rho(A, P)_{\rho_{(A,P)}(S)} &\to \rho(A_S, P_S), \\
r(A, P)_{r_{(A,P)}(S)} &\to r(A_S, P_S)
\end{aligned}
$$

map $\dfrac{a}{\rho_{(A,P)}(s)} \to x$ and $\dfrac{b}{r_{(A,P)}(t)} \to x$. Obviously, one may assume that $s = t$. Then $a(\beta) = b(\beta)$ for each $\beta \in Sper(A_S, P_S)$. As usual, we identify $Sper(A_S, P_S)$ with the generically closed proconstructible subset

$$
\{\alpha \in Sper(A, P); \; \forall \, t \in S : t(\alpha) \neq 0\}
$$

of $Sper(A, P)$. The set

$$C = \{\alpha \in Sper(A, P); \ s(\alpha) \neq 0, a(\alpha) = b(\alpha)\}$$

is constructible and contains $Sper(A_S, P_S)$. By compactness, there are $t_1, \ldots, t_n \in S$ such that

$$Sper(A_S, P_S) \subseteq U = \{\alpha \in Sper(A, P); \ t_1(\alpha) \neq 0, \ldots, t_n(\alpha) \neq 0\} \subseteq C.$$

Defining $t = t_1 \cdot \ldots \cdot t_n \in S$ we obtain $t(\alpha)a(\alpha) = t(\alpha)b(\alpha)$ for all $\alpha \in Sper(A, P)$. This means that $ta = tb$ in $\Pi(A, P)$, i.e., $ta = tb \in \rho(A, P) \cap r(A, P) = (A, P)$. But then the canonical homomorphism $(A_S, P_S) \rightarrow \rho(A_S, P_S) \cap r(A_S, P_S)$ maps $\frac{ta}{ts} \in A_S$ onto x. Thus, the map is surjective, and the proof of the proposition is finished. $\qquad \square$

Given any class X of monoreflectors of $\mathbf{POR/N}$ let X_H be the subclass of H–closed monoreflectors, X_q the subclass of quotient–closed monore-flectors and $X_{qH} = X_{Hq} = X_q \cap X_H$. So far we know that, by restriction, $\nu^\vee$ and $\rho^\wedge$ yield the maps

$$\nu_q^\vee : [\rho \wedge \nu, \rho]_q \rightarrow [\nu, \sigma]_q,$$
$$\rho_q^\wedge : [\nu, \sigma]_q \rightarrow [\rho \wedge \nu, \rho]_q.$$

According to Proposition 21.2, $\nu_q^\vee \rho_q^\wedge$ is the identity of $[\nu, \sigma]_q$. We do not know whether the other composition is the identity map of $[\rho \wedge \nu, \rho]_q$. The map $\nu^\vee$ can also be restricted to

$$\nu_H^\vee : [\rho \wedge \nu, \rho]_H \rightarrow [\nu, \sigma]_H$$

(by Theorem 10.2). It is an open problem whether $\rho^\wedge$ can be restricted in the same way.

We shall now consider the set $[\rho \wedge \nu, \rho]_{qH}$. Our first main result is a characterization of these reflectors:

Theorem 21.4 Suppose that $r \in [\rho \wedge \nu, \rho]$, $r : \mathbf{POR/N} \rightarrow \mathbf{D}$.

(a) The following conditions (i) and (ii) are equivalent:
    (i) r is quotient–closed and H–closed.
    (ii) If $(A, P) \in ob(\mathbf{D})$ and $\alpha \in Sper(A, P)$ then $\kappa_{(A,P)}(\alpha) \in ob(\mathbf{D})$.
(b) If the equivalent conditions of (a) hold then $r = \rho^\wedge \nu^\vee(r)$.

Proof First we prove **(b)**. Let **E** and **F** be the reflective subcategories belonging to $\rho^\wedge\nu^\vee(r)$ and $\nu^\vee(r)$. It is clear that $\mathbf{E} \subseteq \mathbf{D}$, we have to show that they are equal. We shall prove this using condition (a)(ii) and H–closedness of $\nu^\vee(r)$ (which follows from condition (a)(i)). Suppose that $(A, P) \in ob(\mathbf{D})$. From the definition of continuous piecewise rational functions it follows that

$$(A, P) = \rho(A, P) \cap \prod_{\alpha \in Sper(A,P)} \kappa_{(A,P)}(\alpha).$$

Note that $\mathbf{D} \cap \mathbf{TOF} = \mathbf{F} \cap \mathbf{TOF}$, hence Theorem 17.4 and Theorem 17.6 in connection with H–closedness of $\nu^\vee(r)$ and condition (a)(ii) imply that

$$\nu^\vee(r)(A, P) = \sigma(A, P) \cap \prod_{\alpha \in Sper(A,P)} \kappa_{(A,P)}(\alpha).$$

We conclude that

$$(A, P) = \rho(A, P) \cap \nu^\vee(r)(A, P) = \rho^\wedge\nu^\vee(r)(A, P).$$

(a) (i) $\Rightarrow$ **(ii)** Pick any **D**–object (A, P) and let $\alpha \in Sper(A, P)$. Since **D** is H–closed we have $A/\alpha \in ob(\mathbf{D})$. The totally ordered quotient field of A/α is in **D** by quotient–closedness. – **(ii)** $\Rightarrow$ **(i)** If (A, P) belongs to $\mathbf{F} = \mathbf{D} \cap \mathbf{VNRFR}$ and $\alpha \in Sper(A, P)$ then $\kappa_{(A,P)}(\alpha)$ also belongs to **F** (by hypothesis). According to Proposition 17.6 this means that $\nu^\vee(r)$ is H–closed. The proof of part (b) shows that $r = \rho^\wedge\nu^\vee(r)$, hence r is quotient–closed (Proposition 21.3). It remains to prove that r is H–closed. This is much harder. Fortunately most of the work has already been done. We only have to check the hypotheses (i)–(iv) of Lemma 12.11.

(i) is satisfied by the choice of **D** (note that $\mathbf{CPWRFR} \subseteq \mathbf{BIFR/N}$). **(ii)** was shown above and **(iii)** holds by hypothesis. Only **(iv)** requires some thought. So, suppose that $(A, P) \in ob(\mathbf{D})$ and that $0 \leq u \in A$. We define $U = \{\alpha \in Sper(A, P); u(\alpha) > 0\}$ and consider a semi–algebraic function $a \in \sigma(A, P) \cap \prod_{\alpha \in Sper(A,P)} \kappa_{(A,P)}(\alpha)$ such that

- $a|_U \in A_u$; and
- there are some $0 \leq b \in A$ and some $1 \leq n \in \mathbb{N}$ such that $|a|^n \leq \sigma_{(A,P)}(b)$ and $b(\alpha) = 0$ for all $\alpha \notin U$.

It is claimed that $a \in (A, P)$: From Theorem 12.10 and Step 9 in its proof we know that $a \in \rho(A, P)$. But then

$$(A, P) = \rho(A, P) \cap \prod_{\alpha \in Sper(A,P)} \kappa_{(A,P)}(\alpha)$$

(note that $(A, P) \in ob(\mathbf{CPWRFR})$) implies that $a \in (A, P)$. $\hfill\square$

Corollary 21.5 Let $r: \mathbf{POR}/N \to \mathbf{D}$ and $s: \mathbf{POR}/N \to \mathbf{E}$ be reflectors belonging to $[\rho \wedge \nu, \rho]_{qH}$. Then $r = s$ if and only if $\mathbf{D} \cap \mathbf{TOF} = \mathbf{E} \cap \mathbf{TOF}$.

Proof If $r = s$ then there is nothing to prove. Conversely, if $\mathbf{D} \cap \mathbf{TOF} = \mathbf{E} \cap \mathbf{TOF}$ then $\nu^\vee(r) = \nu^\vee(s)$ by Theorem 17.6. Applying Theorem 21.4(b) one gets $r = \rho^\wedge \nu^\vee(r) = \rho^\wedge \nu^\vee(s) = s$. $\hfill\square$

It is an important consequence of Theorem 21.4 (b) that the restriction $\nu_{qH}^\vee: [\rho \wedge \nu, \rho]_{qH} \to [\nu, \sigma]_H$ is injective. By contrast, we do not know whether the restrictions $\nu_q^\vee: [\rho \wedge \nu, \rho]_q \to [\nu, \sigma]_q$ and $\nu_H^\vee: [\rho \wedge \nu, \rho]_H \to [\nu, \sigma]_H$ are injective. Also, it remains open whether $\nu_{qH}^\vee$ is surjective. In fact, this problem is equivalent to the question (raised before Theorem 21.4) whether $\rho^\wedge(s)$ is H–closed for all $s \in [\nu, \sigma]_H$. For, if $\rho^\wedge(s)$ is H–closed then it belongs to $[\rho \wedge \nu, \rho]_{qH}$ (by Proposition 21.3) and $s = \nu^\vee \rho^\wedge(s) = \nu_{qH}^\vee \rho^\wedge(s)$ (Proposition 21.2). On the other hand, suppose that $\nu_{qH}^\vee$ is surjective and that $s \in [\nu, \sigma]_H$. Then there is a unique $r \in [\rho \wedge \nu, \rho]_{qH}$ with $\nu_{qH}^\vee(r) = s$. Because of

$$\rho^\wedge(s) = \rho^\wedge \nu^\vee(r) = r \in [\rho \wedge \nu, \rho]_{qH} \subseteq [\rho \wedge \nu, \rho]_H$$

we see that $\rho^\wedge$ can be restricted to a map between H–closed reflectors.

Another open question is whether H–closedness implies quotient–closedness or vice versa for reflectors between $\rho \wedge \nu$ and ρ. The following result shows that these properties are not completely unrelated:

Corollary 21.6 Suppose that $r, s \in [\rho \wedge \nu, \rho]$ are monoreflectors with $r \le s \le \rho^\wedge \nu^\vee(r)$. Let $\mathbf{D} \supseteq \mathbf{E}$ be the corresponding reflective subcategories.

(a) If r is H–closed and s is quotient–closed then s is also H–closed.
(b) If r is quotient–closed and s is H–closed then s is also quotient–closed.

In both cases we have $s = \rho^\wedge \nu^\vee(r)$.

Proof (a) If s is not H–closed then there exist an $\mathbf{E}$–object (A, P) and some $\alpha = Sper(A, P)$ such that $\kappa_{(A,P)}(\alpha) \notin ob(\mathbf{E})$ (Theorem 21.4). Since s is quotient–closed the localization of (A, P) at $q = supp(\alpha)$ yields

$(A_q, P_q) \in ob(\mathbf{E})$. This also belongs to $\mathbf{D}$, hence so does the residue field $\kappa_{(A,P)}(\alpha) = \kappa_{(A_q,P_q)}(\alpha)$. The reflector $\rho^\wedge \nu^\vee(r)$ does not change any totally ordered field belonging to $\mathbf{D}$, hence $\kappa_{(A,P)}(\alpha) \in ob(\mathbf{E})$. This contradiction finishes the proof of part (a). – **(b)** If s is not quotient-closed then there exist (A, P) and $\alpha \in Sper(A, P)$ with $\kappa_{(A,P)}(\alpha)$ not in $\mathbf{E}$ (as in part (a)). But s being H–closed, $A/\alpha \in ob(\mathbf{E}) \subseteq ob(\mathbf{D})$. Because r is quotient-closed this implies that $\kappa_{(A,P)}(\alpha) = qf(A/\alpha)$ belongs to $\mathbf{D}$. Every totally ordered field of $\mathbf{D}$ is a member of $\mathbf{E}$. This contradicts the choice of (A, P) and α, and the proof of (b) is finished. – In both cases Theorem 21.4(b) applies to show that $s = \rho^\wedge \nu^\vee(s) = \rho^\wedge \nu^\vee(r)$. □

Corollary 21.7 If $r \in [\rho \wedge \nu, \rho]$ is H–closed then $\rho^\wedge \nu^\vee(r)$ is quotient-closed and H–closed. This is the strongest H–closed monoreflector $s \in [\rho \wedge \nu, \rho]$ with $\nu^\vee(s) = \nu^\vee(r)$.

Proof The first claim follows from Corollary 21.6 (a) and Proposition 21.3. The second claim is trivial. □

Given any monoreflector $r \colon \mathbf{POR/N} \to \mathbf{D}$, we introduced the H–closed approximation $r_\infty \leq r$ in Section 13. If r belongs to the image of $\rho^\wedge$ we can show that r_∞ is also quotient-closed:

Corollary 21.8 If $r = \rho^\wedge(s)$ for some $s \in [\nu, \sigma]$ then the H–closed approximation r_∞ belongs to $[\rho \wedge \nu, \rho]_{qH}$.

Proof Note that

$$r_\infty \leq \rho^\wedge \nu^\vee(r_\infty) \leq \rho^\wedge \nu^\vee(r) = \rho^\wedge \nu^\vee \rho^\wedge(s) = r$$

hence $r_\infty = \rho^\wedge \nu^\vee(r_\infty)$ (because $\rho^\wedge \nu^\vee(r_\infty)$ is H–closed and quotient-closed, Corollary 21.7). □

Example 21.9 *H–closed and quotient–closed reflectors in* $[\rho \wedge \nu, \rho]$

(a) The category **CPWRFR** is H–closed and quotient–closed according to Theorem 21.4 (a). This is the only H–closed monoreflective subcategory between **RCR** and **CPWRFR** that contains the entire category **TOF**.

(b) The category **RCR** is H–closed (by definition) and quotient–closed (by Proposition 12.6). Since $\kappa_{(A,P)}(\alpha)$ is a real closed field for every $(A, P) \in ob(\mathbf{RCR})$ and every $\alpha \in Sper(A, P)$ (Proposition 12.4 (e)) this also follows from Theorem 21.4 (a).

(c) Let (A, P) be any intermediate totally ordered ring of $\mathbb{Z}$ and R_0. Then $(\mathbf{A}, \mathbf{P})\mathbf{CPWRFR}$ is H–closed and quotient–closed because it is the intersection of $(\mathbf{A},\mathbf{P})\mathbf{POR/N}$ and **CPWRFR**, which are both H–closed and quotient–closed.   □

We close the section with yet another new characterization of the real closure reflector.

Theorem 21.10 The real closure reflector is the weakest monoreflector $r\colon \mathbf{POR/N} \to \mathbf{D}$ which is stronger than $\rho \wedge \nu$ and for which every residue field $\kappa_{(A,P)}(\alpha)$, $(A, P) \in ob(\mathbf{D})$, $\alpha \in Sper(A, P)$, is real closed.

Proof The real closure reflector clearly satisfies the condition about the residue fields. Conversely, suppose that $\kappa_{(A,P)}(\alpha)$ is real closed for every $\alpha \in Sper(A, P)$ and every $\mathbf{D}$–object (A, P). Since $\mathbf{D} \subseteq \mathbf{CPWRFR}$ (by hypothesis) every $(A, P) \in ob(\mathbf{D})$ can be represented as

$$(A, P) = \rho(A, P) \cap \prod_{\alpha \in Sper(A,P)} \kappa_{(A,P)}(\alpha).$$

The hypothesis that the field $\kappa_{(A,P)}(\alpha)$ be real closed means that $\prod_{\alpha \in Sper(A,P)} \kappa_{(A,P)}(\alpha) = \Pi(A, P)$. But then $(A, P) = \rho(A, P) \cap \Pi(A, P) = \rho(A, P)$.   □

22 Discontinuous semi–algebraic functions

In the previous section we studied the interval $[\rho \wedge \nu, \rho]$ in the lattice of monoreflectors of $\mathbf{POR/N}$. In the present section we pick some H–closed and quotient–closed monoreflector $r \in [\rho \wedge \nu, \rho]$ and consider the interval $[r, \nu \vee r]$. By Theorem 21.4 (b), r is the only reflector in this interval that is weaker than ρ. Therefore any other H–closed reflector in $[r, \nu \vee r]$ must involve discontinuous semi–algebraic functions. We show that the size of the set $[r, \nu \vee r]_H$ depends on the totally ordered field $r(\mathbb{Z})$: If $r(\mathbb{Z}) = R_0$ then $|[r, \nu \vee r]_H| = 2$, if $r(\mathbb{Z}) \neq R_0$ then $|[r, \nu \vee r]_H|$ may be infinite or even $= 2^{\aleph_0}$. The proofs depend on a careful analysis of the discontinuities that can or must occur among the functions in $s(F(1))$, where $s \in [r, \nu \vee r]_H$.

If $r \in [\rho \wedge \nu, \rho]_{qH}$, $r \colon \mathbf{POR/N} \to \mathbf{D}$, then the interval $[r, \nu \vee r]$ always contains a proper class of reflectors. This can be seen using the construction of section 9D. For every ordinal α, let R_α be the class of all real closed fields of cardinality at least $\aleph_\alpha$. For each α the construction yields an intermediate reflector as follows: Let $\mathbf{C}_\alpha$ be the subcategory of $\mathbf{POR/N}$ consisting of all R–algebras, R varying in R_α. Then $\mathbf{C}_\alpha$ satisfies condition 9D.1, hence a reflector $r_\alpha \colon \mathbf{POR/N} \to \mathbf{D}_\alpha$ is defined by $r_\alpha(A, P) = r(A, P)$ if $r(A, P) \notin \mathbf{C}_\alpha$ and $r_\alpha(A, P) = (\nu \vee r)(A, P)$ if $r(A, P) \in \mathbf{C}_\alpha$. It is clear that $r < r_\alpha < \nu \vee r$ and that $r_\beta < r_\alpha$ if $\alpha < \beta$. Thus, the r_α are a proper class and form a chain in $[r, \nu \vee r]$. It was shown in section 10 that none of the reflectors r_α is H–closed. Therefore other methods are required for the analysis of the set $[r, \nu \vee r]_H$.

Let $r \colon \mathbf{POR/N} \to \mathbf{D}$ be a reflector with $\mathbf{D} \subseteq \mathbf{BIFR/N}$. First note that the reflection $r(\mathbb{Z})$ is a totally ordered subfield $(K, T) \subseteq R_0$ (by bounded inversion). Next we consider the reflection of $F(1)$. If $r(F(1))$ contains some discontinuous semi–algebraic function $a \colon R_0 \to R_0$ then more discontinuous functions can be obtained by applying algebraic operations, lattice operations and composition of functions. In particular, we shall see that certain characteristic functions must belong to $r(F(1))$.

Lemma 22.1 Let $a \colon R_0 \to R_0$ be a semi–algebraic function. Define D to be the set of points $x \in R_0$ such that a is discontinuous at x. Then D is a proper semi–algebraic (hence, finite) subset of R_0.

Proof Using the ε–δ definition of continuity, the set D can be described by a first order formula in the language $\mathcal{L}_{po}$, hence D is semi-algebraic. To see that D is proper, pick any point $\alpha \in Sper(F(1)) \backslash R_0$ and note that $a(\alpha) = \frac{b(\alpha)}{c(\alpha)} \in \rho(\alpha)$ with $b, c \in \rho(F(1))$, $c(\alpha) \neq 0$. The set

$$C = \{\beta \in Sper(F(1)); \; b(\beta) = a(\beta)c(\beta) \; \& \; c(\beta) \neq 0\}$$

contains α and is constructible. Then $C \cap R_0$ is an infinite semi–algebraic subset, hence it contains some interval $(u, u + \varepsilon)$. Since b and c are continuous and since $a|_{(u,u+\varepsilon)} = \frac{b|_{(u,u+\varepsilon)}}{c|_{(u,u+\varepsilon)}}$ we conclude that $a|_{(u,u+\varepsilon)}$ is continuous, i.e., $(u, u + \varepsilon) \cap D = \emptyset$. $\qquad\qquad\square$

We distinguish between three different types of discontinuity that a semi–algebraic function $a \colon R_0 \to R_0$ may or may not have. Suppose that a is discontinuous at $u \in R_0$ and consider $a(u^-) = \lim_{x<u} a(x)$, $a(u)$, $a(u^+) = \lim_{u<x} a(x) \in R_0 \cup \{\pm\infty\}$. The discontinuity is said to be of

type I   if $a(u^-) = a(u^+) \neq a(u)$,
type II   if $a(u^-) = a(u) \neq a(u^+)$ or $a(u^-) \neq a(u) = a(u^+)$,
type III   if $a(u^-) \neq a(u) \neq a(u^+) \neq a(u^-)$.

Lemma 22.2 Suppose that $a \in r(F(1))$ has a discontinuity of type I, II or III at $u \in R_0$. If $v \in R_0$ and $K(v) \supseteq K(u)$ then there is a function $b \in r(F(1))$ having a discontinuity of the same type at v.

Proof There is a polynomial $f \in K[X]$ such that $f(v) = u$, i.e., $f - u$ has a root at v. One may assume that this is a simple root. For otherwise, let $g \in K[X]$ be the minimal polynomial of v. Then $g + f - u$ has a simple root at v, and $f - u$ can be replaced by $g + f - u$. The polynomial f is continuous and belongs to $r(F(1))$. Moreover, f can be restricted to a homeomorphism between suitable neighborhoods $(v - \delta, v + \delta)$ of v and $(u - \varepsilon, u + \eta)$ of u. But then the composition $a \circ f \in r(F(1))$ is discontinuous at v if and only if a is discontinuous at a, and the types of the discontinuities agree. $\qquad\qquad\square$

Lemma 22.3 If $a \in r(F(1))$ has a discontinuity of type I (or II) at $u \in R_0$ then for every $v \in R_0$ such that $K(v) \supseteq K(u)$ the characteristic function of $\{v\}$ (or of one of $(-\infty, v]$ and $[v, +\infty)$) belongs to $r(F(1))$.

Proof By Lemma 22.2 we may assume that $v = u$. Without loss of generality assume that $a(u^-) < a(u)$. Then there are $p, q \in \mathbb{Q}$ with $a(u^-) < p < q < a(u)$. We define $b = (a \wedge q) \vee p \in r(F(1))$. There are $r, s \in \mathbb{Q}$, $r < u < s$, such that $b(x) = p$ for $x \in [r, u) \cup (u, s]$ and $b(u) = q$ if the discontinuity is of type I, and $b(x) = p$ for $x \in [r, u)$ and $b(x) = q$ for $x \in [u, s]$ if the discontinuity is of type II. The function $f = (id \wedge s) \vee r \colon R_0 \to R_0$ belongs to $r(F(1))$, hence so does the composition $b \circ f$. In the case of type I, $b \circ f(u) = q$ and $b \circ f(x) = p$ for all $x \neq u$. If the type is II then $b \circ f(x) = p$ for all $x < u$ and $b \circ f(x) = q$ for all $x \geq u$. In any event, $c = \frac{1}{q-p}(b \circ f - p) \in r(F(1))$. If the type is I or II then $c = \chi_{\{u\}}$ or $c = \chi_{[u,+\infty)}$. $\square$

Lemma 22.4 For $u \in R_0$ the following conditions are equivalent:

  (a) There is some $a \in r(F(1))$ that has a discontinuity of type III at u.
  (b) There exist $b, c \in r(F(1))$ such that b has a dicontinuity of type I at u and c has a discontinuity of type II at u.

Proof (b) $\Rightarrow$ (a) According to Lemma 22.3, the characteristic function $\chi_{\{u\}}$ and one of the characteristic functions $\chi_{[u,+\infty)}$ and $\chi_{(-\infty,u]}$ belongs to $r(F(1))$. Then $\chi_{\{u\}} + \chi_{[u,+\infty)}$ or $\chi_{\{u\}} + \chi_{(-\infty,u]}$ is discontinuous of type III.

(a) $\Rightarrow$ (b) There are six different possibilities for the order relation between $a(u^-)$, (u) and $a(u^+)$. It is easy to reduce these six cases to the following two cases:

Case 1 $a(u^-) < a(u) < a(u^+)$
Pick $p, q \in \mathbb{Q}$ such that $a(u^-) < p < a(u) < q < a(u^+)$. Then $a \wedge p$ is discontinuous of type II with $a \wedge p(u^-) = a(u^-) < p = a(u) = a(u^+)$. By Lemma 22.3, $\chi_{[u,+\infty)} \in r(F(1))$. Similarly, $a \vee q$ is discontinuous of type II, and $(a \vee q)(u^-) = a(u) = q < a(u^+)$. Now Lemma 22.3 shows that $\chi_{(-\infty,u]} \in r(F(1))$. Then

$$\chi_{\{u\}} = \chi_{[u,+\infty)}\chi_{(-\infty,u]} \in r(F(1))$$

is a function which is discontinuous of type I.

Case 2 $a(u) < a(u^-) < a(u^+)$

Pick $p, q \in \mathbb{Q}$ such that $a(u) < p < a(u^-) < q < a(u^+)$. Then $a \wedge p \in r(F(1))$ is discontinuous of type I at u, and $a \vee q \in r(F(1))$ is discontinuous of type II at u.   □

Lemma 22.5 Suppose that $a \in r(F(1))$ is discontinuous at $u \in K$. Then for every $v \in R_0$ the characteristic functions $\chi_{\{v\}}, \chi_{[v,+\infty)}$ and $\chi_{(-\infty,v]}$ belong to $r(F(1))$.

Proof Since $K(v) \supseteq K = K(u)$ holds for every $v \in R_0$ it suffices to show that discontinuities of every type occur at u (Lemma 22.2). If a is discontinuous of type III then this follows from Lemma 22.4. Now suppose that a is discontinuous of type I or II and suppose that $a(u^-) < a(u)$. There are $q, r \in \mathbb{Q}$ such that $a(u^-) < q < a(u)$, $r < u$ and $a(x) < q$ for all $x \in [r, u)$. Then the functions $a \vee q$, $f = (id \vee r) \wedge u$ and $g = (id \vee r) \wedge ((2u - id) \vee r)$ all belong to $r(F(1))$, hence, so do the compositions $(a \vee q) \circ f$ and $(a \vee q) \circ g$. One checks easily that $(a \vee q) \circ g$ is discontinuous of type I and $(a \vee q) \circ f$ is discontinuous of type II. The proof is finished with an application of Lemma 22.4.   □

After having obtained some basic information about discontinuities we shall now use it to discuss the intermediate H–closed monoreflectors between $r \in [\rho \wedge v, \rho]_{qH}$ and $\nu \vee r$. We need the following sufficient condition for a reflector to be stronger than ν:

Proposition 22.6 Suppose that $r \colon \mathbf{POR/N} \to \mathbf{D}$, $\mathbf{D} \subseteq \mathbf{BIFR/N}$, is a monoreflector, and let $(K, T) = r(\mathbb{Z})$. If there is a function $a \in r(F(1))$ that is discontinuous at some $u \in K$ then $\mathbf{D} \subseteq \mathbf{VNRFR}$.

Proof It follows immediately from Lemma 22.5 that the characteristic function of every semi–algebraic subset $S \subseteq R_0$ belongs to $r(F(1))$. Thus, $r(F(1))$ is a von Neumann regular f–ring, i.e., $\nu(F(1)) \subseteq r(F(1))$. According to Example 13.7 the reflector ν has arity 1, hence $\nu \leq r$.   □

Theorem 22.7 If $r \colon \mathbf{POR/N} \to \mathbf{D}$ with $\mathbf{RCR} \subseteq \mathbf{D} \subseteq \mathbf{R_0CPWRFR}$ is H–closed and quotient–closed then there is no H–closed monoreflective subcategory **E** that lies properly in between **D ∩ VNRFR** and **D**.

Proof According to Theorem 21.4 (b), no H–closed monoreflective subcategory **E** with **D ∩ VNRFR** $\subseteq$ **E** $\subset$ **D** contains **RCR**. Let $s\colon \mathbf{POR/N} \to \mathbf{E}$ be the corresponding reflector. Then for some $n \in \mathbb{N}$ the reflection $s(F(n))$ contains a discontinuous semi–algebraic function (Theorem 10.6). We use curve selection to show that $s(F(1))$ must contain a disontinuous function. Then, because of $s(\mathbb{Z}) = R_0$, it follows from Proposition 22.6 that $\nu \le s$, hence **E** $=$ **D ∩ VNRFR**, and the proof will be finished.

We have to show that $s(F(1))$ contains some discontinuous function. To show the direction of our arguments and to point out the difficulties, we first treat an easy special case.

Let $t\colon \mathbf{POR/N} \to \mathbf{F}$ be an H–closed reflector with $\rho < t \le \sigma$. There is some $1 \le n \in \mathbb{N}$ with $\rho(F(n)) \subset t(F(n))$. If this is the case for $n = 1$ then we are done. So, assume that $n > 1$. There is a discontinuous function $a \in t(F(n))$. We may assume that a is discontinuous at $0 \in R_0^n$ and that $a(0) = 0$. There is some $\varepsilon > 0$ such that 0 belongs to the closure of at least one of the semi–algebraic sets

$$
\begin{aligned}
M_\varepsilon^+ &= \{x \in R_0^n;\ a(x) \ge \varepsilon\}, \\
M_\varepsilon^- &= \{x \in R_0^n;\ a(x) \le -\varepsilon\}.
\end{aligned}
$$

Without loss of generality, suppose that $0 \in \overline{M_\varepsilon^+}$. The Curve Selection Lemma ([39], Theorem 12.1) supplies a continuous semi–algebraic path $p\colon [0,1] \to R_0^n$ with $p(0) = 0$ and $p((0,1]) \subseteq M_\varepsilon^+$. We extend p to a continuous semi–algebraic function $\hat{p}\colon R_0 \to R_0^n$ by setting $\hat{p}(t) = 0$ for $t < 0$, $\hat{p}(t) = p(1)$ for $t > 1$. There is a homomorphism $f\colon \rho(F(n)) \to \rho(F(1))$ corresponding to the semi–algebraic map $\hat{p}$. Because of $\rho < t$ the functorial morphisms $t(\rho_{F(1)})\colon t(F(1)) \to t(\rho(F(1)))$ and $t(\rho_{F(n)})\colon t(F(n)) \to t(\rho(F(n)))$ are both isomorphisms (Proposition 8.1). Therefore we identify $t(F(1)) = t(\rho(F(1)))$ and $t(F(n)) = t(\rho(F(n)))$. Now f yields the homomorphism $t(f)\colon t(F(n)) \to t(F(1))$. Explicitly, if $c \in t(F(n))$ then $t(f)(c) = c \circ \hat{p}$. We obtain $t(f)(a) = a \circ \hat{p} \in t(F(1))$, which is a discontinuous function by construction. So, if $\rho < t$ then $t(F(1))$ contains a discontinuous function, and the proof is finished in this special case.

There is only one point in this entire chain of arguments which does not immediately generalize to the reflector s, namely the use of the Curve Selection Lemma. Instead of the continuous semi–algebraic curve used above we need a curve $p : R_0 \to R_0^n$ which induces a homomorphism

$s(F(n)) \to s(F(1))$. This is the case only if all the components p_i of p belong to $s(F(1))$. We shall construct such a curve. Let $D \subseteq R_0^n$ be the set of discontinuities of a, i.e.,

$$D = \{x \in R_0^n;\ \exists\, \varepsilon > 0\ \forall\, \delta > 0\ \exists\, y : |x - y| < \delta\ \&\ |a(x) - a(y)| \geq \varepsilon\}.$$

This is a semi–algebraic subset of R_0^n of dimension $m < n$. In the same way, the set D' of discontinuities of $a|_D$ is a semi–algebraic subset of dimension $l < m$. If x is any point of $D \backslash D'$ then $a|_D$ is continuous in a neighborhood of x. We assume that $x = 0$ and that $a(0) = 0$. Since a is discontinuous we find some $\varepsilon > 0$ such that 0 belongs to the closure of at least one of the following sets:

$$\begin{aligned}
M_\varepsilon^+ &= \{x \in R_0^n;\ a(x) > \varepsilon\}, \\
M_\varepsilon^- &= \{x \in R_0^n;\ a(x) < -\varepsilon\}.
\end{aligned}$$

Continuity of $a|_D$ at 0 implies that there is some $0 < \eta \in R_0$ with $|a(x)| < \varepsilon$ for every $x \in D \cap B_\eta(0)$ (where $B_\eta(0)$ is the open ball of radius η about 0). With

$$B_\varepsilon^+ = M_\varepsilon^+ \cap B_\eta(0),\quad B_\varepsilon^- = M_\varepsilon^- \cap B_\eta(0)$$

we note that $B_\varepsilon^+ \cap D = \emptyset$, $B_\varepsilon^- \cap D = \emptyset$ and $0 \in \overline{B_\varepsilon^+}$ or $0 \in \overline{B_\varepsilon^-}$. Without loss of generality, assume $0 \in \overline{B_\varepsilon^+}$. By [19], Proposition 8.1.17, there is a *Nash curve* $\varphi : (-1, +1) \to R_0^n$ such that $\varphi(0) = 0$ and $\varphi((0,1)) \subseteq B_\varepsilon^+$. For each $t \in [0,1)$, let $d(t)$ be the distance between the point $\varphi(t)$ and the set $R_0^n \backslash B_\varepsilon^+$. Then d is a continuous semi–algebraic function and $d(t) > 0$ for $t \in (0,1)$. According to the Łojasiewicz inequality ([19], Corollaire 2.6.7) there are some $0 < c \in R_0$ and some $N \in \mathbb{N}$ such that $ct^N < d(t)$ for all $t \in [0, \frac{1}{2}]$. Let $R_0(T)_{\text{alg}}^{\wedge}$ be the field of algebraic Puiseux series ([19], Example 1.3.6(b)). For every component φ_i of φ, the Taylor series is denoted by $S_i(T) = \sum_{j \in \mathbb{N}} s_{ij} T^j \in R_0[[T]]_{\text{alg}} \subseteq R_0(T)_{\text{alg}}^{\wedge}$ ([19], Corollaire 8.1.5, Lemma 8.1.15); let $P_i(T)$ be the initial polynomial $\sum_{j \leq N} s_{ij} T^j$, $p_i : R_0 \to R_0$ the corresponding polynomial function. Then the order of the power series $S_i(T) - P_i(T)$ is at least $N + 1$, hence

$$\sum_{i=1}^n (S_i(T) - P_i(T))^2 < c^2\, T^{2N}.$$

Returning from the formal setting to functions (cf. [19], p. 149) we conclude that there is some $0 < \delta \in R_0$ with

$$\left(\sum_{i=1}^n (\varphi_i(t) - p_i(t))^2 \right)^{\frac{1}{2}} < ct^N$$

for all $t \in (0, \delta)$. Altogether the polynomial functions p_i determine a polynomial curve $p \colon R_0 \to R_0^n$. By construction of the curve, $|\varphi(t) - p(t)| < ct^N < d(t)$ for $t \in (0, \delta)$, hence $p(t) \in B_\varepsilon^+ \subseteq M_\varepsilon^+$ for each $t \in (0, \delta)$. It is clear that $p(0) = 0$. Thus, p defines a homomorphism $f \colon R_0[T_1, \ldots, T_n] \to R_0[T] \subseteq s(F(1))$ which extends to $\overline{f} \colon s(F(n)) \to s((F(1))$. If $c \in s(F(n))$ then $\overline{f}(c) = c \circ p$. In particular, $\overline{f}(a) \in s(F(1))$ is discontinuous. $\qquad\square$

As a particularly remarkable special case of the theorem we record

Corollary 22.8 The reflectors ρ and σ are consecutive elements in the lattice of H–closed monoreflectors of **POR/N**. $\qquad\square$

We continue to study the interval $[r, \nu \vee r]_H$ for some reflector $r \colon \textbf{POR/N} \to \textbf{D}$ that belongs to the interval $[\rho \wedge \nu, \rho]_{qH}$. But now we assume that $(K, T) = r(\mathbb{Z})$ is properly contained in R_0. We would like to prove the existence of infinitely many H–closed reflectors that belong to the interval. We start by constructing reflectors $s \colon \textbf{POR/N} \to \textbf{E}$ that have arity 1 and for which the functions in $s(F(1))$ can have only prescribed discontinuities.

Let $\mathcal{F}(K)$ be the partially ordered (by inclusion) set of all field extensions of K inside R_0. In $\mathcal{F}(K)$, let $\mathcal{F}$ be a proper nonempty subset such that $L \in \mathcal{F}$, $L \subseteq M$ implies $M \in \mathcal{F}$. We call such a subset of $\mathcal{F}(K)$ an *upper subset*. Moreover, assume that $\mathcal{F}$, as an upper subset, is generated by the finite extensions of K belonging to $\mathcal{F}$, i.e., given $M \in \mathcal{F}$ there is a finite extension $L \supseteq K$ that belongs to $\mathcal{F}$ and is contained in M. Every field belonging to $\mathcal{F}(K)$ carries the total order induced by R_0. Now $s(F(1))$ is defined to be the subset of all semi–algebraic functions $a \in (\nu \vee r)(F(1))$ having the following property: If a is discontinuous at $u \in R_0$ then $r(K(u)) \in \mathcal{F}$. We record a number of elementary facts about $s(F(1))$:

Lemma 22.9 The set $s(F(1))$ is a sub–f–ring of $(\nu \vee r)(F(1))$ containing the image of $(\nu \vee r)_{F(1)}$. $\qquad\square$

By restriction of the codomain, $(\nu \vee r)_{F(1)}$ yields a homomorphism $F(1) \to s(F(1))$ which we denote by $s_{F(1)}$. Eventually it will be shown that $s_{F(1)}$

defines an H–closed monoreflector of arity 1. The first step is to prove

Lemma 22.10 $s_{F(1)} \colon F(1) \to s(F(1))$ is an epimorphism in **POR/N**.

Proof By Proposition 7.16 it suffices to show that $Sper(s_{F(1)})$ is injective or, equivalently, that $Sper(j)$ is surjective (where $j \colon s(F(1)) \to (\nu \vee r)(F(1))$ is the inclusion). The support maps of $s(F(1))$ and $(\nu \vee r)(F(1))$ are both homeomorphisms onto the Brumfiel spectra. The Brumfiel spectra include all minimal prime ideals. Since j is an extension, every minimal prime ideal of $s(F(1))$ extends to a prime ideal of $(\nu \vee r)(F(1))$. It follows that the generic points of $Sper(s(F(1)))$ belong to $im(Sper(j))$. The dimension of $Sper(s(F(1)))$ is at most 1. For, if $\alpha_2 \subset \alpha_1 \subset \alpha_0$ is a specialization chain in $Sper(s(F(1)))$ with α_2 generic then $\alpha_2 = j^{-1}(\beta_2)$ for some $\beta_2 \in Sper((\nu \vee r)(F(1)))$. If $\gamma_2 = s_{F(1)}^{-1}(\alpha_2)$ then the composition

$$\kappa_{F(1)}(\gamma_2) \to \kappa_{s(F(1))}(\alpha_2) \to \kappa_{(\nu \vee r)(F(1))}(\beta_2)$$

is an algebraic extension of totally ordered fields, hence the extension $\kappa_{F(1)}(\gamma_2) \to \kappa_{s(F(1))}(\alpha_2)$ is algebraic as well. Every residue field $\kappa_{F(1)}(\gamma_2)$ is algebraic over $\mathbb{Q}$ or has transcendence degree 1 over $\mathbb{Q}$. The same is true for $\kappa_{s(F(1))}(\alpha_2)$. Therefore no subring can have more than two convex prime ideals. However, $s(F(1))/\alpha_2$ has three convex prime ideals, namely $supp(\alpha_i)/supp(\alpha_2)$ for $i = 0, 1, 2$. This contradiction shows that $Sper(s(F(1)))$ does not have specialization chains of length greater than 1. We conclude that $im(Sper(j)) \subseteq Sper(s(F(1)))$ is generically closed, contains all generic points and is proconstructible. Assume by way of contradiction that $Sper(j)$ is not surjective. Then there is some nonempty closed constructible subset $C \subseteq Sper(s(F(1)))$ with $C \cap im(Sper(j)) = \emptyset$. Since $s(F(1))$ is a reduced f–ring there is a function $c \in s(F(1))$, $0 \le c$, such that

$$C = \{\alpha \in Sper(s(F(1))); \ c(\alpha) = 0\}.$$

Thus, as a function on $Sper(F(1))$, c is positive everywhere, hence $c \in s(F(1))^{\times}$. But then $c(\alpha) \ne 0$ for every $\alpha \in Sper(s(F(1)))$. This contradicts the assumption that $C \ne \emptyset$. Hence, $Sper(j)$ is surjective, and it follows that f is an epimorphism. $\qquad\square$

Lemma 22.11 The reduced poring $s(F(1))$ belongs to the injectivity class of $s_{F(1)}$.

Proof Let $f\colon F(1) \to s(F(1))$ be any **POR/N**–morphism. It is uniquely determined by $f(T)$, where $T \in F(1)$ is the generator of the free object $F(1)$. Since $\nu \vee r$ is a reflector with $(\nu \vee r)_{F(1)} = js_{F(1)}$ there is a unique morphism $g\colon (\nu \vee r)(F(1)) \to (\nu \vee r)(F(1))$ such that $jf = gjs_{F(1)}$. It suffices to prove that $g(s(F(1))) \subseteq s(F(1))$. For, then the restriction $g'\colon s(F(1)) \to s(F(1))$ of g has the property that $f = g's_{F(1)}$.

Pick any $a \in s(F(1))$ and note that the semi–algebraic function $g(a)\colon R_0 \to R_0$ is the composition

$$R_0 \xrightarrow{\;f(T)\;} R_0 \xrightarrow{\;a\;} R_0.$$

Let $u \in R_0$ be a discontinuity of $a \circ f(T)$. If $f(T)$ is discontinuous at u then $r(K(u)) \in \mathcal{F}$ (since $f(T) \in s(F(1))$). Now suppose that $f(T)$ is not discontinuous at u. Then a must be discontinuous at $f(T)(u)$, hence $r(K(f(T)(u))) \in \mathcal{F}$ (since $a \in s(F(1))$). Let $\alpha \in Sper(F(1))$ be the point corresponding to $u \in R_0$, let $\beta \in Sper((\nu \vee r)(f(1)))$, $\gamma \in Sper(s(F(1)))$ and $\delta \in Sper(r(F(1)))$ be the corresponding points. Then the homomorphisms

$$\kappa_{r(F(1))}(\delta) \to \kappa_{s(F(1))}(\gamma) \to \kappa_{(\nu\vee r)(F(1))}(\beta)$$

are isomorphisms. Moreover, this totally ordered field is isomorphic to $r(K(u))$. For, $u \in \kappa_{r(F(1))}(\delta)$ holds trivially, hence $r(K(u)) \subseteq \kappa_{r(F(1))}(\delta)$. On the other hand, the homomorphism $e\colon F(1) \to K(u)$, $T \to u$ (note that e defines $\alpha \in Sper(F(1))$) extends uniquely to $r(e)\colon r(F(1)) \to r(K(u))$. But then $r(e)$ defines $\delta \in Sper(r(F(1))$, hence $r(e)$ factors as $r(F(1)) \to \kappa_{r(F(1))}(\delta) \to r(K(u))$. It follows that $r(K(u))$ may be identified with $\kappa_{r(F(1))}(\delta)$. This implies that $f(T)(u) = f(T)(\delta) \in r(K(u))$, and therefore $r(K(f(T)(u))) \subseteq r(K(u))$. We conclude that $r(K(u)) \in \mathcal{F}$, as claimed.   □

Summarizing these considerations, this is what has been shown so far: The injectivity class of $s_{F(1)}$ determines a subcategory $\mathbf{E} \subseteq \mathbf{POR/N}$. By Theorem 10.2, $\mathbf{E}$ is H–closed and monoreflective in $\mathbf{POR/N}$, its arity is 1 (Section 13). We denote the corresponding reflector by $s\colon \mathbf{POR/N} \to \mathbf{E}$.

Lemma 22.12 The reflector $r \vee s$ is H–closed, it belongs to the interval $[r, \nu \vee r]$ and coincides with neither r nor $\nu \vee r$.

Proof The supremum of the H–closed reflectors r and s is H–closed (by Theorem 10.2). The reflector $r \vee s$ is stronger than r by definition; it is weaker than $\nu \vee r$ since both r and s are weaker than $\nu \vee r$. (Recall that $s(F(1))$ was defined to be a subring of $(\nu \vee r)(F(1))$.) Next we claim that $s(F(1)) \in ob(\mathbf{D})$. The proof is done by checking that $s(F(1)) \in Inj(\{r_{F(n)}; n \in \mathbb{N}\})$ (Theorem 10.6): Let $f\colon F(n) \to s(F(1)) \subseteq (\nu \vee r)(F(1))$ be any homomorphism. Then f extends to

$$\overline{f}\colon r(F(n)) \to (\nu \vee r)(f(1)).$$

Every function $a \in r(F(n))$ is continuous, hence the points of discontinuity of the composition

$$\overline{f}(a) = a(f(T_1), \dots, f(T_n))$$

are among the points of discontinuity of the functions $f(T_1), \dots, f(T_n)$. It follows that $\overline{f}(a) \in s(F(1))$, i.e., $\overline{f}$ may be considered as a homomorphism $r(F(n)) \to s(F(1))$. Thus, $\overline{f}$ is the desired extension of f.

Because $s(F(1))$ belongs to $\mathbf{D}$ it follows that $(r \vee s)(F(1)) = s(F(1))$. The set $\mathcal{F}$ is nonempty by assumption, say $K(u) \in \mathcal{F}$ for $u \in R_0$. Then also $r(K(u)) \in \mathcal{F}$, and this implies that the characteristic function $\chi_{\{u\}}$ belongs to $s(F(1))$ (being discontinuous at u). But $\chi_{\{u\}} \notin r(F(1))$, hence we conclude that $r < r \vee s$. To prove that $r \vee s < \nu \vee r$, note that $\mathcal{F}$ is supposed to be a proper subset of $\mathcal{F}(K)$, i.e., $K \notin \mathcal{F}$. For every $u \in K$, the characteristic function $\chi_{\{u\}}$ belongs to $(\nu \vee r)(F(1))$ (because $(\nu \vee r)(F(1))$ is a member of **VNRFR**), but it does not belong to $s(F(1))$ (by the restriction on the discontinuities in $s(F(1))$). Thus, $(r \vee s)(F(1)) \neq (\nu \vee r)(F(1))$, and this proves the final assertion. $\square$

So far we have been working with a fixed subset $\mathcal{F}$ of the set $\mathcal{F}(K)$ of extensions of (K, T) in R_0. Now the set $\mathcal{F}$ will be allowed to vary. Accordingly, the reflector s discussed above will be denoted by $s_{\mathcal{F}}$ from now on. The purpose is to produce a large number of intermediate H–closed reflectors of r and $\nu \vee r$. The best possible result would be that there are continuously many such reflectors. However, a completely general result of this type requires a much better understanding of the set of reflections $r(L, S)$ (where (L, S) is a totally ordered finite extension of (K, T)) than we possess at this time. Currently our best general result in this direction is

Theorem 22.13 Let $(K, T) \subset R_0$ be a totally ordered proper subfield, let $r \in [\rho \wedge \nu, \rho]_{qH}$ be the reflector **POR/N** $\to$ **(K, T)CPWRFR**. Then there are infinitely many H–closed monoreflectors in the interval $[r, \nu \vee r]$.

Proof As before, let $\mathcal{F}(K)$ be the set of extension fields of K inside R_0. Since K is a proper subfield of R_0 the set $\mathcal{F}(K)$ is infinite. For every upper subset $\mathcal{F} \subseteq \mathcal{F}(K)$ that is generated by finite extensions, let $s_{\mathcal{F}} \colon$ **POR/N** $\to$ **E**$_{\mathcal{F}}$ be the reflector studied above. If $\mathcal{F} \neq \emptyset$ and $\mathcal{F} \neq \mathcal{F}(K)$ then $r \vee s_{\mathcal{F}} \in [r, \nu \vee r] \setminus \{r, \nu \vee r\}$ (Lemma 22.12). Moreover, if $\mathcal{F} \neq \mathcal{F}'$ then $r \vee s_{\mathcal{F}} \neq r \vee s_{\mathcal{F}'}$: Suppose that there is a field $K(u) \in \mathcal{F}' \setminus \mathcal{F}$. The totally ordered field $K(u)$ (with the restriction of the total order of R_0) is a member of **(K, T)CPWRFR**, hence the characteristic function $\chi_{\{u\}}$ belongs to $s_{\mathcal{F}'}(F(1))$, but not to $s_{\mathcal{F}}(F(1))$. To finish the proof we note that the set $\mathcal{F}(K)$ contains infinitely many upper subsets. $\qquad\square$

In one special case we can actually prove the strongest conceivable result:

Theorem 22.14 There are continuously many H–closed monoreflective subcategories **D** between **VNRFR** and **CPWRFR**.

Proof Continuing with the notation in the proof of the previous theorem, we only need to show that $\mathcal{F}(\mathbb{Q})$ has continuously many upper subsets. Let $\mathbb{P}$ be the set of all prime numbers, let $\mathbb{P}' \subseteq \mathbb{P}$ be a subset. We define $\mathcal{F}_{\mathbb{P}'}$ to be the set of all subfields $L \subseteq R_0$ containing some $\mathbb{Q}(\sqrt{p})$, $p \in \mathbb{P}'$. These sets of subfields meet the requirements, and there are $2^{\aleph_0}$ such sets. $\qquad\square$

23 The lattice of H–closed monoreflectors

In this final section we look at the class of all H–closed monoreflective subcategories of **POR/N** and summarize the information we have collected. We start with a brief discussion of the class of all monoreflectors of **POR/N**. The following chart ((23.1) on the next page) shows the inclusion relations between some of the monoreflective subcategories that we encountered in our investigations.

The class of all monoreflective subcategories of **POR/N** is a complete lattice (by Theorem 8.3) with largest element **POR/N** (trivially) and smallest element **SAFR** (Theorem 8.10). The lattice is a proper class (Theorem 9D.3); in fact, between any two reflective subcategories that are depicted in the chart and contain each other there is a chain of monoreflectors consisting of a proper class. The category **RCR** is distinguished in this lattice by each of the following three remarkable properties:

- **RCR** is the smallest monoreflective subcategory of **POR/N** such that all reflections preserve the real spectrum (Theorem 12.12);
- **RCR** is the smallest monoreflective subcategory of **FR/N** such that all reflection morphisms are essential monomorphisms (Theorem 15.5);
- **RCR** is the largest monoreflective subcategory of **CPWRFR** such that every totally ordered residue field is real closed (Theorem 21.10).

Much more is known about the class of H–closed monoreflectors. The diagram (23.2) (p. 255) exhibits the relationship between various H–closed monoreflective subcategories.

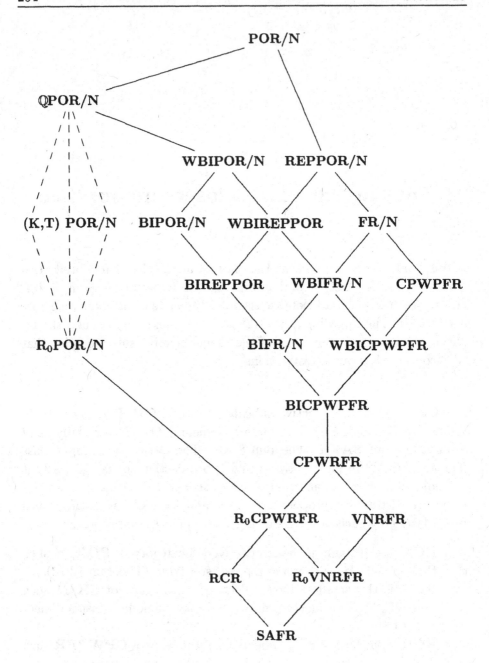

(23.1) Monoreflective subcategories of **POR/N**

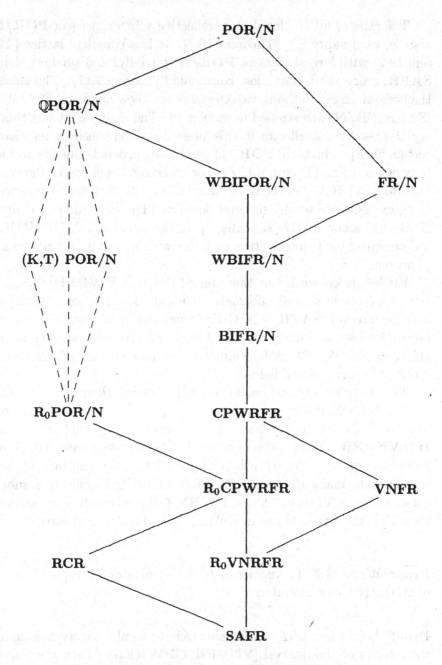

(23.2) H–closed monoreflective subcategories of **POR/N**

The class of all H–closed monoreflective subcategories of **POR/N** is a set of cardinality $2^{\aleph_0}$ (Theorem 10.7). It is a complete lattice (Theorem 10.2) with largest element **POR/N** (trivially) and smallest element **SAFR**, every chain is at most countable (Theorem 10.7). The closer to the bottom we are in the lattice the more we know about it. The interval [**SAFR, FR/N**] was studied in section 19. The main result was that every H–closed monoreflector in this interval is determined by its effect on certain fibre products (in **POR!**) of two totally ordered domains over a totally ordered ring (Theorem 19.8). For reflectors belonging to the smaller interval [**SAFR, CPWRFR**] it suffices to know their restriction to the category **TOD** of totally ordered domains (Theorem 20.1). Finally, an H–closed monoreflective subcategory in the interval [**SAFR, VNRFR**] is determined by its intersection with the category of totally ordered fields (Theorem 17.6).

Far less is known about the interval [**RCR, CPWRFR**]. Using the lattice operations to shift information back and forth between this interval and the interval [**SAFR, VNRFR**] it was shown in Corollary 21.5 that every H–closed and quotient–closed monoreflective subcategory between **RCR** and **CPWRFR** is determined by its intersection with the category **TOF** of totally ordered fields.

The lattice contains pairs of consecutive elements: If $\mathbf{D} \in$ [**RCR, R$_0$CPWRFR**] is H–closed and quotient–closed then there exists no H–closed monoreflective subcategory lying properly between $\mathbf{D} \cap \mathbf{VNRFR}$ and $\mathbf{D}$ (Theorem 22.7). As a special case, **RCR** is an immediate successor of **SAFR** (Corollary 22.8). By contrast, there are continuously many (Theorem 22.14) H–closed monoreflective subcategories between **VNRFR** and **CPWRFR**. These results can be used to show that the lattice is not modular, hence also not distributive:

Proposition 23.3 The lattice of H–closed monoreflective subcategories of **POR/N** is not modular.

Proof Let $\mathbf{D}$ be an H–closed monoreflective subcategory belonging to the interior of the interval [**VNRFR, CPWRFR**]. Then the following diagram depicts a sublattice in the lattice of H–closed monoreflective subcategories of **POR/N**:

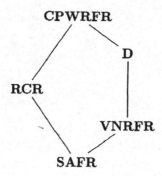

According to [18], Chapter I, §7, the non–modular lattices are characterized by the property that they contain this lattice as a sublattice.
Thus, our lattice is not modular.   □

This summary of our results shows that some parts of the lattice of H–
closed monoreflective subcategories of **POR/N** are quite well understood.
But there are also large areas which we know very little about. Notably
the part close to the top of diagram (23.2) remains mostly terra incognita
and waits for further exploration.

Bibliography

[1] J. Adámek, H. Herrlich, G. Strecker: Abstract and Concrete Categories. John Wiley & Sons, New York 1990

[2] M.E. Alonso, M.F. Roy: Real strict localizations. Math. Z. **194**, 429–441 (1987)

[3] C. Andradas, L. Bröcker, J.M. Ruiz: Constructible Sets in Real Geometry. Springer, Berlin 1996

[4] C. Andradas, J.M. Ruiz: Algebraic and Analytic Geometry of Fans. Memoirs AMS, No. **553**, Amer. Math. Soc., Providence 1995

[5] V. Arnold, G. Shimura: Superposition of algebraic functions. In: Mathematical developments arising from Hilbert Problems (Ed. F. Browder), Proc. Symp. Pure Math. **28**, Part 1, Amer. Math. Soc., Providence 1976, pp. 45-46

[6] E. Artin: Über die Zerlegung definiter Funktionen in Quadrate. Abh. Math. Sem. Univ. Hamb. **5**, 100–115 (1927)

[7] E. Artin, O. Schreier: Algebraische Konstruktion reeller Körper. Abh. Math. Sem. Univ. Hamb. **5**, 85-99 (1926)

[8] P.D. Bacsich: Model theory of epimorphisms. Canad. Math. Bull. **17**, 471–477 (1974)

[9] R.N. Ball, A.W. Hager: Characterization of epimorphisms in archimedean lattice–ordered groups and vector lattices. In: Lattice–Ordered Groups (Eds. A. Glass, C. Holland), Mathematics and its Appl., Vol. **49**, Kluwer, Dordrecht 1989, pp. 175-205

[10] R.N. Ball, A.W. Hager: Epicompleteness in archimedean lattice–ordered groups. Trans. AMS **322**, 459-478 (1990)

[11] R.N. Ball, A.W. Hager: Epicompletion of archimedean l–groups and vector lattices with weak order unit. J. Austral. Math. Soc. **48**, 25-56 (1990)

[12] R.N. Ball, A.W. Hager: Algebraic extensions of archimedean lattice–ordered groups I. J. Pure Applied Alg. **85**, 1-20 (1993)

[13] E. Becker: Euklidische Körper und euklidische Hüllen von Körpern. J. reine angew. Math. **268/269**, 41–52 (1974)

[14] E. Becker, E. Köpping: Reduzierte quadratische Formen und Semiordnungen reeller Körper. Abh. Math. Sem. Univ. Hamb. **46**, 143–177 (1977)

[15] E. Becker, R. Berr, D. Gondard: Valuation fans and residually real closed Henselian fields. Preprint

[16] W.M. Beynon: Duality theorems for finitely generated vector lattices. Proc. London Math. Soc. **31**, 114-128 (1975)

[17] A. Bigard, K. Keimel, S. Wolfenstein: Groupes et Anneaux Réticulés. Lecture Notes in Mathematics, Vol. **608**, Springer, Berlin 1977

[18] G. Birkhoff: Lattice Theory. Amer. Math. Soc., Providence 1973

[19] J. Bochnak, M. Coste, M.-F. Roy: Géométrie algébrique réelle. Springer, Berlin 1987

[20] N. Bourbaki: Algèbre, Chapitres 6/7. Hermann, Paris 1964

[21] N. Bourbaki: Algèbre commutative, Chapitres 1/2. Hermann, Paris 1961

[22] N. Bourbaki: Algèbre commutative, Chapitre 5/6. Hermann, Paris 1964

[23] N. Bourbaki: Algèbre commutative, Chapitres 8/9. Masson, Paris 1983

[24] N. Bourbaki: Topologie Générale, Chapitres 1 à 4. Diffusion C.C.LS. Paris 1971

[25] E. Brieskorn, M. Knörrer: Plane algebraic curves. Birkhäuser, Basel 1986

[26] L. Bröcker: Zur Theorie der quadratischen Formen über formal reellen Körpern. Math. Ann. **210**, 233–256 (1974)

[27] G.W. Brumfiel: Partially Ordered Rings and Semi–Algebraic Geometry. London Mathematical Society Lecture Note Series, Vol. **37**, Cambridge University Press, Cambridge 1979

[28] G.W. Brumfiel: Witt rings and K–theory. Rocky Mountain J. Math. **14**, 733–765 (1984)

[29] G.W. Brumfiel: The real spectrum of an ideal and KO–theory exact sequences. K–Theory **1**, 211–235 (1987)

[30] C.C. Chang, H.J. Keisler: Model Theory, 2nd ed. North Holland, Amsterdam 1977

[31] M.-F. Coste, M. Coste: Topologies for real algebraic geometry. In: Topos–Theoretic Methods in Geometry (Ed. A. Kock), Aarhus Univ. Var. Publ. Ser., No. **30**, Aarhus, pp. 37-100

[32] M. Coste, M.-F. Coste–Roy: Spectre de Zariski étale: du cas classique au cas réel. Univ. Cath. de Louvain, Sém. de Math. pure, Rapport No. 82

[33] M. Coste, M.-F. Coste–Roy: La topologie du spectre réel. In: Ordered Fields and Real Algebraic Geometry (Eds. D.W. Dubois, T. Recio), Contemporary Mathematics, Vol. **8**, Amer. Math. Soc., Providence 1982, pp. 27-59

[34] M. Coste, M.-F. Coste–Roy: Le topos réel d'un anneau. Cahiers de top. et géom. diff. **22**, 19-24 (1981)

[35] M.-F. Coste–Roy: Le spectre réel d'un anneau. Thèse

[36] M.-F. Coste–Roy, M. Coste: Le spectre étale réel d'un anneau est spatial. C.R. Acad. Sci. Paris **290**, 91-94 (1980)

[37] F. Cucker: Fonctions de Nash sur les variètés algébriques affines. These, Université de Rennes, 1986

[38] H. Delfs: The homotopy axiom in semi–algebraic cohomology. J. reine angew. Math. **355**, 108-128 (1985)

[39] H. Delfs, M. Knebusch: Semialgebraic Topology over a Real Closed Field II – Basic Theory of Semialgebraic Spaces. Math. Z. **178**, 175-213 (1981)

[40] H. Delfs, M. Knebusch: On the homology of algebraic varieties over real closed fields. J. reine angew. Math. **335**, 122-163 (1982)

[41] H. Delfs, M. Knebusch: Locally semialgebraic spaces. Lecture Notes in Math., Vol. **1173**, Springer, Berlin 1985

[42] F. Delon, R. Farré: Some model theory for almost real closed fields. J. Symb. Logic **61**, 1121-1152 (1996)

[43] C.N. Delzell: On the Pierce–Birkhoff Conjecture over ordered fields. Rocky Mountain J. Math. **19**, 651-668 (1989)

[44] C.N. Delzell, J.J. Madden: Lattice–ordered rings and semialgebraic geometry, I. In: Real Analytic and Algebraic Geometry (Eds. F. Broglia et al.), de Gruyter, Berlin 1995, pp. 103-129

[45] J. Dieudonné: History of Algebraic Geometry. Wadsworth, Monterey 1985

[46] O. Endler: Valuation Theory. Springer, Berlin 1972

[47] L. Fuchs: Teilweise geordnete algebraische Strukturen. Vandenhoeck & Ruprecht, Göttingen 1966

[48] L. Fuchs: Abelian Groups. Publ. House of the Hungarian Acad. of Sciences, Budapest 1958

[49] L. Gillman, M. Jerison: Rings of Continuous Functions. Grad. Texts in Math., Vol. **43**, Springer, Berlin 1976

[50] A. Grothendieck, J.A. Dieudonné: Eléménts de Géométrie Algébrique. Springer, Berlin 1971

[51] A.W. Hager: $C(X)$ has no proper functorial hulls. In: Rings of Continuous Functions (Ed. Ch. Aull), Marcel Dekker, New York 1985, pp. 149-164

[52] A.W. Hager: Algebraic Closures of l–Groups of Continuous Functions. In: Rings of Continuous Functions (Ed. Ch. Aull), Marcel Dekker, New York 1985, pp. 165-194

[53] A.W. Hager: A description of HSP–like classes, and applications. Pac. J. Math. **125**, 93–102 (1986)

[54] A.W. Hager, J.J. Madden: Algebraic classes of abelian torsionfree and lattice–ordered groups. Bull. Greek Math. Soc. **25**, 53-63 (1984)

[55] A.W. Hager, J. Martinez: Maximum Monoreflections. Applied Categorical Structures **2**, 315–329 (1994)

[56] A.W. Hager, L.C. Robertson: Representing and ringifying a Riesz space. In: Symposia Math., Vol. XXI, Academic Press, London, 1977, pp. 411-431

[57] R. Hartshorne: Algebraic Geometry. Graduate Texts in Mathematics, Vol. **52**, Springer, New York 1977

[58] M. Henriksen: A survey of f–rings and some of their generalizations. In: Ordered Algebraic Structures (Eds.: W.C. Holland, J. Martinez), Kluwer, Dordrecht 1997, pp. 1–26

[59] M. Henriksen, J. Isbell, D. Johnson: Residue class fields of lattice-ordered algebras. Fund. Math. **50**, 107-117 (1961)

[60] H. Herrlich, G. Strecker: Category Theory, 2nd ed. Heldermann, Berlin 1979

[61] D. Hilbert: Ueber die Darstellung definiter Formen als Summe von Formenquadraten. Math. Ann. **32**, 342–350 (1888)

[62] D. Hilbert: Mathematical Problems. In: Mathematical developments arising from Hilbert Problems (Ed. F. Browder), Proc. Symp. Pure Math., Vol. **28**, Part 1, Amer. Math. Soc., Providence 1976, pp. 1-34

[63] D. Hilbert: Grundlagen der Geometrie. Teubner, Stuttgart 1987

[64] M. Hochster: Prime ideal structure in commutative rings. Trans. Amer. Math. Soc. **142**, 43–60 (1969)

[65] R. Huber, C. Scheiderer: A relative notion of local completeness in semialgebraic geometry. Arch. Math. **53**, 571–584 (1989)

[66] J. R. Isbell: Algebras of uniformly continuous functions. Annals of Maths. **68**, 96-125 (1958)

[67] B. Jacob: The model theory of generalized real closed fields. J. reine angew. Math. **323**, 213–220 (1981)

[68] N. Jacobson: Lectures in Abstract Algebra, Vol. III. Van Nostrand, New York 1964

[69] N. Jacobson: Basic Algebra II. W.H. Freeman, New York 1989

[70] P.T. Johnstone: Stone Spaces. Cambridge University Press, Cambridge 1982

[71] K. Keimel: The Representation of Lattice–Ordered Groups and Rings by Sections in Sheaves. In: Lectures on the Applications of

Sheaves to Ring Theory, Lecture Notes in Mathematics, Vol. **248**, Springer, Berlin 1971, pp. 1-98

[72] M. Knebusch: Weakly Semialgebraic Spaces. Lecture Notes in Mathematics, Vol. **1367**, Springer, Berlin 1989

[73] M. Knebusch, C. Scheiderer: Einführung in die reelle Algebra. Vieweg, Braunschweig 1989

[74] T.Y. Lam: The Algebraic Theory of Quadratic Forms. Benjamin, Reading 1973

[75] T.Y. Lam: Ten Lectures on Quadratic Forms over Fields. In: Conference on Quadratic Forms – 1976 (Ed. G. Orzech), Queen's Papers in Pure and Applied Math., No. **46**, Queen's University, Kingston 1977, pp. 1–102

[76] T.Y. Lam: Orderings, Valuations and Quadratic Forms. Regional Conf. Ser. Math., No. **52**, Amer. Math. Soc., Providence 1983

[77] T.Y. Lam: An introduction to real algebra. Rocky Mountain J. Math. **14**, 767–814 (1984)

[78] D. Lazard: Epimorphismes plats. In: P. Samuel, Les épimorphimes d'anneaux. Séminaire d'Algèbre commutative, Paris 1967/68

[79] L. Lipshitz: The real closure of a commutative regular f–ring. Fund. Math. **94**, 173–176 (1977)

[80] G.G. Lorentz: The 13–th Problem of Hilbert. In: Mathematical developments arising from Hilbert Problems (Ed. F. Browder), Proc. Symp. Pure Math., Vol. **28**, Part 2, Amer. Math. Soc., Providence 1976, pp. 419–430

[81] W.A. MacCaull: Positive definite functions over regular f–rings and representations as sums of squares. Ann. Pure Applied Logic **44**, 243–257 (1989)

[82] A. Macintyre: Model–completeness for sheaves of structures. Fund. Math. **81**, 73–89 (1973)

[83] S. MacLane: Categories for the Working Mathematician. Springer, New York 1971

[84] J.J. Madden: Two methods in the study of k–vector lattices. PhD Thesis, Wesleyan Univ., Middletown 1983

[85] J.J. Madden: Pierce–Birkhoff rings. Arch. Math. **53**, 565–570 (1989)

[86] J.J. Madden, J. Martinez: Monoreflections of commutative rings with identity. Preprint

[87] J.J. Madden, N. Schwartz: Separating ideals in dimension 2. In: Real Algebraic and Analytic Geometry (Eds.: M.E. Alonso et al.), revista mathemática Univ. Compl. Madrid, vol. **10**, Madrid 1977, pp. 217–240

[88] J.J. Madden, J. Vermeer: Epicomplete archimedean l–groups via a localic Yosida theorem. J. Pure Applied Alg. **68**, 243-252 (1990)

[89] L. Mahé: On the Pierce–Birkhoff conjecture. Rocky Mountain J. Math. **14**, 983–985 (1984)

[90] A.A. Markov: Insolubility of the problem of homeomorphy. In: Proc. Int. Cong. of Math. 1958 (Ed. J.A. Todd), Cambridge Univ. Press, Cambridge 1960, pp. 300-306

[91] M. Marshall: The Pierce–Birkhoff conjecture for curves. Canad. J. Math. **44**, 1262–1271 (1992)

[92] M.A. Marshall: Spaces of Orderings and Abstract Real Spectra. Lecture Notes in Math., Vol. **1636**, Springer, Berlin 1996

[93] T.S. Motzkin: The Arithmetic–Geometric Inequality. In: Inequalities (Ed. O. Shisha), Academic Press, New York 1965, pp. 205–224

[94] D. Mumford: The Red Book of Varieties and Schemes. Lecture Notes in Math., Vol. **1358**, Springer, Berlin 1988

[95] A. Pfister: Hilbert's seventeenth problem and related problems on definite forms. In: Mathematical developments arising from Hilbert Problems. (Ed. F. Browder), Proc. Symp. Pure Math., Vol. **28**, Part 2, Amer. Math. Soc., Providence 1976, pp. 483–489

[96] M. Prechtel: Endliche semialgebraische Räume. Diplomarbeit, Regensburg 1988

[97] A. Prestel: Lectures on Formally Real Fields. Lecture Notes in Mathematics, Vol. **1093**, Springer, Berlin 1984

[98] A. Prestel: Einführung in die Mathematische Logik und Modelltheorie. Vieweg, Braunschweig 1986

[99] A. Prestel: Model Theory for the Real Algebraic Geometer. Dip. Mat. Univ. Pisa, Pisa 1998

[100] A. Prestel, M. Ziegler: Erblich euklidische Körper. J. reine angew. Math. **274/275**, 196–205 (1975)

[101] S. Prieß–Crampe: Angeordnete Strukturen – Gruppen, Körper, Projektive Ebenen. Springer, Berlin 1983

[102] R. Quarez: The idempotency of the real spectrum implies the extension theorem for Nash functions. Preprint

[103] R. Ramanakoraisina: Sur les schémas réels. Thèse, Université de Rennes, 1983

[104] P. Ribenboim: Théorie des valuations. Université Montréal, Montréal 1965

[105] M.F. Roy: Faisceau structural sur le spectre réel et fonctions des Nash. In: Géométrie Algébrique Réelle et Formes Quadratiques (Eds. J.–L. Colliot–Thélène et al.), Lecture Notes in Math., Vol. **959**, Springer, Berlin 1982, pp. 406–432

[106] G.E. Sacks: Saturated Model Theory. Benjamin, Reading 1972

[107] P. Samuel: Les epimorphismes d'anneaux. Séminaire d'Algèbre Commutative, Paris 1967/68

[108] N. Schwartz: Real Closed Spaces. Habilitationsschrift, München, Januar 1984

[109] N. Schwartz: Real closed rings. In: Algebra and Order (Ed. S. Wolfenstein), Heldermann Verlag, Berlin 1986, pp. 175–194

[110] N. Schwartz: The Basic Theory of Real Closed Spaces. Regensburger Math. Schriften, Bd. **15**, Universität Regensburg, Regensburg 1987

[111] N. Schwartz: The basic theory of real closed spaces. Memoir Amer. Math. Soc., No. **397**, Amer. Math. Soc., Providence 1989

[112] N. Schwartz: Epimorphisms of f-rings. In: Ordered Algebraic Structures (Ed. J. Martinez), Kluwer, Dordrecht 1989, pp. 187–195

[113] N. Schwartz: Eine Universelle Eigenschaft reell abgeschlossener Räume. Comm. in Alg. **18**, 755–774 (1990)

[114] N. Schwartz: Inverse real closed spaces. Illinois J. Math. **35**, 536–568 (1991)

[115] N. Schwartz: Piecewise Polynomial Functions. In: Ordered Algebraic Structures (Eds. J. Martinez, C. Holland), Kluwer, Dordrecht 1993, pp. 169–202

[116] N. Schwartz: Gabriel filters in real closed rings. Comment. Math. Helv. **72**, 434–465 (1997)

[117] N. Schwartz: Epimorphic hulls and Prüfer extensions of partially ordered rings. Preprint

[118] N. Schwartz: The semiring of sums of squares in a formally real field. Preprint.

[119] N. Schwartz: The algebraic topology of real spectra. In preparation

[120] M. Shiota: Nash manifolds. Lecture Notes in Math., Vol. **1269**, Springer, Berlin 1987

[121] H.H. Storrer: Epimorphismen von kommutativen Ringen. Comment Math. Helv. **43**, 373–401 (1968)

[122] L. van den Dries: Artin–Schreier theory for commutative regular rings. Ann. Math. Logic **12**, 113–150 (1977)

[123] V. Weispfenning: Model–Completeness and Elimination of Quantifiers for Subdirect Products of Structures. J. Alg. **36**, 252–277 (1975)

List of categories

We give a complete list of all categories occurring in this work. For most categories we also indicate where they are discussed. Exceptions are categories that are so common that nothing needs to be said (such as the category of sets) and the category of reduced porings (which appears everywhere in this work). In the notation we use some generic prefixes and suffixes. They have the following meaning:

Prefixes

WBI	every object has the weak bounded inversion property
BI	every object has the bounded inversion property
(A,P)	every object is an (A, P)-algebra
TO	every object is totally ordered

Suffixes

R	the objects are rings
R/N	the objects are reduced rings
D	the objects are integral domains
F	the objects are fields

almost real closed fields

> Example 16.9(e), p. 196; Example 17.8(e), p. 207

(A, P)-algebras of continuous piecewise polyno-　　**(A,P)CPWPFR**
mial functions

> Example 13.4, p. 163

(A, P)-algebras of continuous piecewise rational　　**(A,P)CPWRFR**
functions

> Example 21.9(c), p. 240; Theorem 22.7, p. 244; Theorem
> 22.13, p. 250

totally ordered fields   **TOF**

chapter 3, p. 43 f; Example 8.4, p. 82; chapter 16, p. 189 ff;
chapter 17, p. 201 ff; p. 217; p. 233 ff

totally ordered fields with Henselian natural
valuation

Example 16.9(d), p. 196; p. 208

totally ordered fields of cardinality $\leq \kappa$

Example 16.8(b), p. 196

totally ordered integrally closed domains

Example 18.8, p. 214

totally ordered local domains

Example 18.9, p. 214

totally ordered Pythagorean fields

Example 16.9(c), p. 196; Example 17.8(c), p. 207

totally ordered subfields of a fixed totally ordered
field

Example 16.8(c), p. 196

Index

Druck: Strauss Offsetdruck, Mörlenbach
Verarbeitung: Schäffer, Grünstadt

Lecture Notes in Mathematics

For information about Vols. 1–1520
please contact your bookseller or Springer-Verlag

Vol. 1563: E. Fabes, M. Fukushima, L. Gross, C. Kenig, M. Röckner, D. W. Stroock, Dirichlet Forms. Varenna, 1992. Editors: G. Dell'Antonio, U. Mosco. VII, 245 pages. 1993.

Vol. 1564: J. Jorgenson, S. Lang, Basic Analysis of Regularized Series and Products. IX, 122 pages. 1993.

Vol. 1565: L. Boutet de Monvel, C. De Concini, C. Procesi, P. Schapira, M. Vergne. D-modules, Representation Theory, and Quantum Groups. Venezia, 1992. Editors: G. Zampieri, A. D'Agnolo. VII, 217 pages. 1993.

Vol. 1566: B. Edixhoven, J.-H. Evertse (Eds.), Diophantine Approximation and Abelian Varieties. XIII, 127 pages. 1993.

Vol. 1567: R. L. Dobrushin, S. Kusuoka, Statistical Mechanics and Fractals. VII, 98 pages. 1993.

Vol. 1568: F. Weisz, Martingale Hardy Spaces and their Application in Fourier Analysis. VIII, 217 pages. 1994.

Vol. 1569: V. Totik, Weighted Approximation with Varying Weight. VI, 117 pages. 1994.

Vol. 1570: R. deLaubenfels, Existence Families, Functional Calculi and Evolution Equations. XV, 234 pages. 1994.

Vol. 1571: S. Yu. Pilyugin, The Space of Dynamical Systems with the C^0-Topology. X, 188 pages. 1994.

Vol. 1572: L. Göttsche, Hilbert Schemes of Zero-Dimensional Subschemes of Smooth Varieties. IX, 196 pages. 1994.

Vol. 1573: V. P. Havin, N. K. Nikolski (Eds.), Linear and Complex Analysis – Problem Book 3 – Part I. XXII, 489 pages. 1994.

Vol. 1574: V. P. Havin, N. K. Nikolski (Eds.), Linear and Complex Analysis – Problem Book 3 – Part II. XXII, 507 pages. 1994.

Vol. 1575: M. Mitrea, Clifford Wavelets, Singular Integrals, and Hardy Spaces. XI, 116 pages. 1994.

Vol. 1576: K. Kitahara, Spaces of Approximating Functions with Haar-Like Conditions. X, 110 pages. 1994.

Vol. 1577: N. Obata, White Noise Calculus and Fock Space. X, 183 pages. 1994.

Vol. 1578: J. Bernstein, V. Lunts, Equivariant Sheaves and Functors. V, 139 pages. 1994.

Vol. 1579: N. Kazamaki, Continuous Exponential Martingales and BMO. VII, 91 pages. 1994.

Vol. 1580: M. Milman, Extrapolation and Optimal Decompositions with Applications to Analysis. XI, 161 pages. 1994.

Vol. 1581: D. Bakry, R. D. Gill, S. A. Molchanov, Lectures on Probability Theory. Editor: P. Bernard. VIII, 420 pages. 1994.

Vol. 1582: W. Balser, From Divergent Power Series to Analytic Functions. X, 108 pages. 1994.

Vol. 1583: J. Azéma, P. A. Meyer, M. Yor (Eds.), Séminaire de Probabilités XXVIII. VI, 334 pages. 1994.

Vol. 1584: M. Brokate, N. Kenmochi, I. Müller, J. F. Rodriguez, C. Verdi, Phase Transitions and Hysteresis. Montecatini Terme, 1993. Editor: A. Visintin. VII. 291 pages. 1994.

Vol. 1585: G. Frey (Ed.), On Artin's Conjecture for Odd 2-dimensional Representations. VIII, 148 pages. 1994.

Vol. 1586: R. Nillsen, Difference Spaces and Invariant Linear Forms. XII, 186 pages. 1994.

Vol. 1587: N. Xi, Representations of Affine Hecke Algebras. VIII, 137 pages. 1994.

Vol. 1588: C. Scheiderer, Real and Étale Cohomology. XXIV, 273 pages. 1994.

Vol. 1589: J. Bellissard, M. Degli Esposti, G. Forni, S. Graffi, S. Isola, J. N. Mather, Transition to Chaos in Classical and Quantum Mechanics. Montecatini Terme, 1991. Editor: 2S. Graffi. VII, 192 pages. 1994.

Vol. 1590: P. M. Soardi, Potential Theory on Infinite Networks. VIII, 187 pages. 1994.

Vol. 1591: M. Abate, G. Patrizio, Finsler Metrics – A Global Approach. IX, 180 pages. 1994.

Vol. 1592: K. W. Breitung, Asymptotic Approximations for Probability Integrals. IX, 146 pages. 1994.

Vol. 1593: J. Jorgenson & S. Lang, D. Goldfeld, Explicit Formulas for Regularized Products and Series. VIII, 154 pages. 1994.

Vol. 1594: M. Green, J. Murre, C. Voisin, Algebraic Cycles and Hodge Theory. Torino, 1993. Editors: A. Albano, F. Bardelli. VII, 275 pages. 1994.

Vol. 1595: R.D.M. Accola, Topics in the Theory of Riemann Surfaces. IX, 105 pages. 1994.

Vol. 1596: L. Heindorf, L. B. Shapiro, Nearly Projective Boolean Algebras. X, 202 pages. 1994.

Vol. 1597: B. Herzog, Kodaira-Spencer Maps in Local Algebra. XVII, 176 pages. 1994.

Vol. 1598: J. Berndt, F. Tricerri, L. Vanhecke, Generalized Heisenberg Groups and Damek-Ricci Harmonic Spaces. VIII, 125 pages. 1995.

Vol. 1599: K. Johannson, Topology and Combinatorics of 3-Manifolds. XVIII, 446 pages. 1995.

Vol. 1600: W. Narkiewicz, Polynomial Mappings. VII, 130 pages. 1995.

Vol. 1601: A. Pott, Finite Geometry and Character Theory. VII, 181 pages. 1995.

Vol. 1602: J. Winkelmann, The Classification of Three-dimensional Homogeneous Complex Manifolds. XI, 230 pages. 1995.

Vol. 1603: V. Ene, Real Functions – Current Topics. XIII, 310 pages. 1995.

Vol. 1604: A. Huber, Mixed Motives and their Realization in Derived Categories. XV, 207 pages. 1995.

Vol. 1605: L. B. Wahlbin, Superconvergence in Galerkin Finite Element Methods. XI, 166 pages. 1995.

Vol. 1606: P.-D. Liu, M. Qian, Smooth Ergodic Theory of Random Dynamical Systems. XI, 221 pages. 1995.

Vol. 1607: G. Schwarz, Hodge Decomposition – A Method for Solving Boundary Value Problems. VII, 155 pages. 1995.

Vol. 1608: P. Biane, R. Durrett, Lectures on Probability Theory. Editor: P. Bernard. VII. 210 pages. 1995.

Vol. 1609: L. Arnold, C. Jones, K. Mischaikow, G. Raugel, Dynamical Systems. Montecatini Terme, 1994. Editor: R. Johnson. VIII, 329 pages. 1995.

Vol. 1610: A. S. Üstünel, An Introduction to Analysis on Wiener Space. X, 95 pages. 1995.

Vol. 1611: N. Knarr, Translation Planes. VI, 112 pages. 1995.

Vol. 1612: W. Kühnel, Tight Polyhedral Submanifolds and Tight Triangulations. VII, 122 pages. 1995.

Vol. 1663: Y. E. Karpeshina; Perturbation Theory for the Schrödinger Operator with a Periodic Potential. VII, 352 pages. 1997.

Vol. 1664: M. Väth, Ideal Spaces. V, 146 pages. 1997.

Vol. 1665: E. Giné, G. R. Grimmett, L. Saloff-Coste, Lectures on Probability Theory and Statistics 1996. Editor: P. Bernard. X, 424 pages, 1997.

Vol. 1666: M. van der Put, M. F. Singer, Galois Theory of Difference Equations. VII, 179 pages. 1997.

Vol. 1667: J. M. F. Castillo, M. González, Three-space Problems in Banach Space Theory. XII, 267 pages. 1997.

Vol. 1668: D. B. Dix, Large-Time Behavior of Solutions of Linear Dispersive Equations. XIV, 203 pages. 1997.

Vol. 1669: U. Kaiser, Link Theory in Manifolds. XIV, 167 pages. 1997.

Vol. 1670: J. W. Neuberger, Sobolev Gradients and Differential Equations. VIII, 150 pages. 1997.

Vol. 1671: S. Bouc, Green Functors and G-sets. VII, 342 pages. 1997.

Vol. 1672: S. Mandal, Projective Modules and Complete Intersections. VIII, 114 pages. 1997.

Vol. 1673: F. D. Grosshans, Algebraic Homogeneous Spaces and Invariant Theory. VI, 148 pages. 1997.

Vol. 1674: G. Klaas, C. R. Leedham-Green, W. Plesken, Linear Pro-p-Groups of Finite Width. VIII, 115 pages. 1997.

Vol. 1675: J. E. Yukich, Probability Theory of Classical Euclidean Optimization Problems. X, 152 pages. 1998.

Vol. 1676: P. Cembranos, J. Mendoza, Banach Spaces of Vector-Valued Functions. VIII, 118 pages. 1997.

Vol. 1677: N. Proskurin, Cubic Metaplectic Forms and Theta Functions. VIII, 196 pages. 1998.

Vol. 1678: O. Krupková, The Geometry of Ordinary Variational Equations. X, 251 pages. 1997.

Vol. 1679: K.-G. Grosse-Erdmann, The Blocking Technique. Weighted Mean Operators and Hardy's Inequality. IX, 114 pages. 1998.

Vol. 1680: K.-Z. Li, F. Oort, Moduli of Supersingular Abelian Varieties. V, 116 pages. 1998.

Vol. 1681: G. J. Wirsching, The Dynamical System Generated by the 3n+1 Function. VII, 158 pages. 1998.

Vol. 1682: H.-D. Alber, Materials with Memory. X, 166 pages. 1998.

Vol. 1683: A. Pomp, The Boundary-Domain Integral Method for Elliptic Systems. XVI, 163 pages. 1998.

Vol. 1684: C. A. Berenstein, P. F. Ebenfelt, S. G. Gindikin, S. Helgason, A. E. Tumanov, Integral Geometry, Radon Transforms and Complex Analysis. Firenze, 1996. Editors: E. Casadio Tarabusi, M. A. Picardello, G. Zampieri. VII, 160 pages. 1998.

Vol. 1685: S. König, A. Zimmermann, Derived Equivalences for Group Rings. X, 146 pages. 1998.

Vol. 1686: J. Azéma, M. Émery, M. Ledoux, M. Yor (Eds.), Séminaire de Probabilités XXXII. VI, 440 pages. 1998.

Vol. 1687: F. Bornemann, Homogenization in Time of Singularly Perturbed Mechanical Systems. XII, 156 pages. 1998.

Vol. 1688: S. Assing, W. Schmidt, Continuous Strong Markov Processes in Dimension One. XII, 137 page. 1998.

Vol. 1689: W. Fulton, P. Pragacz, Schubert Varieties and Degeneracy Loci. XI, 148 pages. 1998.

Vol. 1690: M. T. Barlow, D. Nualart, Lectures on Probability Theory and Statistics. Editor: P. Bernard. VIII, 237 pages. 1998.

Vol. 1691: R. Bezrukavnikov, M. Finkelberg, V. Schechtman, Factorizable Sheaves and Quantum Groups. X, 282 pages. 1998.

Vol. 1692: T. M. W. Eyre, Quantum Stochastic Calculus and Representations of Lie Superalgebras. IX, 138 pages. 1998.

Vol. 1694: A. Braides, Approximation of Free-Discontinuity Problems. XI, 149 pages. 1998.

Vol. 1695: D. J. Hartfiel, Markov Set-Chains. VIII, 131 pages. 1998.

Vol. 1696: E. Bouscaren (Ed.): Model Theory and Algebraic Geometry. XV, 211 pages. 1998.

Vol. 1697: B. Cockburn, C. Johnson, C.-W. Shu, E. Tadmor, Advanced Numerical Approximation of Nonlinear Hyperbolic Equations. Cetraro, Italy, 1997. Editor: A. Quarteroni. VII, 390 pages. 1998.

Vol. 1698: M. Bhattacharjee, D. Macpherson, R. G. Möller, P. Neumann, Notes on Infinite Permutation Groups. XI, 202 pages. 1998.

Vol. 1699: A. Inoue, Tomita-Takesaki Theory in Algebras of Unbounded Operators. VIII, 241 pages. 1998.

Vol. 1700: W. A. Woyczyński, Burgers-KPZ Turbulence, XI, 318 pages. 1998.

Vol. 1701: Ti-Jun Xiao, J. Liang, The Cauchy Problem of Higher Order Abstract Differential Equations, XII, 302 pages. 1998.

Vol. 1702: J. Ma, J. Yong, Forward-Backward Stochastic Differential Equations and Their Applications. XIII, 270 pages. 1999.

Vol. 1703: R. M. Dudley. R. Norvaiša, Differentiability of Six Operators on Nonsmooth Functions and p-Variation. VIII, 272 pages. 1999.

Vol. 1704: H. Tamanoi, Elliptic Genera and Vertex Operator Super-Algebras. VI, 390 pages. 1999.

Vol. 1705: I. Nikolaev, E. Zhuzhoma, Flows in 2-dimensional Manifolds. XIX, 294 pages. 1999.

Vol. 1706: S. Yu. Pilyugin, Shadowing in Dynamical Systems. XVII, 271 pages. 1999.

Vol. 1707: R. Pytlak, Numerical Methods for Optical Control Problems with State Constraints. XV, 215 pages. 1999.

Vol. 1709: J. Azéma, M. Émery, M. Ledoux, M. Yor (Eds), Séminaire de Probabilités XXXIII. VIII, 418 pages. 1999.

Vol. 1710: M. Koecher, The Minnesota Notes on Jordan Algebras and Their Applications. IX, 173 pages. 1999.

Vol. 1711: W. Ricker, Operator Algebras Generated by Commuting Projections: A Vector Measure Approach. XVII, 159 pages. 1999.

Vol. 1712: N. Schwartz, J. J. Madden, Semi-algebraic Function Rings and Reflectors of Partially Ordered Rings. XI, 279 pages. 1999.